Interpretation of Topographic Maps

Victor C. Miller

Mary E. Westerback
Long Island University—C. W. Post Campus

Merrill Publishing Company
A Bell & Howell Information Company
Columbus Toronto London Melbourne

Dedication

Victor Miller died suddenly on April 1, 1988, at the age of 66. He had a keen wit, a dynamic teaching style, and a unique ability to understand landforms. The maps, block diagrams, and text in this book are intended to excite, delight, and challenge the reader to visualize the "why" and "how" of topographic expression.

In his memory, I would like to dedicate this book to our teachers and students, past and present.

Mary E. Westerback

Published by Merrill Publishing Company
A Bell & Howell Information Company
Columbus, Ohio 43216

This book was set in Palatino.

Administrative Editor: David Gordon
Production Coordinator: Rex Davidson
Art Coordinator: Pete Robison
Cover Art: Courtesy of USGS
Cover Designer: Brian Deep
Text Designer: Cynthia Brunk

Library of Congress Catalog Card Number: 88-61050
International Standard Book Number: 0-675-20919-6
Printed in the United States of America
1 2 3 4 5 6 7 8 9—92 91 90 89

Foreword

Victor Miller came to me as a graduate student from under the aegis of that peerless landscape artist, Armin K. Lobeck, who was then in charge of the Columbia College geology instruction program. We had both learned much from that master of landform interpretation and we both tried to emulate his technique of presenting landforms in perspective drawings that conveyed the essence of the form with only a few strokes. The many delightful block diagrams with which Victor has illustrated this book attest to his success in that endeavor. In my graduate courses in topographic map interpretation, Victor showed a remarkable talent for interpreting from contours alone the rock structures that underlie the surface features and the geomorphic processes of erosion and deposition responsible for shaping those structures into unique landforms.

Victor's doctoral dissertation on the relationships between stream channels and the dip of strata in the folded Appalachians extended his accomplishments in the direction of quantitative fluvial morphometry, then a newly emerging branch of geomorphology. Carrying his acquired skills into the professional interpretation of air photographs, Victor completed his comprehensive treatise *Photogeology,* published by McGraw-Hill in 1961. This work became the reference standard in its field, its value now enhanced by the need to identify and interpret rock structures the world over within the new paradigm of plate tectonics.

Today we are dazzled by the proliferation of false-color LANDSAT images of the earth's solid surface. These marvels of electronic technology may seem to expose to our direct view everything we need to know about the geology and geomorphology of a given study area. But satellite imagery must be checked against the ground-truth, provided in large part by the modern contour topographic map. Few persons realize that a proliferation of new and detailed large-scale topographic quadrangles has paralleled the advances of satellite imagery. Contours contain essential quantitative information on the relief dimension, not directly available from satellite images and air photos. Contour interpretation is what this book is all about. Now, as never before, skill in contour interpretation is an essential tool for exploiting the data we now have available, whether the purpose be to evaluate the natural-resource potential of an area or a host of environmental problems that must be faced.

I could not imagine a team better qualified than Victor Miller and Mary Westerback to produce a textbook of topographic map interpretation. Their accumulated professional experience in field study and academic and applied geology are evident in the material they have selected. Their experience as classroom teachers shows clearly in their personal style of communication with the student. These authors have made a significant contribution to the upgrading of instruction in the earth sciences at a time when standards in science education are sorely in need of decisive improvement.

Arthur N. Strahler

Preface

Topographic maps have long been prominent in the teaching and practice of such branches of earth science as geography, geology, and geological and civil engineering. They are absolutely essential to the several branches of the military. They also serve soil scientists, hydrologists, agronomists, conservationists, and foresters.

The techniques of constructing topographic maps have changed through the generations. Until a few decades ago, survey parties worked in the field with plane table and alidade to measure distances, directions, and differences in elevation. They located and plotted one ground point after another, and these control points served as the bones on which the flesh—the contours—could be draped. In recent times, however, advances in electronics, computers, and photogrammetry have produced topographic maps of almost unbelievable accuracy and detail.

To us, the most fascinating thing about topographic maps is not how they are made, nor where, nor how much their production might cost, nor how long it takes to make them. Neither is their accuracy, despite the obvious need for maximum accuracy and economy. Rather, our fascination with topographic maps lies in how earth scientists have *used* them. They have "read" them for information, as they might read a reference publication.

We may compare such map usage with the ways in which most earth scientists have used aerial photographs and other forms of remote-sensing imagery. In the early days, until the end of World War II, earth scientists simply carried the photographs into the field and used them as base maps. They located their positions on the pictures and marked these points, or ground stations, by making pinpricks through the photograph. They then marked each pinhole on the back side of the photograph with a small circle and number, which corresponded to the numbered entry for that point in the field notebook.

Photograph scale, unlike map scale, varies irregularly from place to place on the picture due to variations in ground elevation. Despite this, many also used the photographs for measuring "fairly accurate" or "approximate" ground distances and directions, and for planning each day's traverses and their approaches to a great variety of field problems. And since a folding refraction stereoscope fits conveniently into a back pocket, it has become routine to examine aerial photographs stereoscopically in the field.

The point is that we use stereophotos in the same way we use maps. From them, we compile data on the nature of an area's terrain—slopes, distances, shapes, and locations. We use the photographs as "maps." We are, in short, *reading* them.

Although topographic maps cannot be viewed in three dimensions, one of the first things students must learn to do is to *visualize* in three dimensions the topography the contour lines depict. Thus, students learn to use topographic maps in the same way that they use stereophotographic "maps." For example, we use such characteristics of topographic maps as scale, distance, direction, contours, and contour intervals to determine valley depth and hilltop height. We also use them to compute slope magnitude (gradient) and to construct topographic profiles, to determine the precise location of a specific point on the ground, and to select the best route to travel somewhere and estimate how long it will take.

Photogeology, the study of geology on aerial photographs, has progressed far beyond the restricted "map" uses of photographs. Geologists, physical geographers, and civil engineers not only extract from photos such topographic data as locations, distances, and slopes; they also use their knowledge of rock types, geological structures, and such processes as weathering, erosion, and deposition to identify as many rock units as possible in an area, to identify fault traces and determine their offsets, to note dips and folds, and often to decipher the sequence of tectonic and geomorphic events through which the mapped area has passed. An experienced photogeologist can reach beyond mere recognition of topographic features and relations and pass into the fascinating realm of *photointerpretation.*

Our purpose in this book is to call your attention—whether you are a student or a professional earth scientist—to the fact that topographic maps hold vast information, far beyond the limited world of location, distance, height, and slope. But to claim this treasure of information, you must first become aware of its existence, then learn what questions to ask and how to find the answers. In time you will learn that the most important question to ask about a map, after noting the assemblage of features that make up the topography, is *why*.

ACKNOWLEDGMENTS

We are grateful for the assistance and encouragement that we have enjoyed from so many, not just in preparing this book, but over the many years during which our thoughts and convictions about map reading and interpretation have become organized and solidified. Those who must remain nameless, because they are so numerous, include our former professors and colleagues, those still living and those departed, and our unheralded former students who served over the decades as sounding boards (if not unsuspecting guinea pigs) for our methods of conveying an appreciation for, and a competence in, the art of map interpretation.

The seeds of this book were planted when co-author Victor Miller began his undergraduate studies at Columbia College nearly a half century ago with the late Armin K. Lobeck, and were brought to flower under the inspiration and guidance of Arthur N. Strahler. There are no words that can convey the gratitude and respect that he felt for these two consummate artists, scientists, educators, and friends.

A quarter century ago, co-author Mary Westerback was a graduate student of Victor Miller. This master teacher inspired her to pursue a career in teaching geology and earth science at the college level. The heritage of knowledge, and the joy of sharing this with others, form the cornerstone of her teaching career. It has been her pleasure to work with her distinguished colleague in writing this book, and in thus sharing this heritage with others.

Our special thanks go to Mr. Sanjeev Kalaswad, advanced doctoral candidate at the State University of New York at Binghamton, for criticism and constructive suggestions during the early stages of writing and organizing this book, and to Dr. Heinrich Toots of the Geology Department of Long Island University, who made useful suggestions in the final version. We appreciate the diligent efforts of our colleagues who reviewed the manuscript: Elizabeth Abbott, University of Florida; Parker Calkin, SUNY at Buffalo; Keith Clarke, Hunter College; Walter W. Doeringsfeld, Phoenix, Arizona; James F. Fryman, University of Northern Iowa; Donald Owen, Indiana State University; Jeff Patton, University of North Carolina; and Katherine Price, DePauw University.

We appreciate the masterful way in which our administrative editor, David C. Gordon, and his associates Peter Robison, art editor, and Rex Davidson, production editor, led us through the production of this book. We also thank Fred Schroyer, for his fine copyedit.

We are grateful to Allison R. Palmer of the Geological Society of America for permission to reproduce their Decade of North American Geology Geologic Time Scale, and to the American Geological Institute for allowing us to use many of their definitions in our glossary .

Finally, we warmly thank our families, especially Kaye, May, and Ivar, for their never-ending support, encouragement, and patience.

VCM
MEW

Contents

Introduction

"READING" VS. "INTERPRETATION"

Anyone can learn how to *read* a topographic map, but that is only a minor part of what this book is all about. Our primary aim is to show you how to *interpret* topographic maps. (You cannot actually interpret a topographic *map*. Instead, you interpret the *topography as it is depicted* on that map.)

For example, we may note on a map that one stream flows in a narrow, steep-sided valley, whereas a few miles away another stream catches our eye as it meanders leisurely along on a broad floodplain that is bordered by well-dissected bluffs. We are *reading* the map's graphic symbols, recognizing streams, valleys, floodplains, meanders, and bluffs. But this is not *interpretation*.

Suppose we ask ourselves, "Why is one valley narrow and steep-sided, and the other flat-floored and wide?" The key word, "*why*," forces *interpretation* of observed topographic and geologic features. Sometimes the topographic and geologic elements that provide the answer may lie beyond the limits of the area depicted on the map. In such cases we may have to use our imaginations.

Let us take another example. Suppose we note numerous closely spaced streams on one part of the map, whereas in another part the streams are more widely spaced, with gentler, more-rounded slopes. If we simply note this difference, we are merely reading the map, observing a topographic contrast. We are not interpreting. However, if we ask, "*Why* does this contrast exist?" then we are stepping into the realm of inquiry and interpretation. (Remember: even if we do answer the question, we are still not actually interpreting the *map*, but rather the topography as the map *depicts* it.)

From a topographic map, we can directly extract only topographic data. But in many cases, it is *geologic* information we seek. Thus, when we "interpret a topographic map," we are trying to see the geology as it is reflected in the topography. We interpret the rocks and the structures from which that topography (landscape) is carved. We also interpret geomorphic agents such as running water, weathering processes, glacial ice, and the like, which did the carving or which tend to cover up the structure.

SCOPE OF THIS BOOK

We have not attempted to survey all of the landscape features created by tectonic forces and agents of erosion. What we are trying to do, through a careful selection of maps, is to teach you how to visualize—to "see," in three dimensions—the landforms depicted by the contours. Once you can visualize the present-day map area, you will be in a position to learn how to visualize the sequence of structural, tectonic, climatic, and geomorphic events that worked hand-in-hand in the past to create this present topography.

This visualization of events, spanning significant geologic time intervals, will in turn lead you to understand *why* these events transpired. Finally, we hope you will develop the capacity to foresee the future, to visualize a sequence of possible events that are yet to come, events set in motion by past and present events and processes.

BACK TO SQUARE ONE?

If you are comfortable with reading topographic maps, fine—enjoy the book. But if you are shaky on the basics, or uncertain about some aspects of topographic maps, we suggest that you turn to the Appendix, "Getting Acquainted with Topographic Maps." There we detail the several elements that constitute a topographic map. These include map scale, direction, contours, contour interval, contour drawing, and the construction of topographic profiles and stream gradients. We also briefly review townships and ranges, as well as latitude and longitude.

The Appendix is strongly exercise-oriented, affording you the opportunity to learn, by *doing* rather than by memorizing, the basics of topographic map reading. For example, once you have actually gone through the procedure of drawing a contour line, you will never have to recall the verbal definition of a contour. You will know what a contour is, how it was drawn, and what it depicts—in other words, how to *visualize* it.

CHAPTER OVERVIEW

In Chapter 1 we review the definitions of important terms we will be using, and point out topographic features and relations that often are tip-offs to the geology of an area. Chapter 2 includes two maps, one illustrating what we mean by reading and recognition, and the other illustrating what we mean by interpretation.

As for Chapters 3–13, let us explain how we have organized the maps and materials we will be working with. We have avoided creating a show-and-tell type of book. We see no value in reproducing a map and captioning it, "Shown here are a cirque, a fault scarp, a drowned coastline (whatever that is), and terraces." If you are a student, then you are trying to learn about maps, and you might not be able to identify some of those features. It might help to modify the caption to "Scarp **A** is a fault scarp.[1] The amphitheater-like depression **B** is a good example of a cirque." But even that falls short of the mark, since only *recognition* is involved.

If we were students, we would certainly want the author to explain what a fault scarp *is,* and to explain *why* that particular scarp is identified as a fault scarp. If the scarp follows the trace of a fault, why is it not a fault-line scarp?[2] What does the author know and see that we do not?

In some instances we have, because of the simplicity of the topography, taken the show-and-tell approach, but for the most part, we have delved into the *why*.

[1,2]Such terms as *fault scarp* and *fault-line scarp* are defined and discussed in Chapter 1.

In avoiding show-and-tell, we ran into a problem: how to organize our maps. How, for example, should we classify a map that shows a glaciated coastal area of folded and faulted strata that are cut by intersecting dikes? Doing our best, we divided many of the maps in this book into the following categories, by chapter: 3—Glaciation, 4—Shorelines, 5—Volcanics, 6—Piedmont Slopes (Pediments and Alluvial Fans), 7—Depressions, 8—Streams, 9—Water Gaps and Wind Gaps, 10—Dips and Folds, and 11—Faults and Fractures.

There are two other chapters that are different in many ways. For want of another name, we call the first of these, Chapter 12, "Anomalies," though we might equally well have chosen "Puzzlers" or "Mismatches." Each area in this group has at least one feature, or group of features, that does not fit the overall characteristics of its area. For example, an odd-looking linear[3] feature may seem "lost" in the midst of an unpatterned scattering of other features, or an oddly positioned or nonoriented flat surface may appear to be out of place in a setting dominated by linear terrain elements. The serious student of map interpretation should find this chapter a lot of fun and a real challenge.

In Chapter 13 we have collected "Problems." These are not all areas of particularly complex geology or geomorphology, though they tend to be. Nevertheless, each one demands considerable attention and well-thought-out detective work. This chapter epitomizes what we think of as interpretation of topography as depicted on a map. If you agree that the anomalies of Chapter 12 are fun, you should find that the problems in this chapter offer one sheer delight after another!

PLEASE NOTE

With the exception of a few shorelines and specific streams to which we refer in the text, the maps in this book consist solely of contours. On each map we have indicated the contour interval (C.I.). Though in most cases the contours are numbered, we urge you, particularly if you are a beginning student, to concentrate on *topographic heights and slopes,* not on elevations. By concentrating on the lines—their directions, spacings, arrangements, and shapes—you will become far more proficient at visualizing the topography as a three-dimensional model. (Please do not develop the irritating and frustrating habit of constantly searching for the elevation numbers.)

For each map, we provide mile and kilometer scales. Some maps have been enlarged or reduced from the original, and this is indicated on each. An arrow indicates the north direction of each map. Note that, although many of the maps are oriented with north toward the top of the page, others are oriented with north toward the *left side* of the page. Thus, all text references to direction are relative to *geographic* north, and not necessarily to the top of the page.

Each map is accompanied by its own text. All ground features to which we refer are identified by letter, both on the map and in the text. As shown in Figures I–1 and I–2, the maps are divided into thirds—north, middle, and south. The features are labeled alphabetically from left to right across the northern third of the map, and continue similarly across the middle third, and finally across the southern third.

---

[3]A topographic feature that is straight or just slightly curved is said to be *linear,* as distinct from a *curvilinear* feature, in which the degree of bending is more pronounced. It is like the distinction between "large" and "fairly large" potatoes, but it is surprisingly workable.

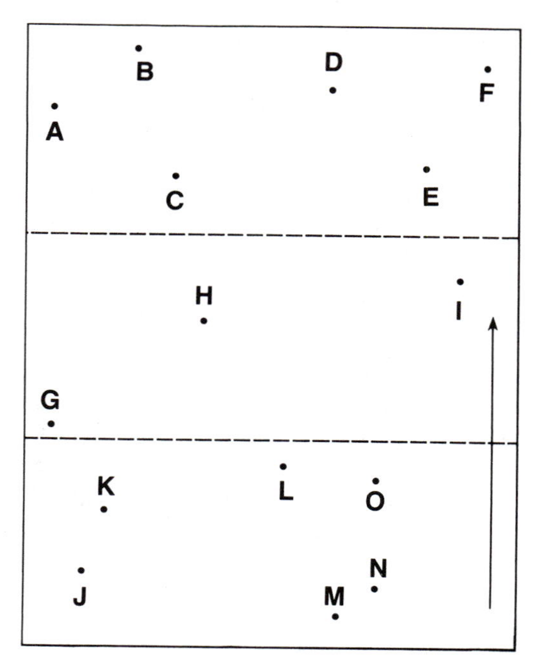

FIGURE I–1
Sketch showing our system for designating ground points referred to in text, when north is toward top of page.

Many of the maps are also accompanied by simple block diagrams. In most instances these block diagrams do not include geologic cross sections, though some do. We have left the block sides blank so you can add the geology where appropriate and possible.

We must stress that the block diagrams are generalized or idealized. Many details are purposely omitted. We have done this to convey the character

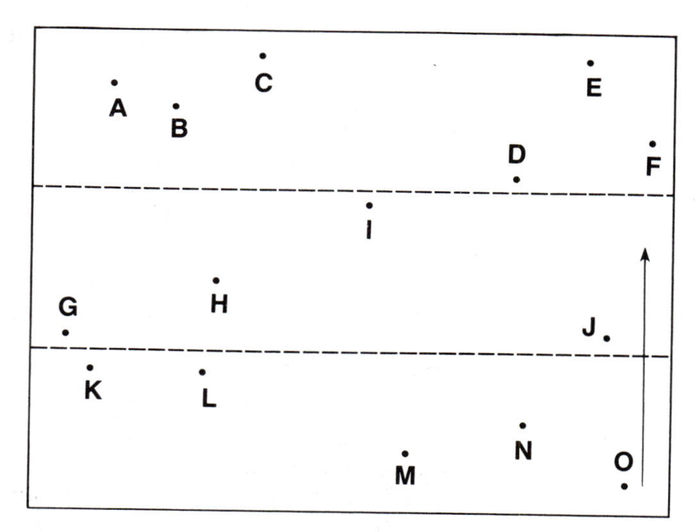

FIGURE I–2
Sketch showing our system for designating ground points referred to in text, when north is toward bound edge of page.

of the structure-topography relation or the contrast between areas, since these are more important than the precise topographic detail. For example, we show the typically rounded eastern topography around Kayjay, Kentucky (Figure 8–5 in back of book) as having a remarkably Arizona Plateau–type appearance, complete with virtually flat upland surfaces and bordering escarpments.

Not only must you, the interpreter of topographic maps, learn how to visualize the topography which the map and its contours depict, you must also acquire the capacity to "see" the simplicity that is almost always masked by that topography's complexity.

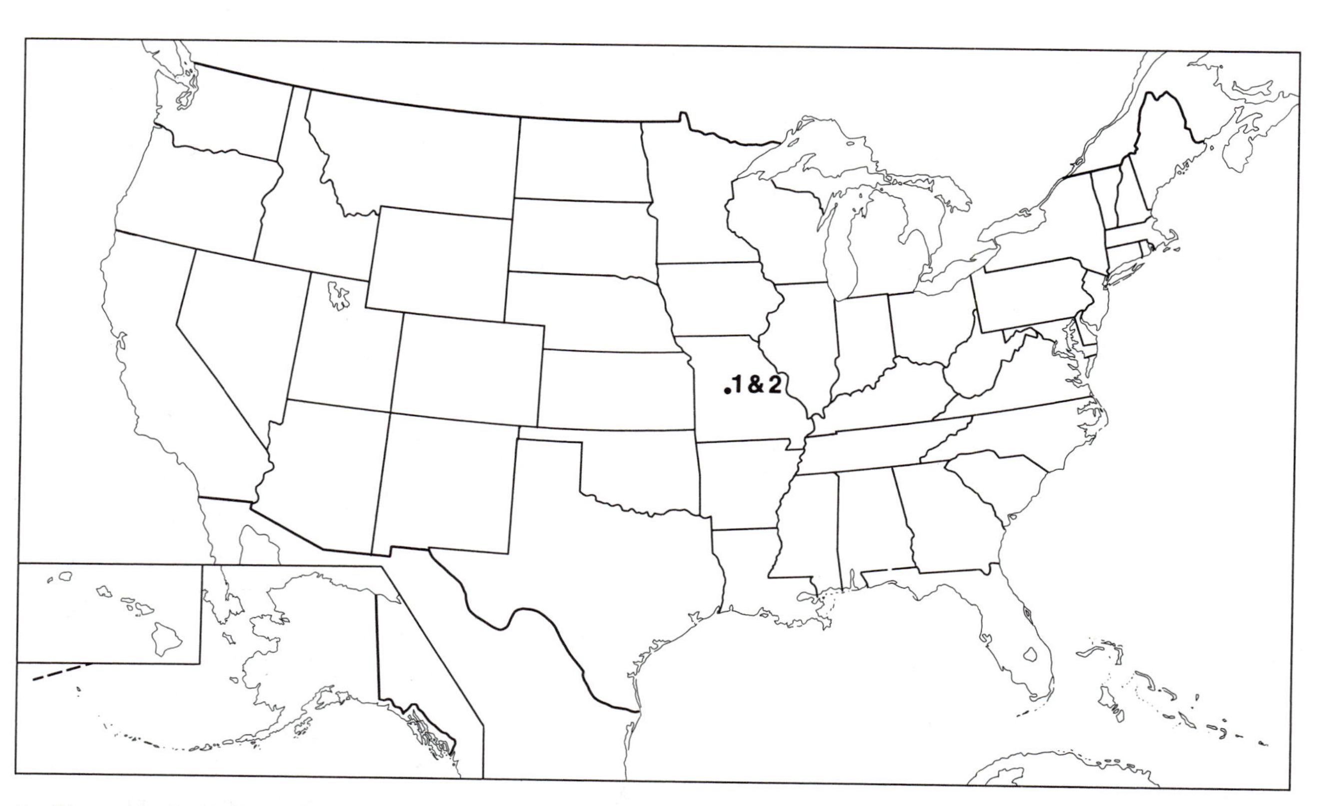

1. Figure 1–9. Bolivar, Missouri.

2. Figure 1–10. Bolivar, Missouri.

1

General Principles

A REVIEW

To interpret the geology or geomorphology of an area from its topography, you must first understand geology and geomorphology, as well as their intimate interrelation. Thus we will review the terms, definitions, processes, and relationships most useful in deciphering geomorphic and geologic information from topographic maps.

Steep Slopes = Snug Contours; V's Point Upstream

First we must stress two fundamental characteristics of topographic maps. The first is that, regardless of a map's contour interval (C.I.), *the steeper the slope, the more closely spaced will be its contour lines*. Conversely, *the gentler the slope, the greater will be the spacing*. (A common mistake by novices is to assume that the contour interval is the horizontal distance on the ground between two contour lines. They may think of a map as having many "contour intervals," simply because its contours are not equally spaced. In the Appendix we illustrate this relationship between slope and contour spacing.)

Equal in importance is the relation between contour lines and stream valleys—the contour V—which is depicted in the Appendix, Figures A-13 and A-14. Remember: *the contour V always points upstream*.

LINEARITY

See Straight, Think Steep

Most topographic features, and hence the contour lines that depict them, are more or less irregular in outline and in slope. So universally true is this, that in the study of topographic maps, geologic maps, and remotely sensed images, you should pay particular attention to any natural feature shown by straight or nearly-straight lines. An excellent rule of thumb: when you see a straight topographic feature, or line of features, especially if it stands in contrast to its surroundings, *think steep*.

A straight line denotes steepness, but not necessarily a steep slope. A low, straight break-in-slope might mark the trace of a steeply dipping fault plane, or a steeply dipping contact of some kind. Steeply dipping beds, joints, faults, and dikes are often expressed topographically by linear and nearly linear scarps, scarplets, benches, ridges, stream segments, rows of depressions (such as sinks), or rows of knobs. The linearity does not tell you what the feature is; it does tell you that something steeply inclined lies along that trend.

Parallelism—Topographic "Grain"

Closely akin to linearity is *parallelism*. Some areas exhibit a definite topographic "grain," due to an assemblage of more or less parallel elements. No individual element may be especially linear, but the *group* is. Numerous geologic factors may cause such an arrangement of parallel features, and these need not be bedrock elements. For example, drumlin swarms (Figure 3–1 in back of book) are remarkably parallel and en echelon. There is nothing bedrockish about a drumlin, and depositional drumlins are not the only glacial features that display parallelism. Many bedrock areas that have been overridden by ice sheets have been scoured and streamlined in the general direction of ice movement. This type of topographic "grain" is well developed throughout great areas of New England, for example.

Selective weathering and erosion along parallel fractures may produce topographic parallelism at a scale that can be picked up by contours. Many metamorphic terrains also exhibit a sometimes ill-defined but detectable parallelism or grain. Some such areas bear a disturbing resemblance to glacially scoured and grained topography. Compare, for example, the overall appearance of area 3 in Figure 8–2 (Virginia Piedmont) with the topographies of Figures 3–4 (Connecticut) and 4–7 (Maine) (all three illustrations are in back of book).

Perhaps the best geologic setting for developing large-scale topographic parallelism is folded strata having different resistances to weathering and erosion. The Valley and Ridge province of the Appalachians is the outstanding fold belt in North America.

DIFFERENTIAL EROSION

Fundamental to all topographic analysis is the principle of *differential erosion*, which usually involves differential weathering. Some rocks are more resistant than others to processes of destruction. Thus we often find that, in a given area, certain rock units form ridges, scarps, and generally high ground, whereas others promote the development of valleys, cliff-sapping slopes, and lowlands. Often closely tied in with differential erosion, and subject to the controls exerted by rock types, are drainage density, slope dimension and steepness, and local relief. In many of the areas illustrated in this book we will have ample opportunity to consider these topographic variables and their relations to the underlying bedrock.

STREAMS AND SLOPES

Much has been written about the classification of streams, drainage patterns, slopes, and the topography associated with faults. Since map interpreters must have a good working knowledge of these classifications, we will review them briefly. Abounding in this book are such terms as subsequent stream, dendritic pattern, rejuvenation, obsequent slope, and resequent fault-line scarp.

Though the work and concepts of W. M. Davis have been a center of controversy, many of his terms are still ideally suited to describing a wide range of geomorphic features and area characteristics. One need not "believe" in Davis to put these terms to good use. He classified stream valleys as young or mature, for example. We tend to concentrate on the streams.

Streams

A *young stream* (stream **A,** Figure 1–1a) is actively downcutting and hence is carving a generally V-shaped valley. You may, of course, also call its valley a youthful, or young, or even a young-looking valley.

An *early mature stream* has recently ceased downcutting but is still cutting laterally. (All streams cut laterally, to a certain extent, whether or not they are also downcutting.) An early mature stream is beginning to meander and carve its floodplain. On the outside of its meander bends it is eroding laterally, and on the inside of its meander bends it is depositing.

A *mature stream* (stream **A,** Figure 1–2a) has cut a floodplain broad enough to accommodate its meander belt. It continues to be a dominantly laterally cutting stream.

A *rejuvenated stream* has either begun to cut down again, or to cut down more vigorously. Rejuvenation does not necessarily indicate uplift and/or tilt. Climate change, change in sea level, and stream capture are but three of the other possible causes for rejuvenation.

Dissection

The terms "youthful" and "mature" are also applied to the degree of dissection of an area. As defined by W. M. Davis, an area that is *youthfully dissected* (or "in youth") was originally undissected, but into which a relatively few streams were able to cut their valleys (Figures 1–1a and 1–1b). Most of the original surface remains as broad, undrained interfluvial tracts. There are quite a few new-looking surfaces that are not original surfaces, but which have resulted from the stripping away of overlying rock layers. Though perhaps technically

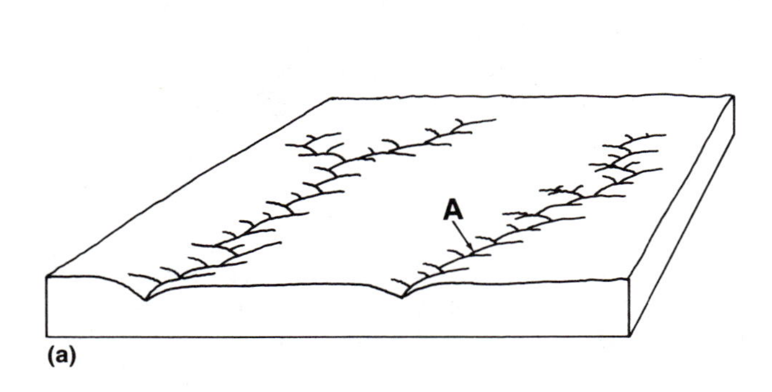

FIGURE 1–1a
Block diagram of a young stream.

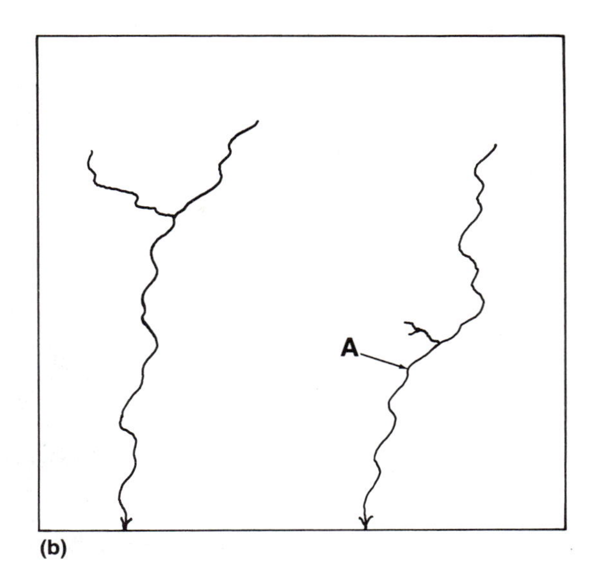

FIGURE 1–1b
Sketch map of area depicted in Figure 1–1a.

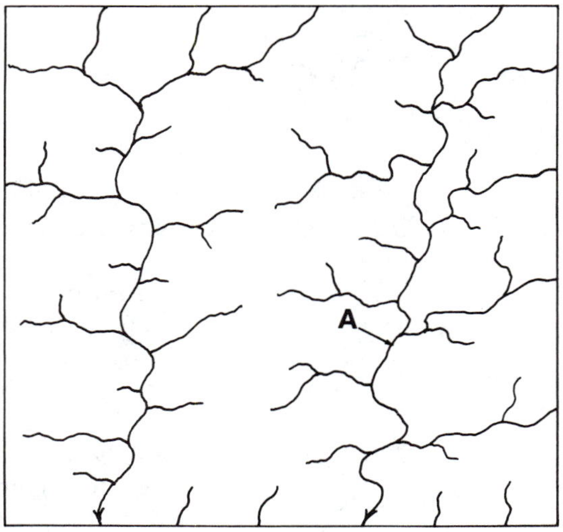

FIGURE 1–2a
Block diagram of a maturely dissected area.

FIGURE 1–2b
Sketch map of area depicted in Figure 1–2a.

incorrect, we have found it extremely useful to describe such a surface as youthfully dissected. Such a surface is only partially scored by valleys or canyons, and maintains intact most of its simple, fairly smooth upland. Usually, but not always, the streams that cut into a youthfully dissected surface are themselves young streams.

A *maturely dissected* area (Figures 1–2a and 1–2b) may be called "in slope." The "original" upland has been destroyed as a surface, but it is still recognizable by the accordance of the divides. As this mature stage passes, the initial accordant interfluves are progressively destroyed and the overall topography is lowered. It approaches a new base level that is defined by the floodplains of the major streams that flow across the area.

For many years, areas that had undergone much destruction and lowering were said to be in "old age." We see no reason to retain this term, since such an area may still reasonably be called "maturely dissected." Further, if we forget about any real or imagined "original" surface, and free ourselves from thinking about stage, we strengthen the term "maturely dissected." It becomes a purely descriptive term, no longer shackled by the concepts of cycle and phase.

STREAM GENESIS

Streams (and their valleys) have also been classified "genetically," though we are not particularly satisfied with the term "genetic." In many instances, a stream's genesis is not pertinent to its classification under this system. However, for the sake of practicality, we will adhere to the terminology of that system, with little modification.

The "Sequents"

A *consequent stream* develops on an original slope of some kind, such as a lava flow, a moraine, a newly emerged coastal plain, or the bed of a former lake. It develops simply because a slope exists, a slope that must be drained when water comes to its surface. A consequent stream need bear no relation to dip direction or any other bedrock structure. In fact, no bedrock need be present at all.

A *subsequent stream* (streams **SST,** Figure 1–3) has extended itself headward along a line, belt, or zone of relative erosional nonresistance. The most common such lines/belts/zones are the traces of steeply dipping faults or joints and the outcrops of steeply dipping nonresistant strata or dikes. These streams admirably illustrate both the principles and the importance of differential erosion.

An *obsequent stream* (stream **OST,** Figure 1–3) flows in a direction opposite to dip direction. An *obsequent slope* (slopes **OSL,** Figure 1–3) is inclined in a direction opposite to the dip direction.

A *resequent stream* (streams **RST,** Figure 1–3) flows in the same direction as dip. It cannot be a stream that developed on an original slope which happened to coincide with dip direction, since that stream would be a consequent stream. (This would be a special variety of consequent stream, since most consequent streams bear no relation of any kind to dip direction—all they require is any original slope, regardless of what may have created it.) Thus, we must think of a resequent stream as one that flows *both* on an erosional slope and in the direction of dip. It is unrelated to any consequent stream or to other streams that may have existed in the area.

A homoclinal ridge, such as a hogback, is drained by both obsequent and resequent streams. The slope inclined in the direction of dip is the *resequent slope* (slopes **RSL,** Figure 1–3). If dip and slope coincide relatively well, such a slope is also called a *dip slope.* Some earth scientists refer to the obsequent slope as the "scarp slope," but we find that a poor term, since all scarps are merely steep slopes, and steep slopes need bear no relation to dip. In fact, scarps form in dipless areas just as well as in dip areas (granite and gneiss terrains, for example).

An *insequent stream* (stream **IST,** Figure 1–3), the last of the "-sequent" types, does not fall conveniently into any of the other categories. It is typically a tributary that seems to have developed where it did for no apparent reason. Since insequent streams do not adjust in any observable way to such controls as dip, strike, or fractures, they simply extend their channels headward in any direction. An area drained by a large number of insequent streams displays a drainage pattern resembling the branches of a tree or a root system. It is called a *dendritic pattern* (Figure 1–2b).

Numerous other stream patterns have been designated, the most common being trellis, annular, radial, angular, rectangular, and parallel. Where appropriate, we will call attention to specific patterns on the maps.

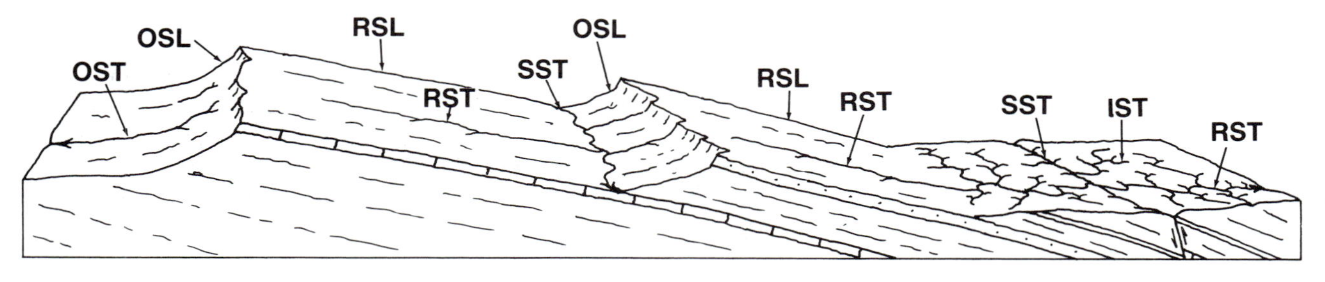

FIGURE 1–3
Block diagram showing genetic streams and slopes (obsequent, resequent, subsequent, and insequent).

Antecedent and Superposed Streams

We recognize two other genetic stream categories, antecedent and superposed streams. Both share one outstanding characteristic—they flow through or across, rather than around, one or more topographic obstacles, such as hills, ridges, and mountains. The difference between them lies in how they developed.

An *antecedent stream* (Figure 9–2 in back of book) is older than the structural event that raised the barrier across its path. The stream continued to cut through the rising obstacle in precisely the same way that a circular saw cuts through a log that is pushed against it.

A *superposed stream* (Figures 9–5 and 9–6 in back of book), on the other hand, has undergone a completely different history. In this case, the obstacle is a topographic surface (mountain ridge, hogback, or whatever) that has been buried by younger rock materials. A stream develops on the top surface of the blanketing rocks, then cuts down through these rocks until it encounters the buried old topography. It is superposed (positioned from above) upon the formerly buried surface, into which it then continues to cut.

Capture by a Pirate

In a surprising number of cases, streams that drain the opposite flanks of a divide have different gradients, discharges, velocities, turbulences, competences, and capacities. In such cases, the steeper, more-active stream is usually able to erode headward more determinedly and rapidly than the gentler stream. The more aggressive stream's headward growth and erosion will thus continue at the expense of the other stream, and the divide between the two will migrate. The headwaters part of the weaker stream's valley will be destroyed, and this piece of real estate will be incorporated into the watershed of the stronger stream.

The aggressive, more rapidly downcutting, headward-eroding stream can approach the flank of an opponent stream, rather than the head. When the stronger stream reaches the gently inclined (and higher) bed of the weaker stream, it captures the entire upstream part of the higher stream, converting it into a tributary. The process of *stream capture* is also called *stream piracy*. The stream doing the capturing is the *captor stream*. The downstream segment of its victim is said to have been *beheaded*. The pronounced bend in the course of the new tributary is designated the *elbow of capture*. We shall be looking at some excellent examples of both capture and imminent capture (e.g. Figures 8–10 and 8–11 in back of book).

Alluvial Fans and Pediments

The *alluvial fan*, a gently sloping fan-shaped surface created by the deposition of stream-carried rock particles at the mouth of a canyon or valley, is a familiar landform. Coalescent fans constitute an alluvial apron, or simply an alluvial slope or alluvial plain. Virtually identical to the fan geometrically, but not genetically, is the *pediment*, which is the product of erosion rather than deposition. The more-extensive pediment complexes are correspondingly referred to as pediment aprons or pediment plains.

It is not easy to distinguish between the fan and the pediment, on maps or in the field. It is often impossible to tell one from the other, though sometimes we can deduce which is which. We devote Chapter 6 to these important landforms, and you will note that alluvial slopes and pediments are also prominent topographic elements on many of the other maps.

FAULTS AND SCARPS

Faults and their associated scarps are both interesting and important. Of all structure-related landforms that the map interpreter encounters, the three most-commonly neglected, omitted, incorrectly identified, or misinterpreted are the fault scarp, the obsequent fault-line scarp, and the resequent fault-line scarp.

Fault Scarps

A *fault scarp* (scarp **FS,** Figure 1–4a) is simply a scarp or steep slope produced as a direct result of displacement along a fault. Many geologists assume that all scarps associated with fault traces are necessarily fault scarps. Unfortunately, this assumption can raise havoc when deciphering an area's tectonic and geomorphic history.

Fault-Line Scarps

A *fault-line scarp* (scarps **RFLS** and **OFLS,** Figure 1–5), like a fault scarp, is also a scarp or steep slope developed along the trace of a fault. Unlike the fault scarp, however, it is not the product of the initial fault displacement, but of prolonged differential erosion. For a fault-line scarp to develop, the original fault scarp must first be destroyed by erosion.

Keep in mind that the designation "fault scarp" does not require the appearance of a clean-cut, newly exposed fault plane. As fault blocks rise, they are immediately subjected to weathering and erosion. Thus, soon after its creation, the fault scarp undergoes dissection, downwasting, and backwasting by these combined processes of destruction. Therefore the front itself, as it is chewed up, also retreats as a topographic feature, back from the trace of the fault (note fault scarp **FS,**Figure 1–4b). Regardless of how severely it is dissected, eroded down, and eroded back, the fault scarp remains a fault scarp. At no time, as long as the structurally higher side remains topographically higher, does this scarp acquire the designation "fault-line scarp." *Remember:*

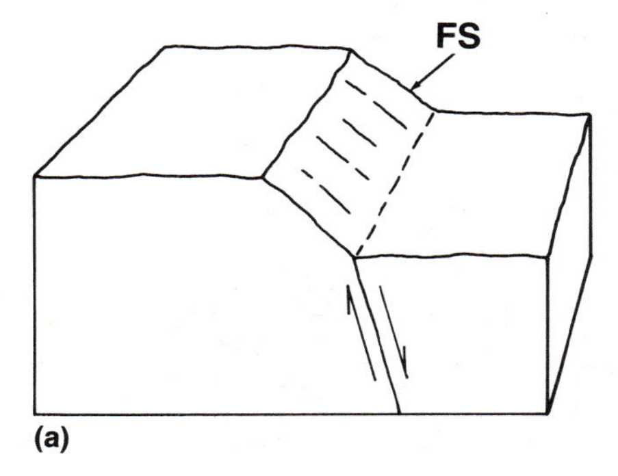

FIGURE 1–4a
Block diagram of a fault scarp.

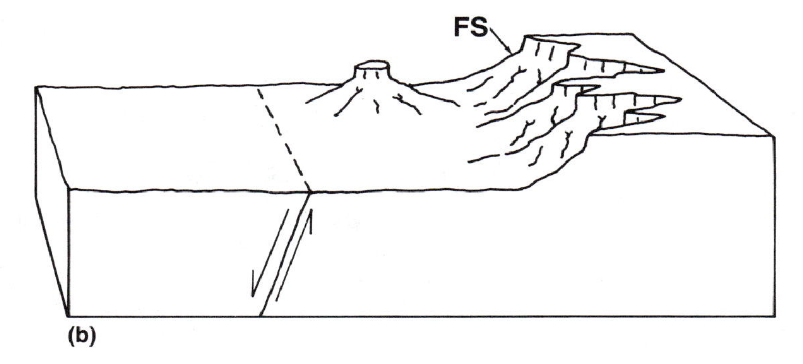

FIGURE 1–4b
Block diagram of a dissected fault scarp.

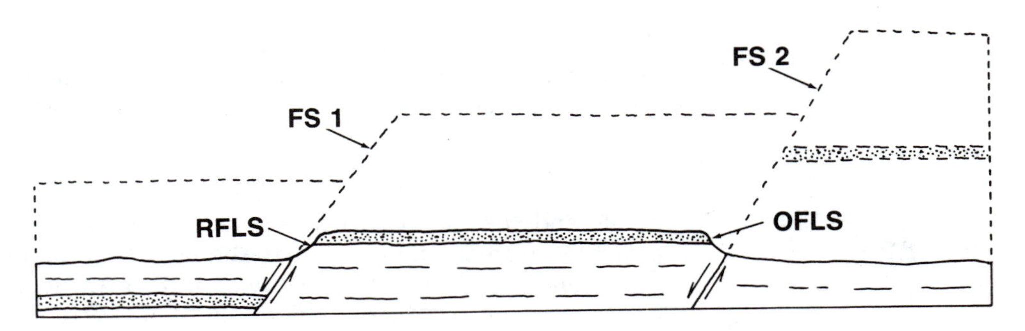

FIGURE 1–5
Cross section of resequent and obsequent fault-line scarps.

once a fault scarp, always a fault scarp, from the instant of creation to final destruction.

We emphasize that the two types of fault-line scarps, obsequent and resequent, are in no way related to obsequent and resequent streams, or obsequent and resequent slopes. Streams and slopes are so classified because of their relations to the direction of local geologic dip. Fault-line scarps, on the other hand, are classified according to which *direction* they face. The *resequent* fault-line scarp (scarp **RFLS,** Figure 1–5) faces in the *same* direction as the original fault scarp (**FS 1**). The *obsequent* fault-line scarp (scarp **OFLS,** Figure 1–5) faces in the direction *opposite* that of the original fault scarp (**FS 2**).

A Hinge Fault—or Something Else?

Confusion over the proper classification of fault-associated scarps can lead to unfortunate conclusions. For example, suppose that along one segment of a north-south-trending fault trace (**FLS 1** to **FLS 2,** Figure 1–6a) a resistant sandstone crops out west of the fault and a nonresistant shale crops out east of it. Some distance to the north, along the same fault trace, the situation is reversed, with a sandstone to the east and a shale to the west. Along this northern section, a fault-associated scarp faces west (sandstone upland overlooking shale lowland). In the south, a similar scarp faces east (sandstone upland overlooking shale lowland). It would be easy to look at this composite topography and postulate a hinge or scissor fault (Figure 1–6b), east block up in the north

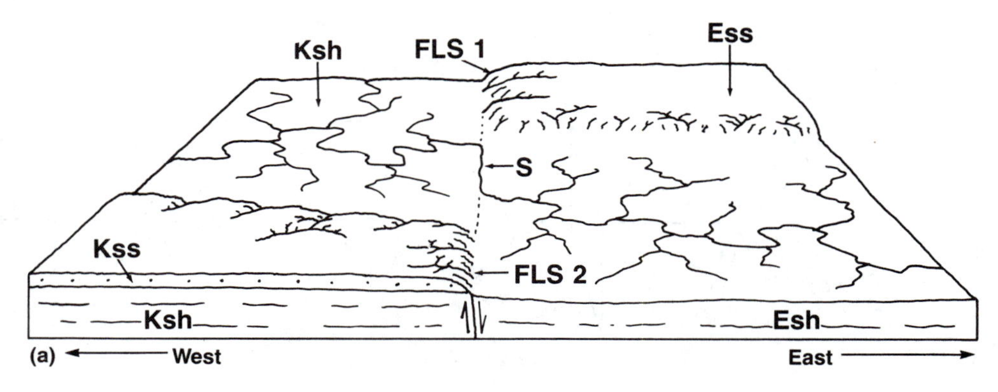

FIGURE 1–6a
Block diagram of faulted area of horizontal strata.

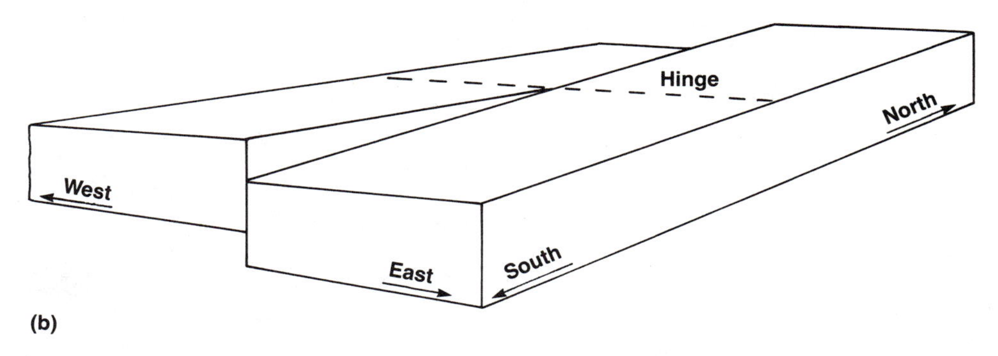

FIGURE 1–6b
Block diagram of a hinge or scissors fault.

section and down in the south section. This interpretation assumes that the scarps are fault scarps.

However, suppose that a geologic map of this area shows that the strata west of the fault are Cretaceous in age (**Kss** and **Ksh,** Figure 1–6a), whereas those east of the fault are Eocene (**Ess** and **Esh**). This would mean that, along the entire fault, the throw is up on the west (the older Cretaceous rocks faulted up against the younger Eocene rocks). It would also rule out the possibility that the west-facing scarp in the north could be a fault scarp, since any fault scarp associated with this fault would have to face east.

The west-facing scarp (**FLS 1**) must therefore be a fault-line scarp. Is it resequent or obsequent? At the time of the faulting, any fault scarp that existed must have faced to the east, since the Cretaceous is faulted up against the Eocene. Since this scarp faces to the west, it must be an obsequent fault-line scarp.

Could the east-facing scarp (**FLS 2**) be a fault scarp, even if scarp **FLS 1** is not? Along this scarp, the older Cretaceous sandstone looks down upon the adjacent younger Eocene shale lowland, so the topography does coincide with the structure as far as up-and-down is concerned. If we knew nothing about the relative ages, and could discern nothing from our maps about the lithologies, we might map this as a "possible" fault scarp. However, once we learned that the rocks in the upland are resistant and those in the lowland are not, we would have to suspect differential erosion's role in creating this scarp.

We have established scarp **FLS 1** as a fault-line scarp, and one with a long history of postfaulting erosion. We can now feel additional confidence in interpreting scarp **FLS 2** as a fault-line scarp. In this case, **FLS 2** is a resequent fault-line scarp, since it faces eastward, as did the original fault scarp.

The only indication of the fault's presence in the central part of the area, where shale lies against shale, is the anomalously linear subsequent segment of the main stream (segment **S,** Figure 1–6a).

Many of the maps in this book depict fault-associated scarps. Some of them can be identified with considerable assurance as either fault scarps or fault-line scarps. Some may be deduced as probably one or the other. Still others elude us; they could equally be one or the other.

ASYMMETRY AND DIP

Topography is often a reliable indicator of dip direction. Always pay particular attention to the *asymmetry* of a ridge, divide, or valley. The differential erosion of gently and moderately dipping strata of unlike resistances often produces

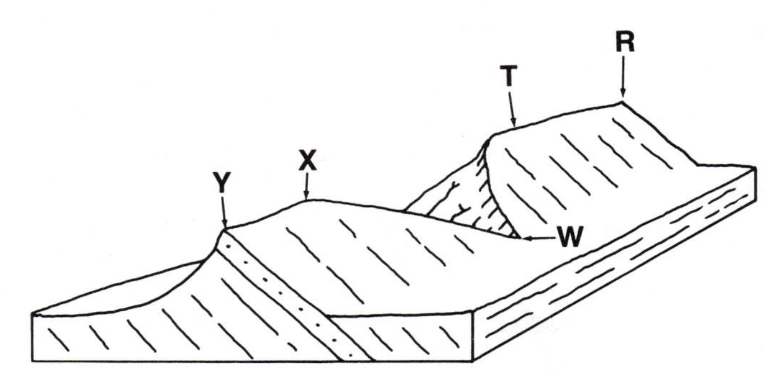

FIGURE 1–7
Block diagram of a hogback and water gap.

asymmetric ridges, divides, and valleys. In such areas the gentler slopes are usually developed on the resequent flanks.

The two asymmetric ridges in Figure 1–3 are special cases, in that their resequent slopes are also dip slopes. Many dip-revealing asymmetric ridges and divides are not so clear-cut as these. Note, by the way, that the valley separating these ridges is equally asymmetric, and hence is just as reliable an indicator of dip.

Taken as a whole, the ridge in Figures 1–7 and 1–8a is relatively symmetric, though the concavity of its obsequent slope makes the upper part of the ridge quite asymmetric. In fact, the slope contrast (concave vs. straight) between obsequent and resequent flanks is itself an asymmetry of form. Another asymmetry is the pronounced contrast in drainage. Compare the fine-textured drainage network in the nonresistant shale that underlies the caprock on the obsequent slope, to the virtual lack of dissection of the dip slope developed on the ridge-forming sandstone. (The drawings are schematic, of course. In real life the sandstone slope would also be drained by streams, but they would be far more widely spaced than those on the obsequent slope.) All of these varieties of asymmetry and obsequent/resequent contrast are illustrated in the maps that we will examine in the following chapters.

Outcrop V and Dip

Another reliable indicator of dip direction is the topographically expressed *outcrop V*. It is formed by the intersection between valley or water gap slopes and the plane of a dipping bed (Figure 1–7). Asymmetry is generally most evident when the dip ranges from extremely gentle to moderate. However, the outcrop V can be detected in many instances even when dip is steep. Suppose we were at point **R** (Figure 1–7), an outcrop of the sandstone that forms the hogback. If we wished to trace the top edge of the sandstone outcrop from point **R** to point **Y**, we would walk along the ridge crest to **T**, and then turn and follow the sandstone down to point **W**. From **W** we would walk back up to **X**, and then along the ridge to **Y**. The contours of Figure 1–8a clearly show the actual outcrop V, which is formed by outcrop lines **TW** and **WX**.

It would be possible to walk down (Figure 1–8a) from **T** to point **Z**, at the west end of the water gap, and then climb back up to point **X**. Our path, along line **TZ** and back along line **ZX**, would also form a V, but it would not be an outcrop V. The reason is that we would not be tracing the sandstone outcrop.

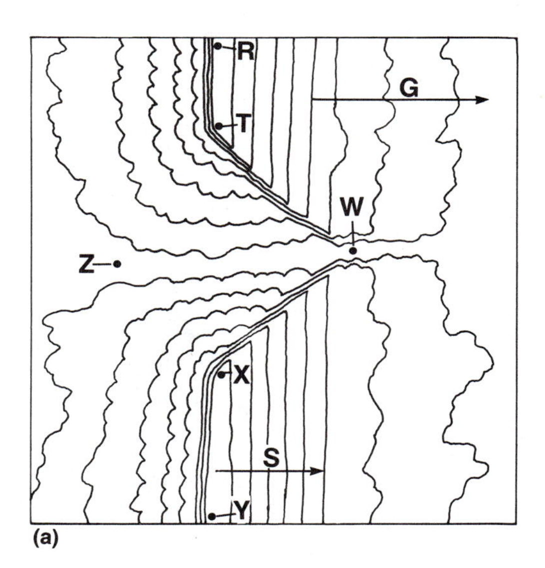

FIGURE 1–8a
Sketch map of a hogback and outcrop V.

Immediately after leaving **T** we would be walking on the underlying shale, on which we would also be climbing until we again reached the sandstone at **X**.

Figure 1–8b shows a narrow, linear, fairly symmetric ridge. Despite that symmetry, dip direction is clearly to the east (right). In this instance the outcrop V is revealed by the linearity of the contours extending along the east flank, and, in contrast, the smooth bending to the east of the contours that run along the west flank. Each sharp point is an outcrop of the steeply east-dipping resistant bed. This topography is merely an exaggerated variety of that shown

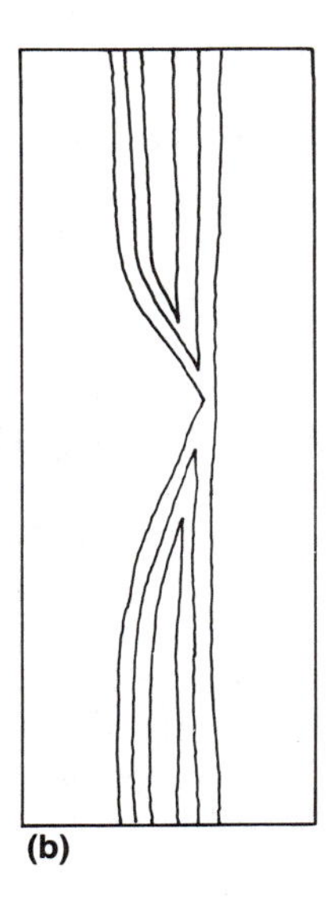

FIGURE 1–8b
Sketch map of narrow linear symmetric ridge and outcrop (steep dip).

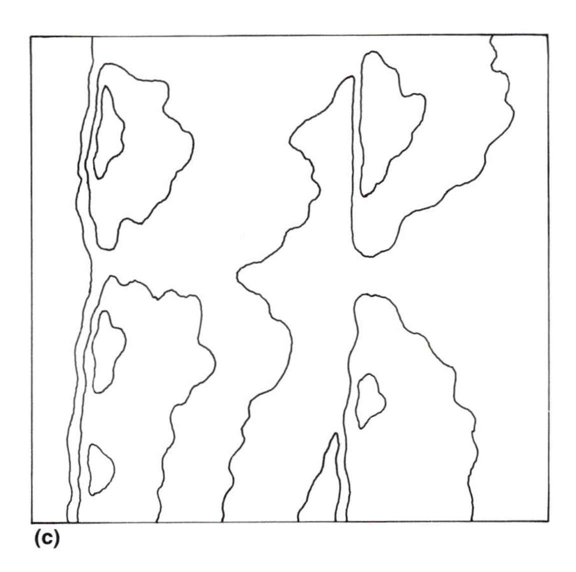

FIGURE 1–8c
Sketch map of cuestas (gentle dip).

in Figure 1–8a. It is the combination of ridge narrowness, symmetry, and linearity that indicates the steepness of the dip.

Gentle Dip, But Linear Slope—A Conundrum?

The dip in the area shown in Figure 1–8c, on the other hand, is so gentle that the outcrop V does not even exist. However, its absence is more than made up for by the pronounced asymmetry, which virtually proclaims the dip to be extremely low. These "ridges," then, are better described as cuestas.

If the dip is so gentle, why are the west-facing "obsequent" slopes so remarkably linear? At the beginning of this chapter we stressed the relation between steep bedrock structure and topographic linearity. How can we "think steep" when dealing with a cuesta, when the essential structural characteristic of a cuesta is the gentleness of its dip?

We are here confronted with the criteria for two mutually exclusive structures, steep dip and gentle dip. This will remain a problem only so long as we restrict our thinking to the bedding planes of the strata. Could not the steeply dipping structures be fault planes? Gently dipping resistant and nonresistant strata that are displaced by steeply dipping faults would be expressed topographically in this manner.

This is an excellent example of a case in which the apparent solution to one problem immediately creates another. Suppose this is an area of tilted fault blocks (it might well be) with linear, steep, west-facing fault scarps (they might well be, since they are virtually undissected as well as linear). Could not the gentle easterly slopes merely be tilted segments of a once nearly horizontal widespread erosional surface, developed on a nonlayered, dipless rock such as granite or schist?

There is another possibility. This might be an area of block-faulted lavas. Numerous areas in the southern part of the Columbia Plateau region (e.g. Figure 11–3 in back of book) resemble the area depicted in Figure 1–8c.

When confronted with an area such as this, keep in mind the possibility that the area is not really as it appears on the map. For instance, an unfortunate combination of map scale and contour interval might make these west-facing slopes appear far steeper and more linear than they actually are.

Despite our unanswered questions about Figure 1–8c, one thing is certain: asymmetry, whatever its cause, is not the product of chance alone. It must have a structural explanation. For now, it is a good idea to look upon Figure 1–8c as a block-faulted "something." Perhaps it is a surface that truncates bedrock structure. Perhaps it is a surface that developed to coincide with and reflect structure. Perhaps it is an area with no surface-expressed bedrock structure (e.g. an alluvial slope). Our next task would be to narrow the possibilities. Further study of the surrounding area might produce helpful information. Reference to literature and geological maps might help.

TOPOGRAPHY: REAL VS. DEPICTED

In most cases, the best way to verify our map interpretation is to go into the field and study the area firsthand. We say "in most cases" because at times we can see more on the map than we can see in the field.

Figures 1–9 and 1–10 tell a story that cannot be told in words. (These figures are in the back of this book; they may be removed for convenience.) Figure 1–9 is part of the Bolivar, Missouri quadrangle. Mapped in the field by R. U. Goode, it was published in 1884 at a scale of 1/125,000 and a contour interval of 50 feet. Note the small rectangular area in the southwest part of the map, particularly its three hills and its drainage to the northwest and east.

That same small rectangular area was remapped (Figure 1–10) by photogrammetric methods in 1959, and revised in 1981. Also designated the Bolivar, Missouri quadrangle, its contour interval is 10 feet, at a scale of 1/24,000.

As these two maps show so beautifully, the topography depicted on a map is just that, a *depicted* topography. The actual topography may be very similar to that depicted on a given map, or it may only vaguely resemble the depicted topography. The closeness of the depicted to the actual depends on such factors as scale, contour interval, time devoted to the mapping, the expertise of the cartographer, and the mapping instruments and techniques.

Remember: when you interpret a topographic map, you may think you are interpreting the topography, but what you are *really* interpreting is the topography as it is *depicted*.

The importance of contour interval is impressively demonstrated by Figures 13–4 and 13–4a in back of book, both 1/250,000 maps of the same Fort Smith, Arkansas/Oklahoma area. The first has a contour interval of 100 feet, and the other a contour interval of 25 meters (82 feet). We can interpret the former much better than we can interpret the latter.

We are now ready to look at some maps. As we progress, we will come across many other criteria that are critical to our interpretations. We hope the perspective afforded by this background material will aid your interpretations.

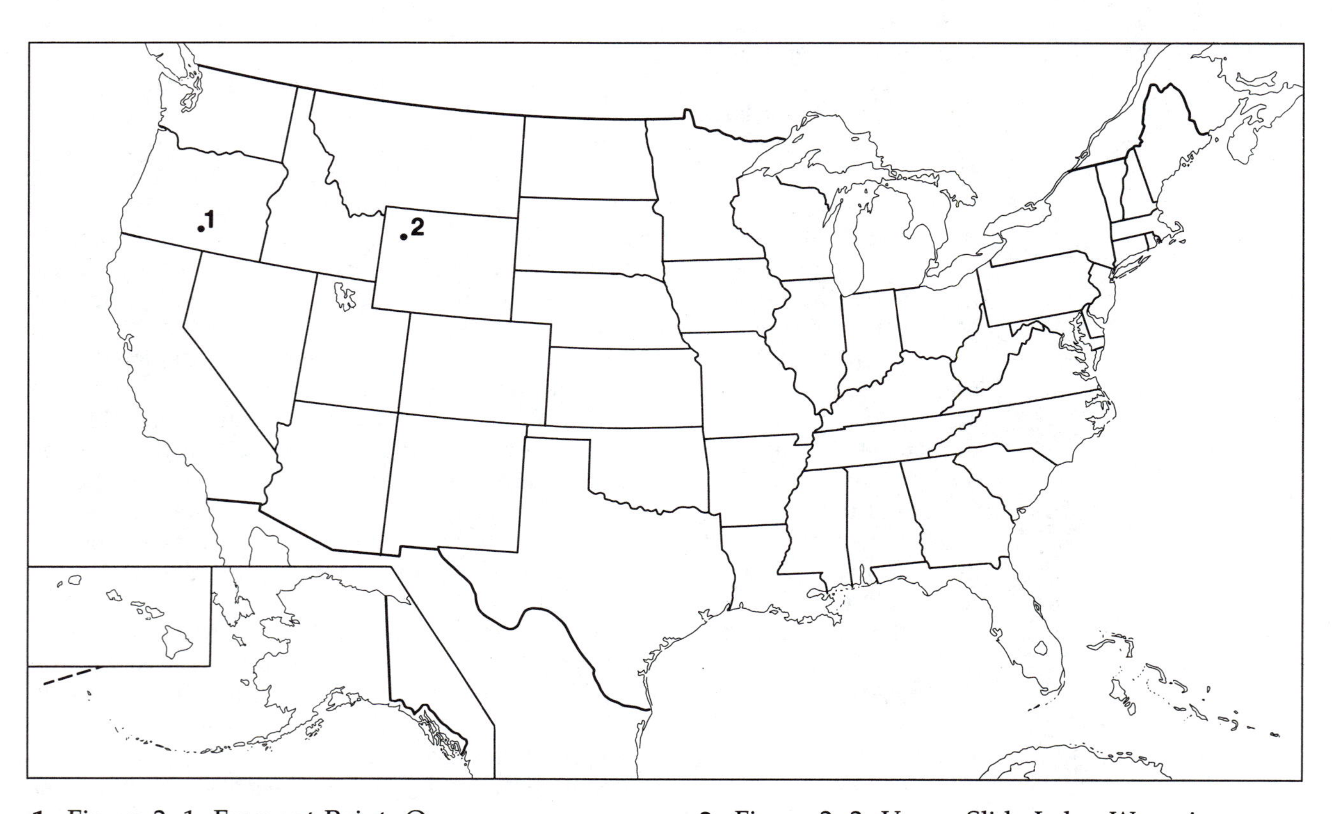

1. Figure 2–1. Fremont Point, Oregon.

2. Figure 2–2. Upper Slide Lake, Wyoming.

2

Map Reading and Map Interpretation

INTRODUCTION

This chapter affords you the opportunity to study excellent examples showing the distinction between *map reading* and *map interpretation*. Please refer to Figures 2–1 and 2–2 in the back of this book (they may be removed for convenience).

In Figure 2–1, in which north is at the top of the page, (note north arrow), you will observe and recognize several topographic features—upland, lowland, valley, slope, escarpment, and slump. Such recognition or identification of features is what we call *map reading*.

On the other hand, in Figure 2–2 (north is toward the left of the page), we do not limit ourselves to identifying objects. Instead, we ask and attempt to answer "why" questions, prompted by observing the features we "read" from the map. We are now *interpreting* this map.

You must firmly establish in your mind the difference between these two processes. Throughout the remainder of this book, we will be doing one or the other, or both.

Fremont Point, Oregon

(Figure 2–1 in back of book, C.I. 20 feet)

According to the *Geological Map of the United States,* this area in south-central Oregon consists entirely of late Tertiary-Quaternary lava flows and tuffs.

The type of ridge asymmetry illustrated in Figure 1–8c is well shown at **A** and along divide **FJ** in this map area. The area lies in the northwesternmost corner of the Basin and Range geomorphic province, a region characterized by block faulting. The main west-sloping upland rises some 3000 feet above the level of the lake.

Though only a very small area is shown here, it is relatively safe to think of this linear 3000-foot-high escarpment as fault-associated. The youth of the bedrock, combined with a virtual absence of stream-cut valleys, attests to relatively recent fault displacement. This rules out an erosional history that would produce such a clearly defined fault-line scarp. Therefore, we must interpret this as a fault scarp.

Is this really map interpretation? Or is it the interpretation of *geology* from the *topography* that is *represented* on a map? Is it not "cheating" to consult a geologic map or some other reference to learn that this area falls in the Basin and Range Province? No! No more than it is cheating when a doctor, called to investigate stomach cramps, asks the patient about the most-recent meal or about any strenuous exercise within the past hour.

So, is this really map interpretation? Yes, as we have defined it. It is map *reading* to observe the gentle west slope leading down from divide **AEFJ,** or the overall steep drop down from the divide to the lake, or the linearity of the scarp and its lack of dissection. It is *interpretation* to tie all these observations together with the reference information, and state how that scarp probably formed.

The outstanding topographic feature on this map, however, is not the entire scarp. It is the glob-like mass that is bordered on the west by arcuate gouge **BEF** in the scarp face, on the north by **BCD,** on the south by **FIK,** and extending eastward nearly to the lake. This is a gigantic area of slump.

Some anonymous wit claimed that the hippopotamus is so named because the first human to catch sight of one exclaimed, "That is a hippopotamus!" When asked how he knew what it was, the person replied, "Because it looks like a hippopotamus!"

This geomorphic monster, with its steep-walled arcuate scar **BEF,** its humps **G,** its depressions **H,** and its overall billowing, saggy-jowled appearance, we simply identify (not interpret!) as a large slump, because it looks like a large slump.

What else could it be?

Upper Slide Lake, Wyoming

(Figure 2–2 in back of book, C.I. 40 feet)

Three completely different types of landforms dominate this area. The most clearly defined, and easiest to identify, is the large floodplain **1** of the only major stream, **Q**. The only items of interest in the floodplain area are the two lakes, which do not appear to be man-made. They are obviously not oxbows. We will investigate their origin.

The second area type, **2,** is represented by two large, dissected uplands that rise north and south of the floodplain. These areas have generally smooth slopes (evenly spaced contours); some steep, some less steep. The slopes tend to intersect along clearly defined lines (e.g. dashed line **U**).

Finally, there are several areas of irregular, hummocky terrain, the largest of which is area **3** in the west. Area **3** displays numerous crenulated contours, mounds (**J**), and depressions (**C**).

Areas **2** are certainly areas of bedrock. In the **2** area south of the river, the prevailing dip is apparently to the northeast, toward the valley floor. This is a good place to pause and look back at the discussion of outcrop V's (Figure 1–7). As traditionally perceived, the outcrop V is a V-shaped outcrop produced by the intersection of a dipping bed and the walls of a valley. These are veritable V's, since their points are directed downward.

There are other outcrop V's that are equally as distinctive, and that are equally as reliable as indicators of dip. However, there does not seem to exist an appropriate term for them. They are not called outcrop V's for the very good reason that they are the V-shaped outcrops of dipping rocks that cross *divides* rather than valleys. These are upside-down V's like the two in Figure 2–3 (**ABC** and **CDE**). A normal outcrop V (**BCD**) also occurs between them.

Note that each upside-down V in Figure 2–3 forms what bears a remarkable resemblance to a triangular facet (Figure 2–4), and even more to a flatiron (Figure 2–5), though these are each completely different in origin and structure

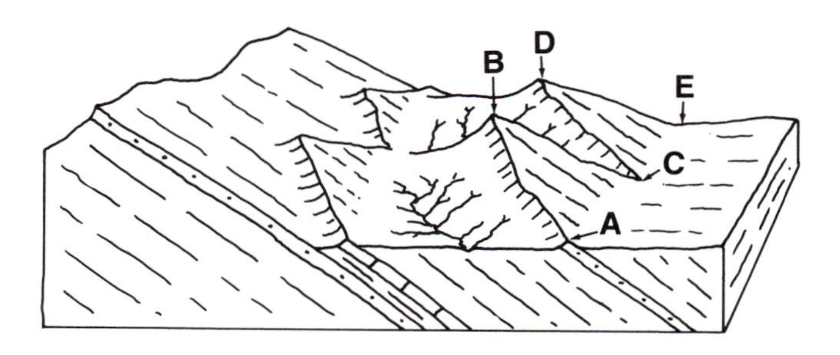

FIGURE 2–3
Block diagram of outcrop V **BCD** and two upside-down outcrop V's **ABC** and **CDE**.

from the features we are discussing. The triangular facet results from the abrupt cutting off of a generally prism-shaped divide by something like coastal erosion, a fault, or a transversely moving glacier. The flatiron is formed by the erosional sculpturing of a resistant layered rock that lies nonconformably upon a crystalline basement.

When we look back at the map (Figure 2–2), we see several features that resemble the upside-down V's depicted in Figure 2–3. For example, both triangular features **P** and **W** seem to reflect northeast dip. Similar dip could also explain the configuration of asymmetric ridge **D,** in the north. The northwest strike of steeply dipping beds may be responsible for small ridge **F,** just to the northeast of lake **b**.

Lines of abrupt contour bends, designated as dashed lines **G** and **H,** appear to combine to form an upside-down outcrop V of steeply dipping beds. They have a dip direction of about N80°E. If this is correct, surface **I** may well be a dip slope. Such strike and dip would help explain the pronounced asymmetry of the slopes that flank area **3,** namely slope **I** (resequent) and the much-steeper slope **K** (obsequent) to the east of area **3**. A similar asymmetry exists between slopes **L** and **N,** which overlook a comparable undulating lowland mass **M,** which is the upper part of slide **E**. The line of sharp contour bends along line **U** may also mark the outcrop trace of another resistant bed.

We must pause to make a most important point. We have repeatedly used the terms "may be," "resemble," and "appear to." We could have said "is," "are," "indicate," and "denote," with little fear of being challenged—after all, who is supposed to be the authority? Yet, we chose less-certain terminology for the important reason that we do not know for certain the meaning of these topographic features, relations, and contrasts. Uncertainty is the constant companion of the interpreter of the topographic map (or any map), or the aerial photograph, the radar image, or the satellite image.

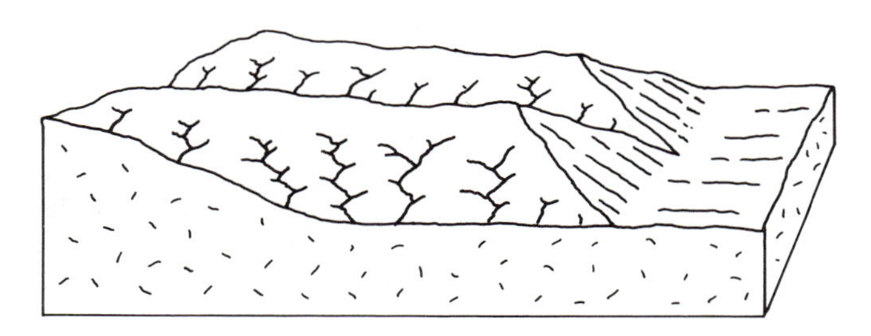

FIGURE 2–4
Block diagram of triangular facets.

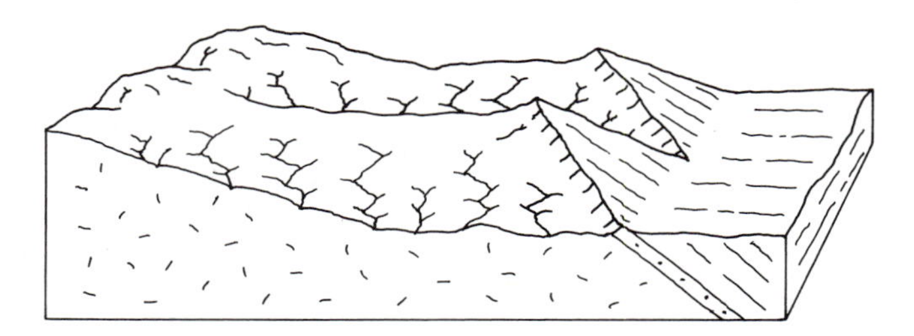

FIGURE 2–5
Block diagram of flatirons.

Anyone attempting to interpret geology from its *representation* on a sheet of paper or a film is dealing with *uncertainty*. Our job is to do all we can to bring the unknown and doubtful forward into the most, or more, probable. Occasionally we can attain the certain, but if you are going to interpret, now is the time to accept the certainty that at least part of the time you are going to be wrong. And, be not too surprised to learn that some of our interpretations may be subject to reconsideration!

Area **3** is obviously a large landslide or rockslide that broke away from the headward part of the broad valley (around **T**) and crashed down-valley onto the floodplain (note the crenulations in the 7400-foot contour south of **D**). Slide material can be identified as far south as point **S**. Our guess is that the source area for this particular mass of debris (around **S**) is mostly the east-facing slope **R,** immediately to the west. Though it is the largest landslide in this map area, **3** is far from the only one. In fact this entire area appears to be unstable.

Although the existence of lake **a** (east of the foot of slide **3**) is apparently explained, the question of why there are two lakes is not so easily disposed of. It would be convenient if we could attribute lake **b** to another slide but since no slide is available at the right place, we must look in other directions. One possibility is that the lakes were originally one large lake. With the passage of time, the river built a delta out into the lake, a delta that eventually extended across to the north shore, sealing off the upstream part (lake **b**) from the downstream part (lake **a**).

Finally, consider the other slides. How many are there? In addition to slide **3** and slide **S,** slide **E** has spread northward across the valley floor, but not quite to the opposite side. **B** resembles a separate small slide, and **A** might be another. There seems to be a small slide at **V,** in the south.

There are two features that we believe to be much-older slides. Their many diagnostic irregularities have been modified, smoothed off, and rounded by weathering and erosion, but they still have their "slide-looking" bulges and scars. We have labeled them **O** and **X**.

The river shown is the Gros Ventre River, which continues to the west into Jackson Hole, where it joins the Snake River. Slide **3** is named Upper Slide and lakes **a** and **b** are collectively called Upper Slide Lake. The famous Kelly or Gros Ventre Slide lies but a few miles to the west. If you have an interest in slides, it would pay you to research the literature and learn more about the geology of this area. Study its structure and stratigraphy, and the fascinating ways in which both structure and lithology are involved in its instability.

1. Figure 3–1. Palmyra, New York.
2. Figure 3–2. Hyannis, Massachusetts.
3. Figure 3–3. Greenfield, Massachusetts.
4. Figure 3–4. Pound Ridge, Norwalk North, Norwalk South, and Stamford, Connecticut.
5. Figure 3–5. Marmot Mountain, Montana.

3

Glaciation

INTRODUCTION

Moraines, eskers, drumlins, tarns. Horns, cols, kames, and kettles. Fiords, arêtes, roches moutonnées. Freezing, melting, plucking, and gouging.

The list seems endless, and endless would be the maps required to illustrate every landform resulting from the transformation of snow into ice, the creation of glaciers, and the erosion, transportation, and deposition that glaciers perform.

When you study maps of stream-cut topography, understand that, in most parts of the world, streams continue to play an important role. They are still "doing their thing." Thus we can see in the topography what they are now doing, interpret what they have done in the past, and anticipate what they probably will do.

However, we cannot do these things with glaciers, unless we travel to very remote high latitudes and high-altitude places where glaciers are still active. Instead, in almost every area where we might encounter the works of glacial ice, whether in alpine troughs or in regions once-capped by magnificent ice sheets, we can only deal with fossil evidence. The glaciers themselves are long gone. We can only try to figure out what they and their meltwaters did, what they removed, and what they deposited. We must determine whether the glaciers actually created the present topography, or merely modified some earlier subaerial topography.

We have assembled but five maps in this chapter, though we have included as many other glaciated areas elsewhere in the book. Instead of illustrating a plethora of glacial landforms, we elected to demonstrate a few of the mystery-solving techniques with which we approach the many fascinating glaciated-area problems. In addition to this chapter's five maps, we direct your attention to several others (all are in the back of this book)—Figures 4–1 (Pleistocene shorelines), 4–7 (partly submerged rugged glaciated coast), 8–10 (stream capture in glaciated Catskill Mountains), 10–6 (periglacial area), and 13–15/16 (morainal belt in New Jersey–New York area).

Palmyra, New York
(Figure 3–1 in back of book, C.I. 20 feet)

Though it might resemble a swarm of salmon at spawning time, frantically fighting their way up a tumultuous rapids, this is part of a drumlin area in western New York. The drumlins' parallelism, their asymmetry (steep stoss, gentle lee), and their streamlined outlines and longitudinal profiles combine to identify them as drumlins, and to indicate ice movement from north to south. Yet, it is not the drumlins per se that make this area fascinating.

The perplexing features in this area are the flat-floored, east-west lowland belts that branch and rejoin their way across the map area in a most disturbing way. What are they? How and when did they come into existence? Are they stream valleys, which they certainly seem to be at first glance? And if they are, why do they branch? And why do they extend at right angles across the grain of the drumlins?

Note the clean way the valleys cut across the drumlins (e.g. at **B, E,** and **J**), truncating them as neatly as a bread knife slices a loaf of crusty French bread. Also note the unusual uniformity in the width of these valleys. Should we not more correctly think of these "valleys" as channels? Large arcuate lake **C** is certainly a strong argument against picturing the "valleys" as having originally accommodated narrow ribbons of water similar to those that now meander across their floors. In addition, some valley segments, such as **A** and **D,** are devoid of streams.

Perhaps the answer more likely lies in the events of the Pleistocene. This area was covered several times by thick ice sheets, and the drumlins are testimonials to the last of these. During the ablation phase of the last sheet, incredible amounts of meltwater were released. We can visualize the water escaping across this area, first feeling its way around and between and among the drumlins. As the water's courses became established, it flushed down these sluiceways.

Since this is one of the first maps we are studying, let us ask: what have we been doing so far—reading or interpretation? We have been doing both.

Recognizing these upside-down-teaspoon hills as drumlins is hardly high-caliber interpretation, though it is a step beyond reading. Reading, of course, would include noting the drumlins' size, shape, and orientation. Noting the size, form, distribution, and orientation of the east-west belts is also basic map reading. Classifying them as meltwater channels, however, is interpretation.

Finally, consider the streams that currently populate this area. How should we classify them? For example, note streams **F, G, H,** and **I,** in the southwest. Streams **F, G,** and **H** are all parallel and quite linear, whereas stream **I** is much more sinuous. Are the first three streams of one type, and the fourth of another? Before we can answer this question, we must learn something about their origins. What do we have to work with?

With the exception of the outwash channels, the entire topography of the map area was created by the overriding glacial ice, was it not? Thus, it was not just the drumlins that were left behind by the glaciers, but the lowlands separating them as well. These four streams, and all the others that lie beyond the outwash channels, must simply have developed on this undulating postglacial terrain. All of them must therefore be consequent streams.

The numerous linear streams are linear, not because they developed (as subsequent streams) along nonresistant bedrock bands or fractures. They are linear because the lowlands on which they formed were, and are, linear lowlands separating the parallel drumlins. Stream **I** is not linear because the lowland it developed on was not linear. It is more irregular, more noncontrolled-looking, simply because it lies among several drumlins, and not between two parallel drumlins.

Even the large, interweaving outwash channels are *consequent*. They developed between and among the drumlins, making their way from higher ground to lower. They also bear no relation to underlying structure or bedrock lithology (which we will not describe, since such details are not pertinent to our interpretation).

Hyannis, Massachusetts

(Figure 3–2 in back of book, C.I. 10 feet)

Two of the most "different" types of terrain are karst areas and moraines. Unfortunately, they look very much alike. Fortunately, they rarely occur in the same area.

There is no such thing as a truly "typical" anything, at least in topography. What would be a typical valley, a typical divide, or a typical mountain? What about a typical moraine?

Nevertheless, this is an excellent map to illustrate the topography of a moraine. In this case, it consists of a belt of undulating, hummocky terrain, replete with hillocks and closed depressions. Had we indicated that this area was in central Kentucky, we might have been able to use this same map to illustrate karst!

So, what can we interpret? We cannot do much interpretation here, but we can make some important observations. Hyannis is situated in the biceps-part of the muscle-flexing Cape Cod arm which extends into the Atlantic Ocean from southeastern Massachusetts. In this area, known to have undergone multiple glaciations, it would be unreasonable to "interpret" terrain such as this as resulting from the solution of limestone. What we have here, then, is a belt of land which we can designate confidently as a moraine. The belt is the well-defined hill/depression zone that extends east-west across the northern part of the map area.

The swampy lowland to the north is the area that had to be occupied by the glacier while it was dropping the material that formed the moraine. The meandering channels, particularly the assemblage at **A,** look like, and are, tidal meanders (Cape Cod Bay lies just to the north).

The several northern lakes (lakes **B**) lie in kettles. If these depressions are indeed kettles, and if the morainal belt is indeed morainal, how can we explain the large lakes **C** to the south of the moraine? The broad, low surface **D** must be an outwash plain, built of the finer debris released by the melting of the glacier and carried away from the moraine by outwash streams.

The depressions occupied by lakes **C** are apparently also kettles, local areas protected by ice blocks while outwash streams were depositing the material that formed the broad outwash plain **D**. Do they indicate that the glacier must have advanced to at least the south edge of the map area, and not just to the latitude of the moraine?

Though lakes **C** must occupy depressions, the nearby contours **EG** and **H** are not hachured because they extend, as normal unclosed contours, beyond the confines of the map area. Stream **F** indicates external drainage for the low area (defined by contour **EG**), which is occupied by the several lakes **C**.

Greenfield, Massachusetts

(Figure 3–3 in back of book, C.I. 10 feet)

This is part of the Connecticut River Valley, the alluvial terraces of which William Morris Davis so magnificently described and analyzed in his *Geographical Essays*. It is ideal for demonstrating the need to exercise caution in explaining the obvious.

The two major terraces of the area are clearly defined. Surface **DIL** is the higher terrace, and hence the older. Terrace **AHR** is lower and younger.

Note that terrace **DIL** is remarkably well preserved, being only slightly dissected along its western edge (**BJS**). Contours **E** and **K** show that terrace **DIL** slopes very gently to the south-southwest. This may be the sole clue needed to suggest why the eastern edge of the terrace (**FMW**) is virtually undissected, in contrast to the dissection along **BJS**. All drainage of surface **DIL,** channeled or not, must be toward the south-southwest. Thus, only the rainfall and snowmelt that occur directly on the eastern border slope **FMW** are available to attack that border slope, but the southwest border slope **BJS** must accommodate runoff from the entire terrace area. The porosity and permeability of the alluvial material that constitutes terrace **DIL** must have aided in protecting this large surface.

The several terrace levels attest to the synchronized downcutting and lateral shifting of the Connecticut River. Note that terrace surface **AHR** continues southward as the surface labeled **V**. Within the map area, hill **Q,** which lies between the river and surface **V,** rises at least 200 feet above **V**. Hill **Y** also rises at least 300 feet above **V**. It is obvious that the river's lateral corrasion created surface **AHR** as far south as point **T**. But it is not apparent what stream is responsible for surface **V**. If the Connecticut River did cut surface **V,** how could the river have then shifted westward to its present position to the *west* of hill **Q**?

To the north and west of hill **Q,** stream **UP** appears to be adjusted to its master stream, the Connecticut River. At the time surface **AHR** was being cut, stream **UP** was probably also adjusted to the Connecticut. Could surface **V** somehow be related in origin to stream **UP,** rather than to the Connecticut? It might be easier to imagine the diversion of this smaller tributary to its present course around the north flank of hill **Q,** than to envision shifting the master stream from area **V** to its present course.

However, stream **UP** originates in lake **N** and is really not a very significant stream. It may have played a minor role in shaping that part of the **AHRV** surface that lies to the east of point **T,** but if some stream in the immediate area did contribute to the building of surface **V,** we must not overlook the possibility that it was tributary **X**. The fact is that the answer to this problem does not lie within the confines of this map. We would have to consult areas to the east and south before we could further speculate.

Is there any way to determine when and how surface **DIL** was formed? Did the Connecticut River deposit sediments to that level before it began to cut the terraces? Perhaps depressions **G, O** (occupied by lake **O**), and **N** (occupied by lake **N**) can help us with these questions. This is not a limestone area, and the depressions certainly cannot be the products of stream or ice erosion. Streams do not cut such undrained depressions. Glaciers do not erode holes in uncemented sediments and leave the remaining surfaces almost perfectly preserved (and with gentle, down-valley slopes). Finally, it would be hard to visualize this river—or any other—creating surface **DIL** by lateral corrasion, without notice of these adjacent depressions. These depressions have all the characteristics of kettles.

If we assume that depressions **G, O,** and **N** are kettles, we may propose that surface **DIL** is the top of outwash deposits. These deposits were dropped in the lowland surrounding a group of stagnant ice blocks. Surface **DIL** may thus be assigned an ablation-stage age (during final glacial retreat).

In conclusion, note the contrast between the smoother, gentler, more-streamlined northern slope of hill **C** and that hill's steep, irregular, almost jagged south slope. Such asymmetry, which is opposite to that of a drumlin, is common to bedrock hills that have been overridden. In this case, they were overriden from north to south by glacial ice: roches moutonnées.

Pound Ridge, Norwalk North, Norwalk South, and Stamford, Connecticut

(Figure 3–4 in back of book, C.I. 10 feet)

This area, which lies in southwestern Connecticut, was overridden by Pleistocene ice sheets that came down from the north. It is a region of extremely complex geology. Though it is impossible for a geologist to study this map "cold" and identify the rocks and structures straightaway, it is still possible to draw some meaningful generalizations.

The entire map area exhibits a pronounced linear north-south grain. The landforms appear so streamlined that it is safe to attribute at least some of the grain to glaciation. However, there are certain large belts that stand considerably higher than others, and features of such magnitude cannot be written off as glacial in origin.

The main lowland can be traced north-south across the entire area, from **A** to **D** to **G**. In the south it is joined, from the west, by another low belt, **CF**. It also seems to branch to the southeast along low trend **HI**. Lowland **BE**, in the northeast, is quite pronounced as well. These lowland belts very probably have been carved from less-resistant rocks by normal subaerial differential erosion. However, the several upland expanses, being underlain by more-resistant bedrock, have better withstood the attack of weathering and erosional processes. Additional differential scouring may have been accomplished by the overriding glaciers.

We should expect ground moraine throughout this area. It may be significant that the topography in the lowlands is more irregular, hummocky, and unpatterned than it is throughout the upland tracts. In the uplands, parallelism is far more pronounced, and an overall smoothed-off surface is suggested by the sweeping roundness of the contour bends. These differences might be explained by the thicker, more-varied glacial deposits in the lowlands than in the higher areas.

An abundant feature throughout this region is the ubiquitous stone fence, although you cannot see them on the map. Many of the boulders used in making these fences are striated. Their surfaces were scratched and gouged as they were dragged across exposed bedrock by the overriding ice sheets.

Can you imagine what this area would look like if it were partially submerged? The coastal area depicted in Figure 4–7 will give you a good idea.

Marmot Mountain, Montana

(Figure 3–5 in back of book, C.I. 40 feet)

The V-shaped contour bends that cross south-flowing stream **VU** indicate that the valley, although steep, is a stream-cut valley. In contrast, the arcuate bends in the contour lines that define the head **C** of valley **CD** clearly depict a rounded, scooped-out amphitheater, a landform readily identified as a cirque. Several other cirques are located to the northeast of divide **AW**: cirques **B**, **M**, and **O**. Alcove **P** also appears to be a cirque, though its contours are not nearly as well rounded and smooth. Note that a tarn **N** occupies a depression in cirque **O**.

Downstream from cirque **O**, the trough becomes decreasingly U-shaped, gradually acquiring the V profile (at **E**) of a nonglaciated valley. Note the similar downstream transition from cirque **P** to valley-segment **Q**.

The northeast flank of ridge **FS** and lowland **JL** to the northeast were clearly scoured out by a large glacial mass. Four glacial lakes (**G, H, I,** and **K**) occupy depressions in this area.

It is interesting that this area was subjected to glaciation and that not all of the area was covered with ice. But the *most* interesting fact is that south- and southwest-facing slopes almost completely escaped glaciation. Maximum glaciation took place on the northeast-facing slopes. The smoothly curved contours just south of point **F** suggest some south-facing glaciation. However, the entire slope below line **RS** is scored by stream-cut V's. In fact, if you draw a line **RSTW**, you will delineate the north edge of a large zone of nonglaciation.

The reason for such asymmetric glaciation is well known. South- and southwest-facing slopes receive more direct insolation than north- and northeast-facing slopes. Thus they are better heated during the warmer hours of the day (after noon).

1. Figure 4–1. Larimore and Emerado, North Dakota.
2. Figure 4–2. Redondo Beach, California.
3. Figure 4–3. Wilton, Virginia.
4. Figure 4–4. Rehoboth, Delaware.
5. Figure 4–5. Mathews, Virginia.
6. Figure 4–6. Honokane, Hawaii.
7. Figure 4–7. Pemaquid Point, Maine.

4

Shorelines

INTRODUCTION

This chapter exemplifies our decision not to give this book a show-and-tell format. You will note that we have not included any examples of deltas, though there are many beautiful maps of the Mississippi River delta that we might have selected. There are no examples of partially submerged dissected folds, and no examples of partially submerged drumlins, such as those in Boston Harbor. There are no examples of the virtually limitless variety of coastal features.

Instead, the maps we did select demonstrate many possible *why* questions that can be asked about coastal areas. They challenge us to look back in time for the most reasonable and logical sequences of events to answer those questions.

Larimore and Emerado, North Dakota

(Figure 4–1 in back of book, C.I. 10 feet)

Although the contour interval is only 10 feet, throughout most of this map area the contours are relatively widely spaced. The only exceptions are along the valleys of the three main streams, **M, D,** and **S,** where the valley sides are so steep that the contours are tightly crammed together on the original maps from which this was traced. We have omitted most of the contours from this map because we do not need them, and because they are so closely spaced that they would not print clearly. Note that we have also omitted all contour numbers.

This is a remarkably undissected area, having but the three streams (**M, D,** and **S**) of any consequence. If these streams and their valleys could be removed (i.e. filled in), the topography would consist almost entirely of a variable, vaguely stair-step, gentle slope to the northeast. This is depicted in the highly schematic Figure 4–1a, a sketch of part of the area viewed from the southeast.

How can we state that the prevailing slope is toward the northeast, without knowing the contour numbers? Why could not contour **P** be higher than contour **O,** rather than lower? Why could not the slope be in the direction of the solid arrow (near center of map), rather than in the direction of the dashed arrow?

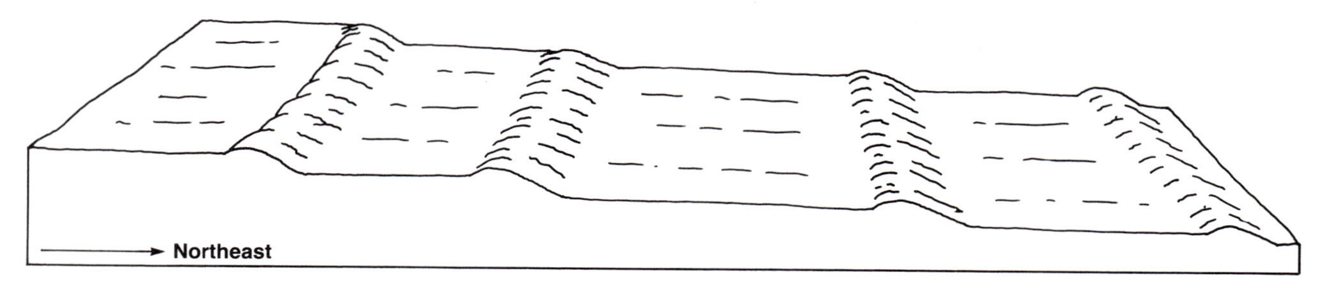

FIGURE 4–1a
Highly schematic block diagram showing topographic characteristics of area depicted in Figure 4–1, viewed from southeast.

If stream **S** had not cut its valley into the gently sloping surface, we might have some difficulty telling the direction of the slope without looking at the numbers. But contour segments **Q** and **R** provide the answer. They run along the side slope of the valley cut by stream **S**. So, we know that contour **R** must overlook stream **S,** and therefore contour **Q** must overlook contour **R**. Since **Q** and **R** are segments of contours **O** and **P,** respectively, the gentle slope must be toward the northeast (dashed arrow).

Note that the **Q-R** relation is duplicated at many places along the sides of stream valleys **M** and **S** (e.g. at **N** and **U**). The relation between stream **MS** and this regional slope tells us that the stream flows to the east. This direction of flow is confirmed by contour V's such as the one at **J**.

Throughout the eastern two-thirds of the area (i.e. east of line **BY**), the only closed contours are those such as **C, E,** and **V**. These identify linear "ridges," which trend parallel to the regional contour trend. A similar "ridge" is shown at **W,** but its contour is not quite closed. As shown in Figure 4–1a, these ridges tend to lie along the outer (eastern) edges of the flat stair-step surfaces.

The slopes in the area to the west of stream **D** are not as smooth and uniform as those to the east. Still, we can detect even in that western area a general topographic parallelism (e.g. contours **X** and **L**) and parallel low, linear ridges (e.g. **K** and **A**).

During late Pleistocene time, a 50-mile wide arm of gigantic Lake Agassiz extended southward from Canada, along the entire North Dakota–Minnesota state line. The area shown on this map is but a small fragment of the lacustrine plain that now characterizes that entire region. Such lacustrine areas, replete with their numerous strand (beach) lines, are obvious. Even if we did not know the map area's location and the identity of the Pleistocene lake, we would be able to recognize the lacustrine topography, just as we can recognize landform types such as barchans and cinder cones without knowing their locations.

What are the several linear low "ridges"? They are simply former beach ridges. Then, into what classification do we assign the various streams? This is more challenging.

The streams of the entire map area are all oriented southwest-northeast or northwest-southeast. All of the former flow to the northeast (e.g. streams **M, S,** and **Z**), down the regional slope. Since in the present era this is most assuredly an original slope, such streams must be consequent streams.

The second group of streams, all of which are tributaries to the northeast-flowing consequents, flow parallel to the beach ridges. Stream **D** is the largest. Others include streams **T, F,** and **a**. Note that they not only flow parallel to the beach ridges, but lie behind (to the west of) them. Instead of occupying valleys created by their own erosion, these streams flow in linear lowlands that were formed by the building up of the beach ridges. Thus, they are streams created by the establishment of drainage in these "original" linear low belts. They too are consequent streams. In fact, all the streams of any importance appear to be consequents.

In Chapter 8 we will consider stream capture. You may wish to cover that material before deciding whether stream **G** (in the northeast) is in danger of being captured by stream **I**, at point **H**.

Redondo Beach, California
(Figure 4–2 in back of book, C.I. 25 feet)

Like the topography of Figure 4–1 (Larimore and Emerado, North Dakota, in back of book), this area consists of a series of stair-step rises, or terraces. The relief of each step here is much greater, and so is the total relief—about 1250 feet, compared to about 250 feet in Figure 4–1. In addition, the topography here is considerably less linear, more irregular, and much more completely dissected. Finally, it lacks the beach ridges that characterize the Dakota area.

This is a coastal area having a history of periodic crustal uplift. This strongly suggests that the several terraces result from a corresponding number of cycles of wave planation and shoreline retreat. The present coastal cliffs may thus be thought of as the next terrace face, which is awaiting uplift above sea level. Surfaces **A, B, G,** and **D** are well-defined terraces. The upland surface **E** may simply be the top of a rise that the waters never reached. On the other hand, it may be the remnant of a once-extensive wave-cut surface that antedated the cutting of terrace **D**.

This is another opportune point to ask whether we have interpreted this map (its topography) or merely recognized the features. We have recognized the features as uplifted terraces. But had the area's location been unknown, had the present shoreline not been shown, and had the ocean not been identified, then interpretation would have been required.

After all, stair-step topography such as this is found in innumerable dissected plains and plateau areas throughout the world—e.g. the eastern part of the Figure 10–1 area (Cambridge, Kansas, in back of book). All that is needed is a sequence of horizontal-to-nearly-horizontal strata consisting of alternating more-and-less-resistant layers. Differential erosion of such a sequence produces just such a topography.

Having identified these terraces as uplifted wave-cut surfaces, we may now classify the streams, such as **C** and **F,** which descend the slopes and empty into the ocean. Since these are the first streams to have formed on these slopes (these *initial* slopes) they must all be consequent streams.

South-facing slope **HK** drops all the way from the level of surface **D** down to level **J** of surface **A**. Why is this slope so steep and so uniform? There are virtually no terrace remnants in this southern area (with the exception of surface **L,** which is a small residual fragment of terrace level **B**). Where are the terraces in this area? Their absence suggests that the attack by waves and currents along this stretch of coast was more intensive than it was to the northwest. As a consequence, each uplifted wave-cut bench in this steep area was destroyed prior to the uplifting of the next.

If events continue into the future as they seem to have occurred in the past, all or much of terrace **J** may be destroyed prior to the next uplift. However, some of terrace **A** on the west may remain unconsumed at that time.

Stream **I** and its companions are also consequent streams.

Wilton, Virginia
(Figure 4–3 in back of book, C.I. 10 feet)

Terraces constitute important elements of the topography in this map area. They are more fragmented than the terraces in Figure 4–1 (Larimore and Emerado, North Dakota) and Figure 4–2 (Redondo Beach, California) (both in back of book).

The sinuous blank area that crosses the map diagonally from west to east is an arm of the Atlantic Ocean. We have designated it simply as "embayment." The most-pronounced terrace fronts (scarps) are **D** and **B**. Figure 4–3a is a sketch of the area viewed from the southeast. Note in this figure that scarps **D** and **B** face each other across the embayment.

Several other features associated with the embayment merit close attention. For example, the embayment broadens from west to east, toward the nearby Atlantic. At many places the embayment's shorelines are smoothly arcuate. The best of these arcuate stretches is **JL;** others are at **A, E,** and **G.** Around these arcuate shoreline bends, the slopes that drop to water level are steep. They are like the undercut slopes around the outside on a stream's meander bends. At other places, such as **F** and **I,** the slopes are much gentler. They are like the bluffs away from which such a stream has laterally migrated.

These features help us identify the embayment as the drowned valley of a meandering stream. Island **K,** situated in the center of swing **JL,** must have been a modest topographic high inside a large meander. Possibly this meander had been cut off, to the south of **K,** prior to submergence.

This map raises an excellent question: what process or processes created the terrace levels and the embayment-facing terrace scarps? If this area were to be uplifted, or the ocean lowered, surfaces **H** and **M** would be transformed into a new, lower terrace level overlooking the stream that would reform along the floor of the new valley (the present embayment). The new valley would in turn be bordered by new terrace faces, which are now the embayment's bluffs and slopes (e.g. **A, F,** and **G**). However, if we turn the clock back, instead of imagining what might happen in the future, we can visualize a broad valley, its floor at level **HM,** bordered by bluffs **B** and **D.**

Whereas Figure 4–2 (Redondo Beach, in back of book) depicts a coastal area, this is a coastal *plain* area. It has already undergone both emergence and partial submergence. Upland surface **C,** in the northeast, may not be the original land surface created when this area rose above sea level. The map has no way of telling us.

Whatever the origin of surface **C,** we can visualize, immediately after uplift, the creation of an east-flowing consequent stream approximately along

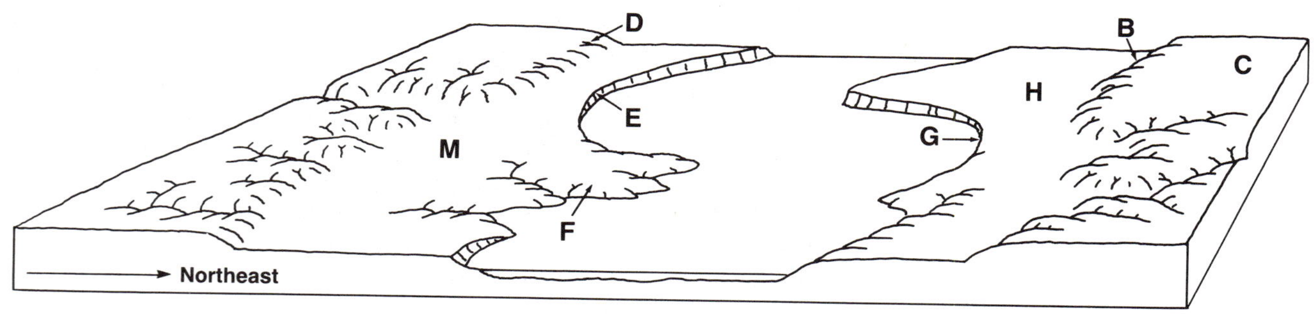

FIGURE 4–3a
Block diagram of part of area depicted in Figure 4–3, viewed from southeast.

the line of the present embayment. We may then imagine its cutting down to level **HM,** attaining a graded profile adjusted to its new base level. A stillstand would have permitted extensive coastal erosion to the east of here, with concurrent lateral cutting along the valley of this master stream. This cutting created an ever-widening floodplain, an erosional surface or strath that eventually extended from bluff **D** across to bluff **B**.

Rejuvenation, due to crustal uplift or eustatic lowering of sea level, with the lowering of base level, would have renewed downcutting by the stream. This would have caused attainment of a graded profile and the lateral carving of a second, lower floodplain or strath (between bluffs **A** and **E**) set well below the former, wider valley bottom (surface **HM**).

Finally, the area was partially submerged, just enough to drown the younger valley and to leave the remains of surface **HM** standing as a broad strath terrace.

Note how much more dissected is the upland area (i.e. above scarp **B**), compared to the more-extensively preserved strath terrace (surface **H**). This may be explained by the fact that, during the entire interval in which surface **H** was being planed off, the upland area was attacked by the combined processes of weathering and erosion.

How does one classify the streams in this area? The original main stream and its principal tributaries must be consequent streams. The many minor tributaries, streams that have simply branched off haphazardly in all directions, are probably insequent streams.

Rehoboth, Delaware

(Figure 4–4 in back of book, C.I. 10 feet)

This area has a total relief of a little over 30 feet. Like the area shown in Figure 4–3 (Wilton, Virginia, in back of book), it is also a dissected coastal plain which has been partially submerged. Well-defined meander scars, particularly **D, E, I,** and **J,** attest to the sinuosity (meandering) of the main stream that once flowed across this area to the sea.

Submergence of such a dissected low topography produces an extremely irregular shoreline, such as segment **AG**. You can imagine that a similarly irregular shoreline once extended southeasterly from point **H,** past **K,** to well beyond the east border of the map. Shoreline erosion of such promontories (**H**) and the redistribution of the materials dislodged by longshore drift and beach drift, resulted in the building of barrier beach **KC**. This beach was and is a bulwark against further attack on once-exposed shoreline **AG**.

If the shoreline south of **K** has not yet attained stability, it will be further eroded back (westward), accompanied by a westward migration of beach **KC**. Very little inland migration of the beach is needed to completely separate embayment arm **F** from embayment **B**. If crustal and eustatic stability continue, the only other processes that will dominate this area will be erosion on the land and deposition in the embayments. Numerous marshy tracts indicate that the latter is well under way.

Mathews, Virginia

(Figure 4–5 in back of book, C.I. 5 feet)

This coastal area is so low and flat that numerous ditches have been dug to drain the marshy ground. Some of the ditch lines (e.g. **L** and **F**) have been drawn as solid heavy lines. Other ditches are indicated by geometric contour designs (e.g. **K** and **G**).

Like the areas depicted in Figures 4–3 (Wilton, Virginia) and 4–4 (Rehoboth, Delaware) in back of book, this area has experienced partial submergence. Embayments **A** and **C** are drowned valleys, as is lake **M**. **M** is a former embayment that has been completely sealed off by the barrier beach that built across its mouth. Water area **H,** on the other hand, is more a lagoonal body, bordered on the west by the mainland and on the east by barrier beach **I**.

The linearity of shoreline **BD,** in the north, suggests erosion. Linear shoreline **JN,** in the southeast, apparently is the product of both erosion and baymouth deposition. The near-attachment of beach **I** to the mainland at **J,** and its terminus in open water at **E,** indicate prevalent longshore drift toward the north.

Honokane, Hawaii

(Figure 4–6 in back of book, C.I. 40 feet)

The critical importance of contour interval in topographic-map interpretation is convincingly brought home by this map. We had originally intended to trace off every fifth contour to produce a map having a contour interval of 200 feet. As you can see, had we used that interval, the valley floor between points **F** and **J** (both on the 200-foot contour) would have appeared as an empty space suggestive of a virtually flat floodplain or similar featureless surface. Even with the smaller C.I. of 40 feet, the lowland floor still appears to be relatively flat.

But just to the north of **J,** a low rise or divide (**I**) extends across the "mouth" of the valley. It does so in the same way that a barrier beach or baymouth bar extends across an embayment or estuary (e.g. Lake **M,** in Figure 4–5, Mathews, Virginia, in back of book). This important landform would not have appeared on the map had we used the 200-foot contour interval. We would have been unable to see the distinct possibility that this is a former arm of the ocean which has been filled in by sediments delivered from the south.

In its upstream segment, the valley displays the usual V-shaped profile typically produced by a downcutting stream. The published quadrangle map indicates that the main stream in this valley does not reach the ocean, but terminates in a low marshy area immediately southwest of barrier **I,** near point **H**. Could it be that we are seeing a large valley that, like those depicted in Figures 4–3 (Wilton, Virginia) and 4–4 (Rehoboth, Delaware), has been partially drowned?

Study the linearity of shoreline segment **KR,** and the clear-cut truncation of its several headland spurs (the cliff at **R** is about 700 feet high). Are these the products of faulting? In other parts of the world, they might be. But this is coastal Hawaii. Here, the most-probable explanation for these features is extensive coastal erosion.

In fact, we can estimate how far these headlands have been eroded back, by extending (drawing) slope **PN** to the north. Even if that slope had been uniform (it may well have been concave), it would have extended to point **O**. (You can convince yourself of this by constructing a profile along line **PN,** and extending it north to **O**.)

This is Hawaii and these are volcanic rocks. At one time the area east of **E** and south of **O** must have resembled area **ECA,** with its upper slope developed on lava flows and possibly some pyroclastic materials. Streams that formed on that slope, like present stream **D,** were all consequent streams.

Deeply entrenched streams **H, L, M, Q,** and **S** are thus much-older consequent streams which have had sufficient time to cut their deep, steep-sided valleys.

Why are there deep valleys and cliffed headlands (divides) in the central and eastern areas, but continuous gentle slopes to the sea in the **ECA** area? There is a linear 360-foot-high cliff at **G**, and a very low linear stretch at **B**. But elsewhere in the western area, the shoreline is irregular and has apparently been eroded back very little if at all.

Here is one way to answer the question: "The reason the western area does not resemble the central and eastern areas is that it has not been cut into by streams or eroded back by coastal erosional processes." However, the trouble with such an answer is that it answers nothing. It simply restates the facts. We must still ask "why?"

If area **ECA** is the same age as the rest of the map area, it might somehow have been protected against coastal erosion and concurrent stream downcutting. Since there is nothing to indicate such protection, we must search elsewhere.

If area **ECA** were much younger than the remaining area, there might not have been sufficient time for downcutting and cliffing to have taken place. In fact, area **ECA** "looks like" a relatively young area. The other area "looks like" a much-older area. Now we have a hypothesis (not a proof) that fits the topography. It appears that, from time to time, as the western area has received great amounts of volcanic material, the area to the east has not, leaving its older surface exposed to prolonged attack and destruction.

Pemaquid Point, Maine

(Figure 4–7 in back of book, C.I. 10 feet)

This is part of the coastal belt of New England. It has undergone both intensive continental glaciation and partial submergence, the latter at least partly the result of isostatic response to the weight of the ice sheets.

The general topographic grain of this area, and its streamlined appearance, is similar to that in the undrowned southwestern Connecticut area depicted in Figure 3–4 (in back of book). It is no more possible to make detailed, specific geologic interpretations from this map than it is from Figure 3–4. But differential erosion is a very dominant factor in carving bedrock of various resistances. So, we propose that the topographically high belts (promontories and islands) are underlain by relatively resistant rocks, and that less-resistant units have been eroded and scoured more completely from the lower (submerged) areas.

The linearity of shoreline trend **AD** and the steepness (arrow **C**) of its scarp suggest the possibility of fault control. In fact, the steepness of scarp **C** and the gentleness of all slopes to its west (e.g. arrow **B**) combine to form a profile asymmetry. This kind of asymmetry is commonly associated with tilted fault blocks in areas of more-recent tectonism, such as the Basin and Range province of the American West. In New England, however, it is unwise to do more than record impressions and propose interpretations.

1. Figure 5–1. Medicine Lake and Timber Mountain, California.
2. Figure 5–2. S P Mountain, Arizona.
3. Figure 5–3. Grand Canyon, Arizona.
4. Figure 5–4. Ship Rock, New Mexico.
5. Figure 5–5. Spanish Peaks and Walsenburg, Colorado.

5

Volcanics

INTRODUCTION

Through the ages volcanic activity has fascinated, mystified, and terrified mankind. Concern about the inscrutable underworld is reflected in the names of deities associated with volcanism—Vulcan (god of fire), Pluto (ruler of the underworld), and Pele (a Polynesian demigoddess said to appear in human form).

Individuals who are far removed from the awesome forces that are ever present in active volcanic areas are not concerned with these legendary gods and goddesses. But in the Hawaiian Islands and other areas of active volcanism, there are those who, even today, seriously regard these temperamental deities.

In 1824, Kapiolani, Christian wife of a Hawaiian chief, defied tradition to demonstrate that Pele did not possess power. She picked and ate sacred berries without first making an offering to Pele; she suffered no ill effects, even though she was standing at the edge of a lava lake (Vitaliano, 1973). Despite this, more than 150 years later, Pele is still more than a superstition to some Hawaiians. During the 1955 eruption at Kapoho, inhabitants chanted and made offerings to Pele. This practice still occurs from time to time at Halemaumau.

The visitor center at Hawaii Volcanos National Park has hundreds of samples of rock and black sands which were returned by visitors who feared that removal of such taboo materials would bring them bad luck.

We too are intrigued by volcanism, and especially by volcanic landforms. We have resisted the temptation to include in this book a comprehensive collection of topographic maps covering the broad spectrum of landforms associated with volcanic and plutonic activity. Consequently, we restrict this chapter to five map areas: three feature volcanic landforms (e.g. cones and flows), and two depict plutonic landforms (e.g. dikes and plugs). In the first group, the contrast between volcanic and surrounding nonvolcanic terrain is striking and important. In the latter, the role of differential erosion is critical.

Medicine Lake and Timber Mountain, California

(Figure 5–1 in back of book, C.I. 40 feet)

Recent volcanics, like the hippopotamus, can often be recognized simply because they look like volcanics. The volcanics in this area are excellent examples.

The most obvious of the individual landforms is small cone **B**. It is almost perfectly circular, with uniform side slopes and crater. Hill **A** also boasts a small crater.

The extremely irregular topography is represented by minutely wiggly contours and numerous hillocks and miniature depressions. They proclaim area **E** to be a recent lava flow which emanated from source area **F**. Smoother, but still well-defined, is flow **D**. Another large flow, apparently older, extends eastward as bulge **J**. Contour crenulations in area **H** and its southeasterly projection **I** suggest a fairly young flow that spread eastward. Part of it probably drained away along the southwest edge of older flow **J**. A similar local tongue is indicated at **G**.

Hill **C,** to the west, appears to mark the center of an older major source area.

So much for show-and-tell. Now we shall get to interpretation.

S P Mountain, Arizona

(Figure 5–2 in back of book, C.I. 40 feet)

Aerial photographs of lava flow **H** must appear in more geology and physical geography textbooks than any other volcanic feature. It is part of the volcanics field north of Flagstaff, Arizona, not far from Sunset Crater National Monument and the majestic San Francisco Peaks. The flow has an extremely irregular surface, well-defined edges (e.g. **G** and **L**), and it extends northward from the base of an almost perfectly circular volcano (cinder cone **P**). These characteristics combine to make its identification (recognition) as a lava flow a simple matter.

The *AAPG Geological Highway Map of the Southern Rocky Mountain Region* shows the area to the north and west of the volcanics to be underlain by Paleozoic carbonates. These are the same strata that form the rim rock of the nearby Grand Canyon. Note the contrast between the contour crenulations in the lava area, and the contour smoothness in the area to the north and west.

Note that some volcanic cones (for example, **D** and **W**) are irregular in form. Some of this irregularity almost certainly is the result of extended erosion; these appear to be much-older cones. However, the peculiar shapes of cones **M** and **V** are not so readily explained by erosion alone. The western half of cone **M** and all of cone **V** display a pronounced north-south linearity. This linearity remarkably parallels the major east-facing scarp **AE,** which dominates the northwestern part of the map area.

In this map area there are several other linear features. All of them parallel scarp **AE**. They are (1) the lower, but clearly defined west-facing scarp **BC;** (2) the western border **I** of lava flow **H;** and (3) what seem to be several linear "valleys" (e.g. **KJ**) cut into the surface of flow **H**. We have not indicated some other similar features, which you can identify with little difficulty.

Why are these features linear and parallel? Do they have any genetic relationship? Perhaps the key is that scarp **AE** ends abruptly to the south at point **N,** where it encounters the igneous material. No one can look at this map and flatly state what formations lie to the north and west of flow **H**. But one thing is apparent: whatever the bedrock, the topography to the north and west is quite different from that of the volcanics.

Scarps **AE** and **BC** are best explained as fault-associated scarps. Their sharpness and overall clean-cut appearance suggest that they are fault scarps. However, suppose the strata that form the scarps are very resistant, and are directly overlain by beds that are very nonresistant (e.g. limestone beneath a silt-shale sequence). The rapid stripping away of the nonresistant upper beds, especially in the down-dropped area between the two faults, could account for the "new" appearance of the scarps.

Incidentally, note the north-flowing stream (defined by contour bends) that has developed along the base of scarp **BC**.

Why is scarp **AE** terminated at the edge of the volcanics (point **N**)? Whether scarp **AE** is a fault scarp or a fault-line scarp, one thing about it is (almost) certain: it extends to the south beneath the volcanics. These volcanics, particularly cone **P** and its flow **H**, are young and well preserved. Not enough time has elapsed, since the covering of the fault trace to the south of **N**, for scarp **AE** to have been produced by differential erosion. Thus we may conclude that scarp **AE**, as well as the associated fault displacement, is prevolcanic. This interpretation is supported by the fact that two lobes of flow **H** (**F** and **O**) flowed westward into the lowland at the foot of scarp **AE**. (See Figure 5–2a, a sketch of these relationships viewed from the northwest. Position the map with south at the top to best see the relation between map and sketch.)

Sketch a line connecting points **S** and **T**, passing through cones **U** and **V**. Now sketch a second line connecting **Q** and **R**, a line which coincides with the western edges of these two cones. This parallelism strongly suggests the possibility of fault influence over the "nonconical" forms of these cones.

Small linear valley **KJ**, on the upper surface of flow **H**, has an unnatural-looking geometric shape, particularly since it is located on a flow, where irregularities abound (note the many small closed contours). Some fracturing along this line may have facilitated rapid erosion (subsequent). Another such valley lies about a quarter-mile west of point **L**.

Finally, we must consider linear edge **I** of lava flow **H**. The topography to the west of **I** appears to be nonvolcanic. Is this linear scarp truly linear, in the sense that the other linear features are? Although different in appearance from scarp **AE**, it is far more linear than the east-facing edge of flow **H** (e.g. around **L**). In addition, it is almost parallel to scarp **AE** and valley **KJ**.

Is it reasonable to suspect fault control along border scarp **I**? We have recent igneous rock overlooking a lowland, which is underlain by much-older

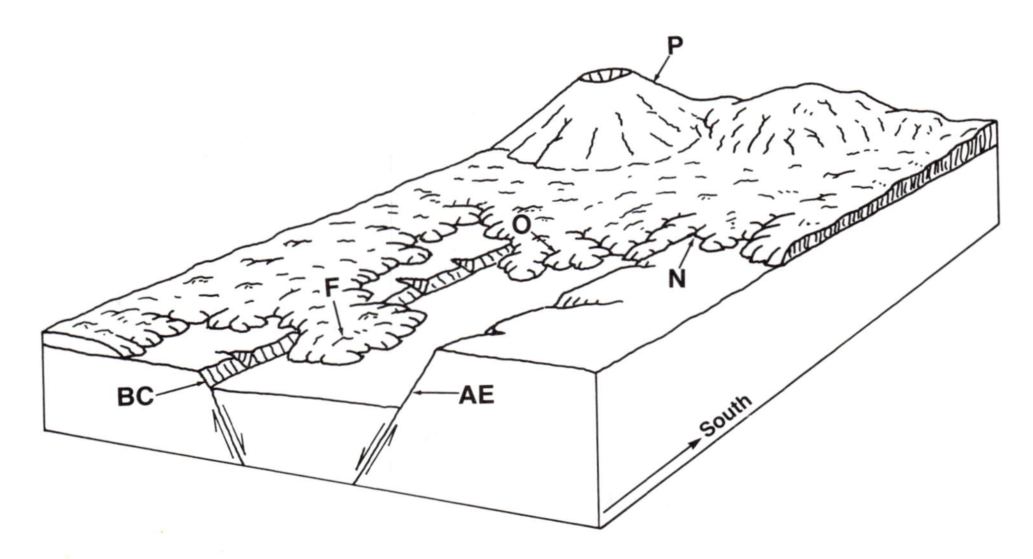

FIGURE 5–2a
Block diagram of part of the area depicted in Figure 5–2, viewed from the northwest.

sedimentary strata. How can we "fault" a lava flow up above much-older bedrock? Besides, there is no topographic indication of fault offset in the sedimentary area to the north of scarp **I** (or is there?).

It is difficult to accept the linearity of scarp **I** as the product of chance. Yet it obviously is not the result of differential erosion along a fault, since the lava flow is much too recent. Nor is it a fault scarp. Perhaps the initial advance of north-flowing lava **H** was restricted along its western edge by an east-facing scarp which *at that time* extended along this general line. But how that led to the present topography is difficult to envision.

Grand Canyon, Arizona
(Figure 5–3 in back of book, C.I. 200 feet)

The approximate scale of this map is 1:250,000, about one-fourth that of Figure 5–2 (S P Mountain, Arizona, in back of book). The contour interval of this map is five times as great, 200 feet, as compared to 40 feet.

Scarp **D,** 600 to 1000 feet high, extends southward along the northwestern side of the map area. It is abruptly terminated, as a linear/curvilinear feature, where it runs into a more-bulbous, massive terrain at point **E**. Linear scarp **K,** 10 miles southeast of **E,** could be a continuation of scarp **D**.

As in the Figure 5–2 area, the geology of this area consists of flat-lying resistant strata (here and there broken by steeply dipping faults), overlain in part by volcanic rocks. Scarp **D,** then, is another fault-associated scarp. It is, in fact, the famous Hurricane Cliffs of northwestern Arizona.

This is a region of contrasts. The most obvious is between broad areas dominated by extremely gentle slopes (e.g. area **CH**) and narrow bands of very steep slopes and cliffs, such as **D, I,** and **O**. Another is the contrast between those scarps and other, gentler, more-irregular upland borders, such as **L, M, J,** and **F**. Finally, there are extensive areas, such as **CH,** that contrast sharply with the large central highland bounded (clockwise) by points **K, J, F, A, B, L,** and **M**. This highland is populated by numerous large and small hills and mountains, between and among which the terrain is quite irregular and hummocky.

This central highland, with its many hills, humps, bulges, and blobs, is the main volcanic area. Lava has flowed, cascaded, and crept over it. In many places lava has flowed down over the prevolcanic cliffs that border the upland, replacing their sheer drops with gentler, more-irregular slopes such as **L, M, J,** and **F**. This blanketing of preexisting cliffs accounts for the asymmetry of valley **GN**. Slope **LM** is the lava slope, and scarp **O** is the unshrouded cliff developed on the resistant sedimentary strata.

Just as scarp **AE** of Figure 5–2 ends at point **N,** south of which it is covered by lava, in this area the Hurricane Cliffs (**D**) are lost at **E**. Here they encounter lavas that flowed down over them from source areas in the highland to the east. Refer back also to the relation of lava tongue **O** to scarp **BC,** as depicted in Figure 5–2a.

Ship Rock, New Mexico
(Figure 5–4 in back of book, C.I. 20 feet)

Ship Rock has for generations been well known to American geographers and geologists. Surface and aerial photographs, as well as sketches and topographic maps of this feature, appear in many textbooks. As a result, the geology of Ship Rock is no secret. This is the resistant core or plug of an ancient volcano, from which the flesh has been stripped by prolonged weathering and erosion.

The central spine looms as a gigantic tower over the surrounding desert lowland. Radiating out from this core is a series of dikes, also more resistant than the country rock. The latter is a sequence of flat-lying, fine- to-medium-grained clastics.

Two miniature, elongate closed contours, **G** and **H,** would mean nothing on almost any other map, or at almost any other location in this map area. But their orientation and location, along an extension of the main dike ridge **F,** reveal them to be small, isolated dike-ridge segments that have not quite been reduced to the general lowland level.

Does this mean that, at **G** and **H,** the dike is less resistant than it is at **F,** where it stands much higher? Probably not, since we may assume the intrusive rock to have the same composition at **F** as at **G** and **H**. More probably, the dike simply thins to the south, away from the central core. Such thinning would permit more-rapid destruction along this more-distant segment of the dike, than along the thicker segment to the north.

Since the identity of the central mountain and its radiating ridges is not something over which we must struggle, let us turn to a few more-interesting, if less-impressive, features. For example, what are slopes **C, D,** and **E,** which lead gently away from dike ridge **F**? All around the mountain and ridges are similar slopes, including **A** and **B**.

The entire area surrounding the core and dikes is apparently well dissected. This tells us that this is an erosional surface. Certainly such a surface, with its myriad channels and divides, could not be the product of deposition. Erosion is also indicated by the fact that, everywhere, the ground adjacent to mountain and ridges slopes away from these prominent features. The plug and dikes do not project up through material that has been deposited around and over them. They have been uncovered by the removal of the less-resistant country rock that once encased them. These sloping surfaces—erosional surfaces—must be pediments, since they cannot be alluvial fans or aprons.

Suppose, however, that you were to object to such an interpretation (and it is an interpretation, since we had to make a choice between alternatives, a choice based on criteria we can discern on the map). Suppose you discovered that all the material in the lowland consists not of Cretaceous strata (as indicated on the *AAPG Geological Highway Map of the Southern Rocky Mountain Region*), but of Quaternary or late Tertiary alluvium. What then?

For the sake of discussion, let us assume that the lowland is underlain by unconsolidated fill, deposited material that at one time partially buried the dike ridges. Would this make slopes **A** through **E** *depositional* rather than erosional? Would they then not be pediments, but alluvial fans?

No. They would still be pediments, for the simple reason that the present lowland topography, including slopes **A** through **E,** would still be the product of erosion, albeit erosion of unconsolidated fill. A pediment is a gently sloping

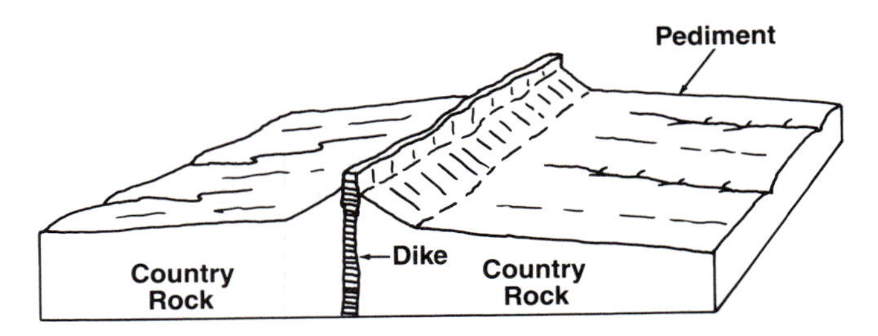

FIGURE 5–4a
Block diagram of a dike ridge, similar to that depicted in Figure 5–4.

erosional surface, leading down from the base of a scarp or steeper slope. It matters not whether that erosional surface has been cut in Precambrian schist, Mesozoic siltstone, or Pleistocene morainal deposits.

Yes, these are pediments. Figure 5–4a shows the relation between pediment slopes and a dike ridge such as ridge **F**.

Spanish Peaks and Walsenburg, Colorado
(Figure 5–5 in back of book, C.I. 250 and 125 feet)

The solid contour lines are drawn at an interval of 250 feet. In a few places we have added 125-foot contours (dashed lines) to show important minor topographic features and relations.

A narrow, linear ridge or wall-like feature need not be an etched-out igneous dike that is more resistant than its country rock. Three other possibilities are (1) a resistant, steeply dipping sedimentary unit, (2) a fault or fracture which has been cemented with minerals more resistant than the bedrock, or (3) a clastic dike of greater resistance than the bedrock. The one thing all have in common is that the ridge is the result of differential erosion.

Conversely, if any of these geologic features should consist of material *less* resistant than the surrounding bedrock, then it would be etched out as a narrow, linear depression, valley, or trench. Finally, should the feature (e.g. dike or cemented fault) have a resistance equal to that of the bedrock, it might have no topographic expression at all.

The above leads us to a point that all interpreters of topographic maps, aerial photographs, and remote-sensing imagery must bear in mind: there is rarely, if ever, such a thing as a truly "typical" or representative landform. Even the ubiquitous alluvial fan is often mistaken for a pediment slope, or vice versa (see Chapter 6).

The Spanish Peaks area of south-central Colorado is famous for its swarms of radiating and intersecting dikes, just a few of which are shown here. Many of the dikes in this map area can be depicted by contours spaced at a 250-foot interval (solid lines), but some would pass undetected without the addition of intervening 125-foot-interval contours (dashed lines). For example, note how critical to interpretation are contours **A, B,** and **F** in the north. And without dashed contour **C,** a small dike ridge adjacent to ridge **B** could escape detection altogether, although the trained eye might note contour bends **D** and **E**.

Observe the unique four-armed starfish patterns (**L** and **P**). They are the points of ridge intersection. Compare these to the sketch in Figure 5–5a, which

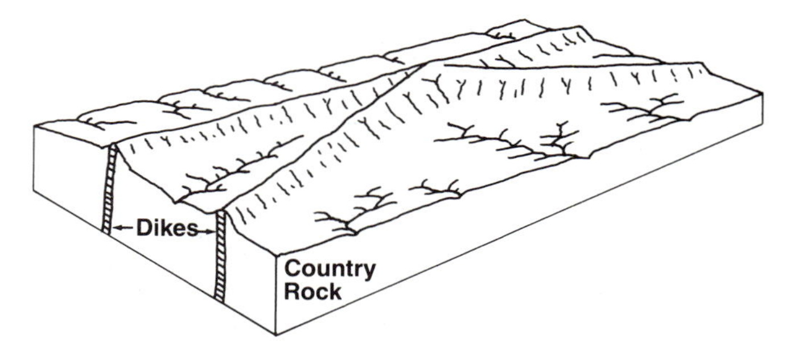

FIGURE 5–5a
Block diagram of intersecting dike ridges. Compare to those depicted in Figure 5–5.

show intersecting dike ridges. But note that this does not *prove* that the ''starfish'' on the map are intersecting dikes.

For example, is it not possible that ridge **JN** is a steeply dipping, resistant sedimentary rock, such as a conglomerate or sandstone, and that ridge **IM** is a resistant, cemented fault plane? But in an area teeming with radiating and intersecting dikes, it is more logical to assume that these are igneous features, etched out by differential erosion.

In areas such as this, it is easy to ''see'' geology that does not exist, and equally easy to deny what you see. For example, hairpin contour bend **Q** indicates that a dike ridge extends eastward, at least to point **R**. But what about **T**? Does the dike extend eastward to, and possibly beyond, point **U**? It might very well do that, particularly since **T** is so ''directed'' and geometric in comparison to nearby **S**. It is acceptable to map the dike as definite to, and just east of, point **U**. The same thinking would apply to point **O,** which lines up well with ridge **IM**.

Finally, what about features **G** and **H** in the northeast? They are as sharply defined as closed contour **K,** and we believe that **K** is a dike ridge. Are **G** and **H** less likely to be dikes because of their orientations? Think about it.

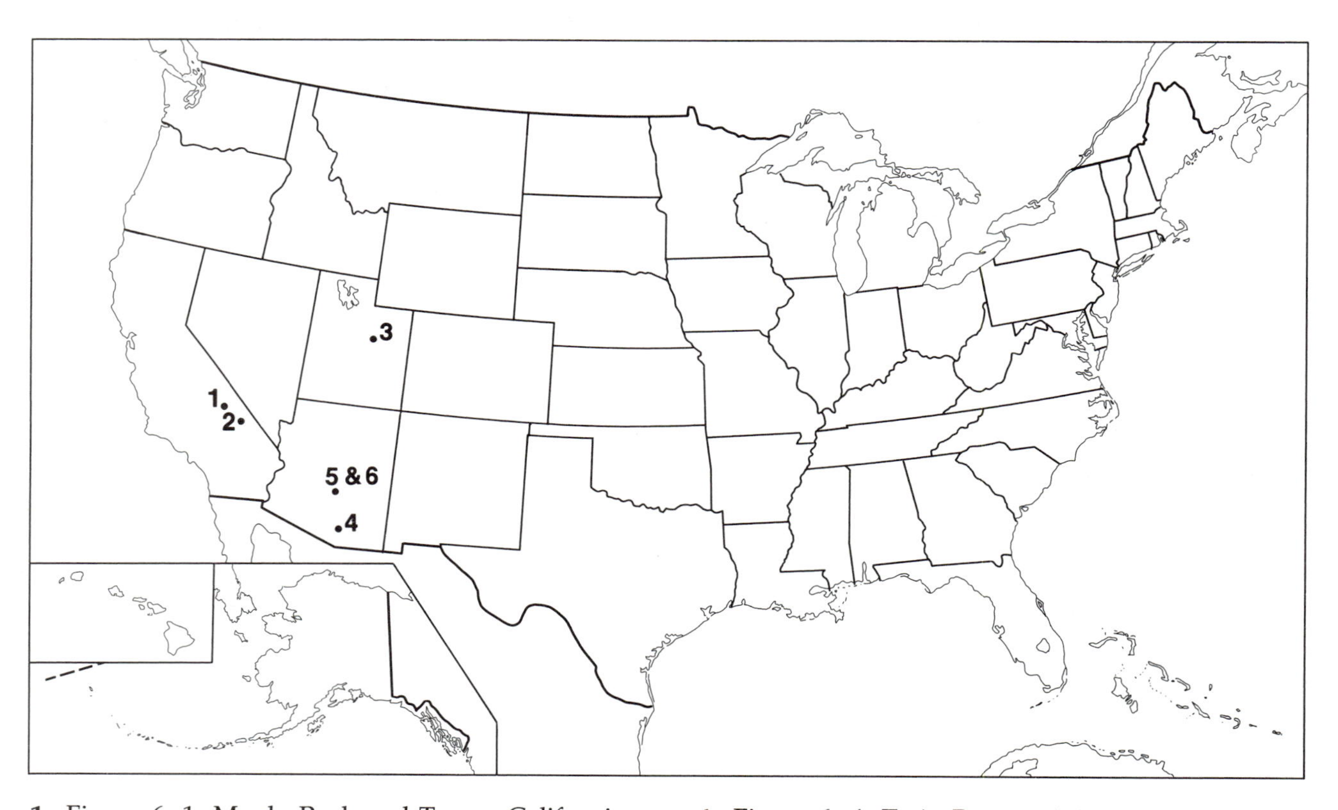

1. Figure 6–1. Manly Peak and Trona, California.
2. Figure 6–2. Avawatz Pass, California.
3. Figure 6–3. Wellington, Utah.
4. Figure 6–4. Twin Buttes, Arizona.
5. Figure 6–5. Sacaton, Arizona.
6. Figure 6–6. Sacaton, Arizona.

6

Piedmont Slopes

INTRODUCTION

Two of the most ubiquitous landforms in arid and semiarid areas are the piedmont slopes known as the *alluvial fan* and the *pediment*. Slopes that consist of coalescent fans are known as *alluvial aprons* or *plains*. Slopes that consist of coalescent pediments are called *pediment aprons* or *plains*. In ordinary conversation and professional writing, even large pediment slopes are usually referred to simply as "pediments." But on the other hand, we restrict the term "fan" to a single fan-shaped landform, a slope radiating away from the mouth of a canyon or valley.

The pediment is an extremely important landform, though its significance is seldom understood or appreciated. Pediments are created and they are destroyed. Many pediments, erosional surfaces that truncate bedrock structures, are capped by various thicknesses of rock debris. This blanket, which lies unconformably upon the underlying bedrock, effectively masks the bedrock geology.

It is virtually impossible for the average geographer or geologist to distinguish a capped pediment from a fan. Their appearances are virtually identical. There are certain criteria that we can apply to deduce which is which. In some instances we can feel reasonably certain; in others, doubt will linger. And there will always be slopes that defy classification, at least from map study alone.

Probably the most important pediment, at least to the map interpreter, is the one that has been partially destroyed, the so-called *pediment remnant*. These generally flat, gently sloping surfaces, capped or not, stand as bench-like or terrace-like landforms. All too frequently they are either mapped and forgotten, or not considered worth understanding.

Manly Peak and Trona, California
(Figure 6–1 in back of book, C.I. 40 feet)

The floor of the eastern lowland is occupied by a playa. The map of the surrounding area shows that the lowland in the southwest is also part of an enclosed (undrained) basin. This is all we know about this map area, except that it lies in the Death Valley portion of the Basin and Range geomorphic province, a region of large-scale block faulting. All range/basin borders are not necessarily fault traces. But all ranges are structurally positive, and all basins are structurally negative. Unlike the Valley and Ridge province of the American East, for example, the Basin and Range province is not an area characterized by *obsequent topography.*[1] Here there are no structurally high lowlands, and no structurally low mountains or ridges.

These observations are fundamental to our consideration of the pediment/fan question in the map area. Observe the slopes leading down to the floors of the basins from the feet of the mountains, such slopes as **M, O, E,** and **S**. Are they what they appear to be, namely fans? Or are they pediments, instead? On what evidence can we base our answer?

First, the basins are enclosed. They have no avenue of escape for any stream which may enter them. No mechanism is available to transport out of the basins the debris that washes into them from the adjacent mountains. They are areas of accumulation. This fact alone indicates that the slopes are fans.

Second, mountain fronts that have sustained prolonged erosion will undergo both downwasting and backwasting. Such scarp retreat is not accomplished uniformly along linear fronts, such as those in this area (particularly scarp **FT**). Instead, as time progresses the front becomes increasingly irregular. Remnants (*inselbergs*) are left behind, standing as sentries in front of the retreating scarp, overlooking the ever-expanding pediment surfaces that are being created as the mountain mass is being destroyed.

But there are no isolated hills in the fringe areas marginal to these mountain fronts. You will find not a single closed contour in the playa to the west of a line drawn from **F** to **T,** a line separating the mountain on the east from the basin floor on the west. You will find no inselberg-like hills.

These two facts are sufficient. A lowland area that is (1) filling with detritus from the surrounding highlands, and that is (2) bordered by mountain fronts that are too straight and regular to have experienced much backwasting, is not going to be an area fringed by broad, sweeping pediment slopes. We therefore identify the individual fan-shaped features as *alluvial fans*. These include **G, H,** and **L,** and **Q** across the basin from them, as well as previously designated slopes **M, O, E,** and **S**. The remaining gently inclined slopes such as **J,** which is not fan-shaped, may be classified as *alluvial slopes*.

An interesting situation is revealed in the vicinity of divide **A,** in the northwest. Note that slope **I,** which may be partly pediment, rises gently northward to a point nearly 100 feet below the divide. To the north, however, drop **B** down to depositional slope **C** is well over 400 feet. (Slope **C** is actually the upper extent of slope **E**.) Erosion along this steeper, over-400-foot-high scarp is important to the future of this area.

As scarp **B** erodes southward, divide **A** will be gradually destroyed. However, the destruction of divide **A** will not stop the destruction taking place along scarp **B,** nor will the southward backwasting of that scarp be halted. This backwasting will continue, and divide **A** will migrate to the south, into area **I**. It will not be long before the upper parts of what is now slope **I** will be removed, and the fragments of its weathering and erosion will be washed northward down scarp **B** and deposited atop slope **CE**. Thus, the northernmost end of what is now the south-sloping floor of the western basin is destined to be altered significantly as it is "invaded" from the north.

Compare and contrast front **FT** east of the playa and front **DR,** which faces it across the basin. Front **FT** is much more clean-cut, steeper, and

---

[1]In the terms "obsequent stream" and "obsequent fault-line scarp," the word "obsequent" indicates that a geomorphic element is *opposite* to a structural element. Thus, we propose the term "obsequent topography" to describe a general or regional topographic/structural discordance. An example is an area characterized by synclinal mountains and anticlinal valleys and lowlands, or by topographically low horsts and topographically high grabens.

higher—more "new-looking." There can be little doubt that line **FT** is the trace of a large, steeply dipping fault. This is a fault scarp along which displacement has taken place so recently that the front is merely scored by "megagullies." It is also cut by a few larger valleys, but even they are not deeply incised, and they are spaced at intervals of a half-mile or more. Destruction is just getting under way along this front.

Front **DR,** on the other hand, is much gentler and far more irregular. It apparently has been subject to weathering and erosion for a considerable time, far more than scarp **FT**. It would be nice to state flatly that scarp **DR** is also a fault scarp, one along which displacement took place sufficiently long ago to permit its more-advanced dissection and downwasting.

However, it is also possible that slope **DR** is a remnant of a downtilted block, perhaps a block that is upfaulted along its western side (near **O** and **J**). Thus **DR** would not be a fault scarp, and we must exercise caution in labeling it. It appears to be a fault scarp, and it probably is. But in mapping this area, we would indicate line **DR** to be the trace of a "questioned" fault. We would show it with a dashed line instead of a continuous line. We would also insert a few question marks between the dashes. On the other hand, we would map line **FT** as a definite fault trace, and show displacement up on the east, down on the west.

Only the northern end of the western basin is included in this map area. Note that the axis of that basin (line **JN**) lies much closer to the eastern border. Note also that the west-facing slopes (e.g. **O**) along the eastern edge of this basin are steeper and more linear than the east-facing slopes that form the western margin of the basin.

We have already observed that eastern scarp **O** might be a fault scarp. Note that the range's divide (**KP**) lies parallel to, and closer to, scarp **O** than to line **DR**. The fact that this range, like the adjacent basin to the west, is asymmetric suggests that the ridge is an east-tilted fault block, perhaps an east-tilted horst.

Before we leave this map, let us ask a few more questions. On what do we base our statement that the western basin is "asymmetric"? Since we can see very little of the highlands to the west of the basin, are we not really comparing the longer, gentler alluvial slopes along **M** with the shorter, steeper alluvial slopes on the **JO** side? This basin is *topographically* asymmetric, but does that necessarily indicate that it is also *structurally* asymmetric? Could this contrast in slopes merely reflect different conditions under which deposition is taking place along the two sides of the basin?

Questions such as these remind us that, though maps tell us many things, in too many cases they simply tantalize us by permitting us to "see" questions and problems for which they cannot provide the answers.

Avawatz Pass, California

(Figure 6–2 in back of book, C.I. 40 feet)

If any topographic map is ideal for the show-and-tell approach, surely this is it. The caption might read: "This is a desert area of block fault mountains, surrounded by well-defined alluvial fans."

The large dissected area in the southeast is certainly mountainous. It is delimited on the north (arcuate line **JLO**) by what could be a fault scarp. And the classical fan-shaped slopes **K** and **N** definitely look like alluvial fans.

But look northwestward from **J**. What is "mountain" and what is "fan" in this area? Is the border between mountain and fan along line **JF**? (There

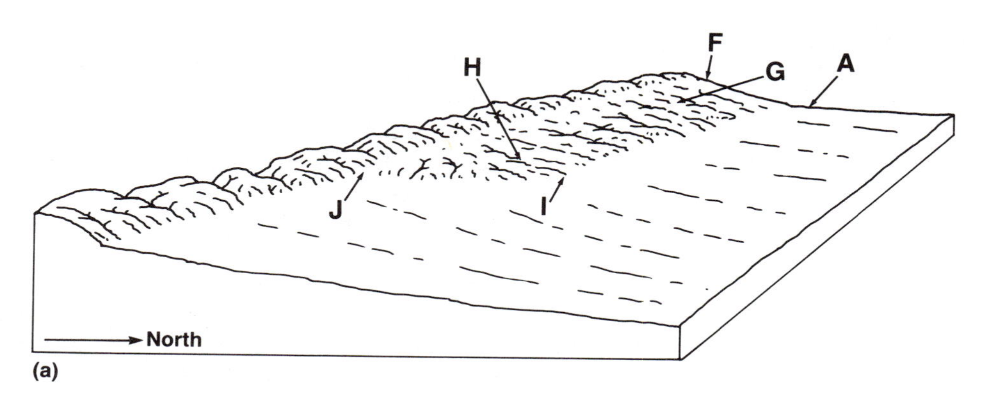

FIGURE 6–2a
Block diagram of part of the area depicted in Figure 6–2, viewed from the east.

appears to be a break in slope, from steeper on the southwest to gentler on the northeast.) Or is the border along line **JIA,** where there is a second break? The prevailing slope in intermediate area **GH** is extremely gentle toward the northeast, as is that of the "fan" area to the northeast of **JIA**. Figure 6–2a is a sketch of this confusing area, as viewed from the east.

Area **GH** has an overall gentle slope to the northeast. Is it not odd that area **GH** is as intricately dissected as the mountainous area? Note the contrast between the amplitude of the contour crenulations in area **GH** and the tiny wrinkles or squiggles in the contours that define "fan" **K,** for example.

Suppose major stream **CB** in the north were to be rejuvenated. This would lower the base level (which is stream **CB**) of all the streams now flowing across "fans" **K** and **N**. They would eventually also be rejuvenated. They would cut down into their many closely spaced channels, dissecting the once nearly smooth "fan" surfaces and transforming them into a gently sloping, intricately dissected expanse—an area remarkably like area **GH**.

What, then, is this strange area **GH**? Along its northeastern edge (line **JIA**) it drops down to the surface of the regional "fan" or "alluvial plain." Could area **GH** be a dissected fan or alluvial slope? This is possible, but why does it stand above the present "alluvial" surface? What is the nature of its boundary with this lowland? What is the nature of its border with the higher, apparently bedrock area to the southwest of line **JF**?

We are sorry to disappoint you, but we cannot answer these questions from the map alone. We have taken the first steps; possibly the answers can only be found in the field. Perhaps we could discover the answers on aerial photographs. The point is that, in the search for information, it is impossible to obtain answers unless questions are asked. It is important to understand that you have not failed, if you raise a question that you cannot promptly answer. The failure is not even to see the question!

Though there are many questions in this map area, we shall consider just one more. Note, in the southwest, the large "alluvial" lowland which slopes (arrows) toward point **M**. All the drainage lines within this lowland similarly converge toward **M**. Just to the north of **M** they meet to form a single north-flowing stream. It flows through mile-and-a-quarter-long water gap **ME** (which is cut through a 600-to-900-foot-high mountain spur), to point **E**. **E,** in turn, is the apex of another, much-lower "fan" which extends for miles out across the lowland to the north. Figure 6–2b is a sketch of this area, viewed from the southeast.

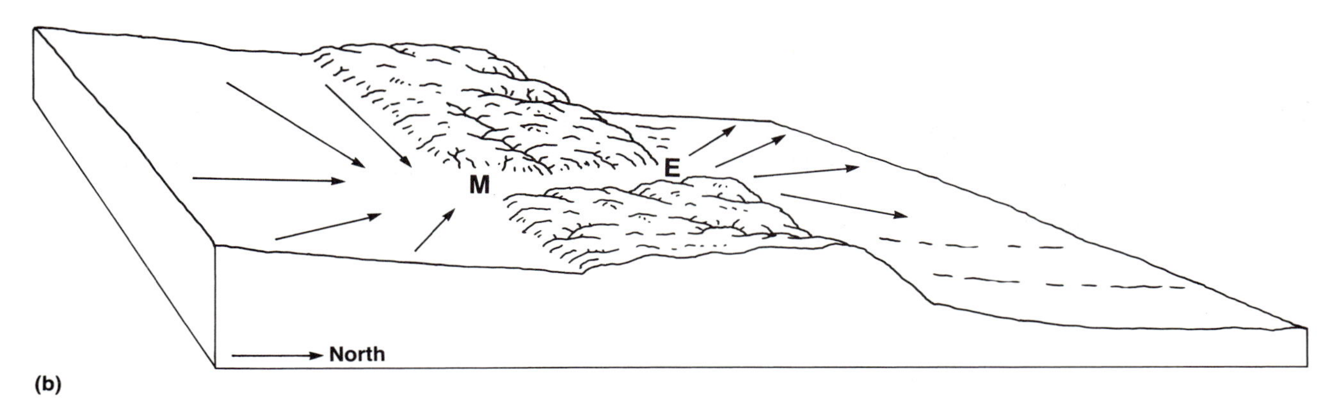

FIGURE 6–2b
Block diagram of part of the area depicted in Figure 6–2, viewed from the southeast.

There are only three conceivable ways that water gap **ME** could have been produced:

1. A north-flowing stream draining the north flank of the mountain spur (a stream such as the one nearby that emerges at **D**) might have eroded headward through the spur. It might have captured the drainage in the vicinity of **M,** drainage that had formerly flowed to the west. If this is what happened, the slopes (arrows) now converging toward **M** would have to be erosional slopes, created by erosional readjustment to the new base level provided at **M** by the head of the gap-cutting captor stream.
2. Prior to the uplift of the mountain spur, a stream from the south flowed north to **M,** and then to **E,** and beyond. Tectonic activity that created the present spur took place over sufficient time, at a sufficiently slow rate, to permit this north-flowing stream (even if intermittent or ephemeral) to maintain its course and to cut the gap: an *antecedent* stream.
3. The mountain spur, the lowland north of **E,** and the lowland south of **M** were created by a normal sequence of tectonic and geomorphic processes. Then the entire area was covered by a thick accumulation of sediments. On the top surface of this blanket, sloping north, a consequent stream developed along line **ME** (but well above the level of the present mountain spur).

   For some reason this stream was rejuvenated. As it cut down, it eventually encountered the crest of the buried spur, through which it continued to slice. The rock (fill) to the north and the rock (fill) to the south of the buried spur were less resistant than the rocks of the spur itself. Thus, normal erosional processes were able to remove these less-resistant rocks more efficiently and rapidly. This recreated (resurrected or exhumed) the mountain spur and its flanking lowlands.

   The only really new feature produced by this sequence of events was water gap **ME**. It was cut by the *superposed* north-flowing stream, which had originally developed atop the pile of debris that had buried the former topography.

These, then, are the three alternative possibilities. Was gap **ME** produced by (1) *headward erosion,* by (2) an *antecedent* stream, or by (3) a *superposed* stream?

If gap **ME** were cut by headward erosion, the slopes converging toward **M** would have to be erosional slopes, would they not?

If gap **ME** were cut by an antecedent stream, those same converging slopes would also have to be erosional.

And, if gap **ME** were cut by a superposed stream, would not the slopes still have to be erosional?

Before we leave this perplexing southwest area, note one final detail. The floor of water gap **ME** is remarkably wide and flat. Surely the stream that cut this gap was not, and still cannot, be that wide! Why, then, is this not a V-shaped valley or notch with a properly narrow floor?

Is it possible that it was once deeper, and that it has been partially filled with detritus washed in from the lowland to the south? Regardless of which alternative explains the actual cutting of the gap, the converging slopes south of **M** (arrows) would have to be aggradational (alluvial). Since the present floor of gap **ME** would also have to be aggradational (alluvial), the entire surface that slopes northward from **E** would also have to be depositional (alluvial).

There are many questions in this area. We have asked but a few, set forth a few hypotheses, and undoubtedly created no little confusion. We have done what map *interpreters* do.

Wellington, Utah

(Figure 6–3 in back of book, C.I. 250 and 125 feet)

Note that this map has two contour intervals. North of break-in-slope **CD**, we use a 250-foot interval because of the extreme topographic relief. In the south we need the 125-foot interval (dashed line contours) to show topographic details that the larger interval cannot show.

A basic knowledge of the regional geology is essential to interpreting and understanding the geomorphology of the lowland overlooked by the main south-facing escarpment, the Book Cliffs. The *AAPG Geological Highway Map of the Southern Rocky Mountain Region* indicates that the cliffs consist of a thick sequence of extremely gently north-dipping Cretaceous and Tertiary strata. The gentle south slopes that lead away from the base of the cliffs (e.g. fan-shaped slope area **ACEM** and lower slope area **FIQ**) form part of a broad lowland. It is occupied by the Price River, which eventually joins the Colorado River south of the map area.

The less-resistant strata that underlie this lowland are conformable beneath the cliff-forming beds. They also dip to the north. Therefore, the south-sloping surfaces cannot be dip slopes. Either they are alluvial deposits strewn upon the bedrock, or they are pediments carved in the bedrock (or carved in a sequence of unconsolidated lowland fill). Thus, they are either erosional or depositional; they cannot be structural. So far, so good.

Now let us compare and contrast slopes **ACEM** and **FIQ,** viewed from the southeast in Figure 6–3a, a highly generalized sketch. **ACEM** is the more smoothly fan-shaped of the two slopes. It is a slightly dissected surface. But slope **FIQ** is a complex of slopes. It is not nearly as simple a topographic unit as **ACEM**.

You can see that the present **ACEM** is all that is left of a once-more-extensive surface. Note how at numerous places it drops off, down to the lower surface, as along steep border slopes **L, N,** and **P.** Is not small cuesta-like feature **O** an isolated erosional remnant or outlier of surface **ACEM**? To the east, another such isolated, cuesta-like landform is designated as feature **J.**

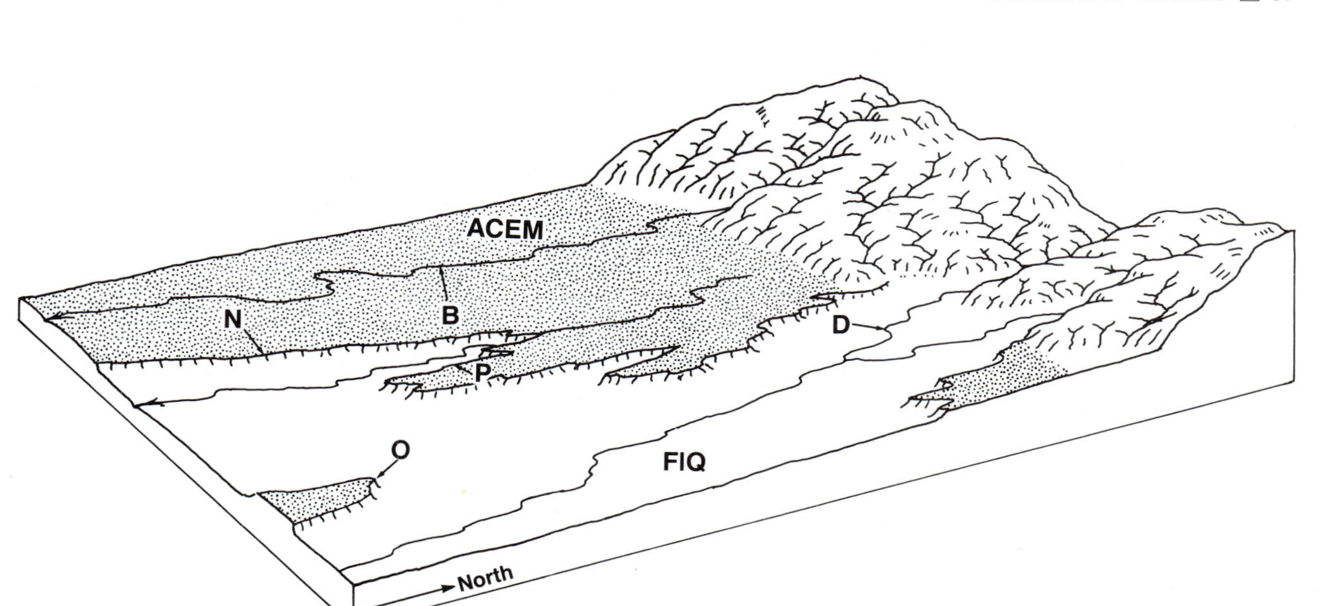

FIGURE 6–3a
Block diagram of part of the area depicted in Figure 6–3, viewed from the southeast.

In fact, triangular area **JKH** appears to be a much-larger remnant of an **ACEM**-like surface. Feature **J** is merely the small apex of the larger remnant.

Apparently, at one time the broad lowland south of the Book Cliffs consisted of a gentle regional slope, leading down from the cliffs to the major stream in the area, the Colorado. **ACEM, O,** and **JKH** are the remains of that slope.

Let us now review the facts, and draw a conclusion. This regional slope lies between a major scarp (the Book Cliffs) and a major externally draining river. The slope is not situated on the flank of an undrained basin, like those which abound throughout the Basin and Range province of the West (e.g. Figure 6–1, Manly Peak and Trona, California, in back of book). Thus, it is highly probable that this former integrated slope complex was the product of pedimentation, and not alluvial aggradation. Therefore we interpret this upper surface—now represented by remnant slope **ACEM,** feature **O,** and area **JKH**—as a *pediment plain*.

If it is a pediment plain, it may be capped by a layer of gravel/sand detritus, as many pediments are. This topographic map cannot tell us whether there is a blanket of debris. But the presence or absence of an alluvial cap is of little consequence; a pediment is a pediment, capped or not.

Strictly speaking it is incorrect to call the top surface of the blanket "the pediment," unless the pediment itself is deeply buried. In other words, if the topography is that of a pediment and not that of a fan, we still designate it as a pediment.

There is only one way this pediment plain could have been dissected and partially destroyed. The streams that flowed across it, southward from the cliffs to the river, must have been rejuvenated. There could have been several causes—perhaps a climatic change, or a lateral northward shift of the master stream closer to the base of the cliffs, or even some kind of stream capture. But for whatever reason, the dynamics of the south-flowing streams were altered, and regional downcutting began.

The larger stream systems, such as **D** and **G,** must have been capable of far more destruction than the smaller streams (e.g. **B**). Hence the pediment plain extending from the mouths of canyons **D** and **G** was more rapidly and completely destroyed than the pediment surface in the vicinity of stream **B**. Stream **B's** headwaters do not extend very far back into the face of the plateau.

Apparently streams **D** and **G** have now attained a state of adjustment, and have virtually ceased downcutting. Another generation of pediment slopes (e.g. arrows) has been created, all inclined toward the local base levels provided by these streams.

We may not be absolutely certain that the upper slopes, such as **ACEM,** are pediments. However, it is almost impossible for the newer, lower slopes not to be.

Twin Buttes, Arizona
(Figure 6–4 in back of book, C.I. 50 feet)

The main mountain mass **K** is Jurassic granite. Hills **F** and **E** are Jurassic quartzite, Precambrian granodiorite, and Paleocene granodiorite. The broad lowland to the north and east of these highlands is underlain by the Paleocene granodiorite, which also makes up hill complex **H** and low hills **A** and **B**. More-distant hill **C,** in the northeast, is Precambrian granodiorite. The southwestern lowland area, like the other lowland area, is underlain by the Paleocene granodiorite. The entire map area is drained externally by the north-flowing Santa Cruz River system. (Geology from *Geological Map of the Twin Buttes Quadrangle, Southwest of Tuscon, Pima County, Arizona.*)

This area presents little challenge in choosing between fans and pediments. Since it is externally drained, this is not an area of thick accumulation. Consequently, the gentle slopes radiating away from the hill/mountain cores must be pediments. Above them rise, as isolated sentinels, inselbergs **A, B, C, D, I,** and **J.**

An interesting geomorphic feature is low divide **L.** In a way, it connects outlying hill **H** to main mountain **K,** just as a sand bar connects a tombolo to a mainland. Study this area in Figure 6–4a, a view from the northeast. Along divide **L** the head of pediment slope **G** (arrow **G**) meets the head of opposite pediment slope **M** (arrow **M**). You could walk up slope **G** to **L,** and then

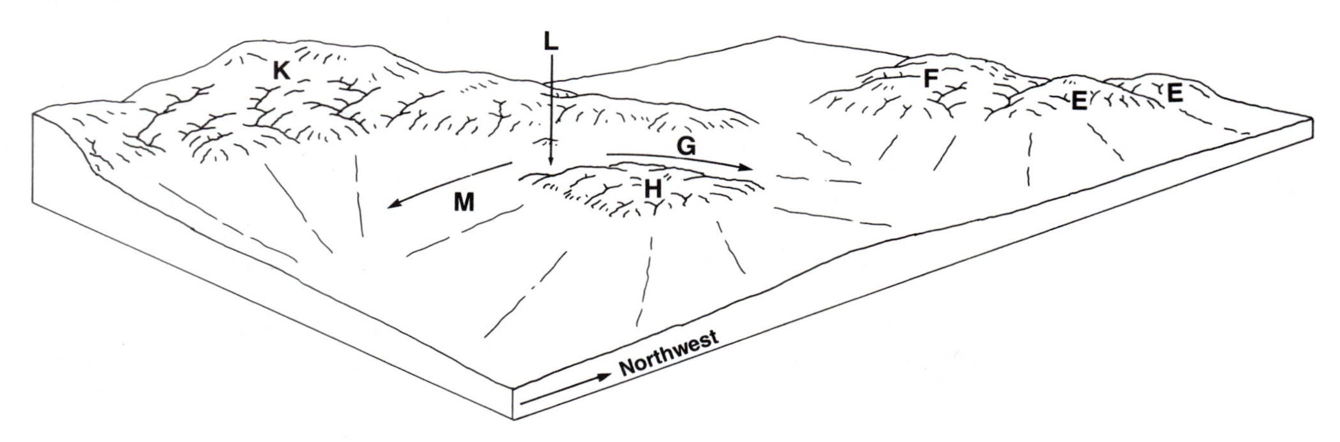

FIGURE 6–4a
Block diagram of part of the area depicted in Figure 6–4, viewed from the northeast.

continue down slope **M,** without encountering any steep slopes or scarps. This low divide, through which passage is so easy, is called a *pediment pass*.

Pediment passes are not the most widely discussed or highly publicized of landforms. At first glance they might appear worthy only of passing mention. However, as we will see when we study Figure 6–5 (in back of book), a map of the Sacaton Mountains of Arizona, there are pediment passes, and then there are pediment passes. . . .

Sacaton, Arizona

(Figure 6–5 in back of book, C.I. 10 feet)

At first glance, this area appears to be merely a series of hills and low mountains, surrounded by slopes that could pass for either pediments or alluvial aprons. Look at outliers such as the larger hills **G, C, A,** and **B,** plus smaller hills **J, L,** and (in the north) **E**. These suggest that considerable parts of the slopes are pediments, the products of extensive erosional processes, and that these isolated hills are inselbergs.

Like the area shown in Figure 6–4 (Twin Buttes, in back of book), this is part of the Basin and Range province, a region in which pediments abound. It is reasonable to conclude that the gentle slopes in this area are, at least in part, pediments. However, we did not select this map solely to illustrate pediments.

The specific landforms that make this such a fascinating area are not the large pediment slopes, but several remarkable pediment passes, especially **D, M, W,** and **X**. There are other passes in the area, but these are the ones that merit our attention. Figure 6–5a shows topographic elements in the central part of the map area, viewed from the southeast.

Look back to pediment pass **L** in Figure 6–4 (in back of book). At that point you can cross from one side of the divide to the other without climbing any steep slopes or scarps. The head of one pediment simply ends at a flat divide, beyond which you can continue down the gentle surface of the opposite pediment slope (slopes **G** and **M**).

But when we examine pediment pass **M** in this area (Figure 6–5), we find a perplexing situation. Approach this pass from the west, moving up the gentle pediment slope (arrow **K**) to point **M**. Now look to the northeast, out across pediment slope **O** (arrow **O**). You are also looking *down* upon the head of slope **O**. The closely spaced contours at **N** reveal a step down from pass surface **M** to the head of pediment slope **O**. Pediments **K** and **O** do not meet at an accordant elevation, as do pediment slopes **G** and **M** in Figure 6–4.

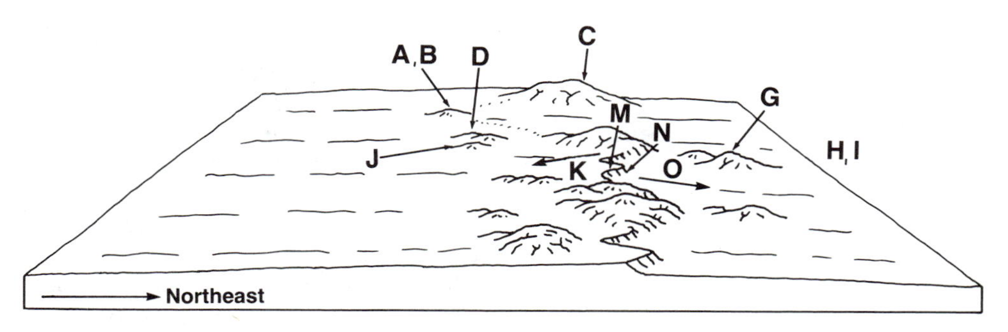

FIGURE 6–5a
Block diagram of part of the area depicted in Figure 6–5, viewed from the southeast.

Pediments are partly formed by parallel retreat of the steeper marginal slopes of adjacent uplands. For example, nearby pediment **R** (arrow **R**) is growing headward (southwestward) by the erosional retreat of mountain front **Q**. Similarly, pediment **S** (arrow **S**) is growing headward (southward) by the erosional retreat of front **V**, and so on. Pediment slope **O**, therefore, is still growing headward. At **P** its headward growth is being accomplished by the parallel erosional retreat of front **P**. Its headward growth at **N**, however, is special. Here it is the head part of pediment **K** that is being destroyed, as steep-slope **N** retreats to the southwest.

Pediments **O** and **K** are thus not yet in a state of equilibrium or adjustment. Divide **M**, which separates them, is migrating to the southwest. This migration will continue until the divide has shifted to the vicinity of the black circle near **K**. Then the opposing pediments will finally attain equilibrium, and this pass will resemble pass **L** of Figure 6–4.

At pass **W** an even more-violent situation exists. The divide at the head of pediment slope **Y** (arrow **Y**) overlooks pediment **T** (arrow **T**) from a height of well over 100 feet.

In contrast, look at pass **X**. It is reasonable to assume that at one time the main mountain divide stood at or near point **a**. Suppose the divide that now exists at **W** developed at **a**. It is easy to imagine its westward migration, accompanied by the considerable growth of pediment **b** (arrow **b**), headward along what is now slope area **Z** (arrow **Z**), to present divide **X**.

Apparently, passes **D** and **F** are no longer discordant but are at or are approaching equilibrium. Hills **U**, **I**, and **H** are all inselbergs.

Construction of a northeast-southwest profile through pass **M**, a north-south profile through pass **W**, and another extending through pass **X** and slope **b** also would be time well spent.

Now, be sure to read the next section for a fascinating comparison of two maps showing this same area.

Sacaton, Arizona

(Figure 6–6 in back of book, C.I. 50 feet)

This is a portion of an earlier map of the same area depicted in Figure 6–5 (in back of book). The two maps were made not only at different times, but by completely different cartographic techniques. Figure 6–5 is modern, made by photogrammetric methods from aerial photographs taken in 1963. Prior to publication, contouring accuracy was field-checked in 1966. Publication scale is 1/24,000.

Figure 6–6, on the other hand, was surveyed in 1904–06, a time when the aerial photograph was not even dreamed of as a realistic mapping tool. Undoubtedly the survey team employed the telescopic alidade, plane table, and stadia rod. Publication scale is 1/62,500.

To facilitate comparison between the two maps and methods, we have enlarged the older map (× 1.7) for presentation here at a scale closer to that of Figure 6–5. The contour interval of the 1966 quadrangle is ten feet, whereas that of this map is 50 feet.

It is true that the accuracy of topographic detail shown on this old map is subject to question. But the difference between the two maps should be evaluated from another standpoint. When we visualize the conditions under which the earlier cartographers must have worked—their means of transportation, and their survey instruments—we are forced to see Figure 6–6 as the truly remarkable piece of work it is.

This map would be absolutely worthless in the hands of the earth scientist intent on gathering data for a study in quantitative geomorphology (e.g. drainage density, basin geometry, and the like). Yet, the significant features pertaining to the pediment passes are right before us, just as they are in Figure 6–5. It is just as possible to interpret the passes, their history, and their future, by using this map, as it is by using the much more-precise map in Figure 6–5.

In fact, the more we compare the maps, the more we are struck by the fact that the lack of distracting detail on this map actually brings out the general relations more clearly. Sometimes, it would seem, older *is* better!

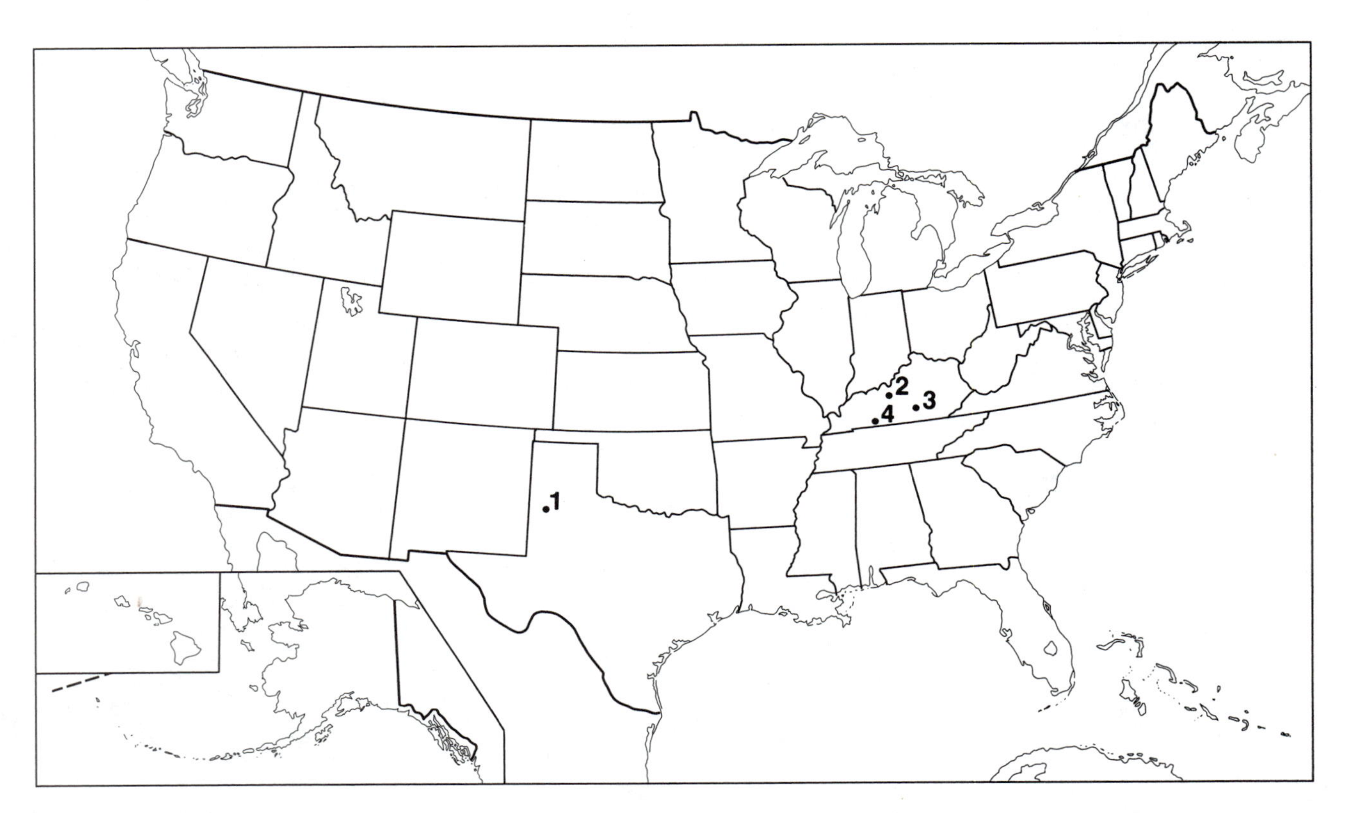

1. Figure 7–1. Anton, Texas.
2. Figure 7–2. Mammoth Cave, Kentucky.
3. Figure 7–3. Monticello, Kentucky.
4. Figure 7–4. Trenton, Kentucky.

7

Depressions

INTRODUCTION

There are several ways in which a depression can be created:

1. A local area may subside because of compaction, or from the removal of rock material beneath. For example, lowering of the water table can cause compaction; and solution can remove rock material.
2. Material may be deposited around a central area (e.g. by glacial meltwater or by wind), leaving the area as a low surrounded by the higher deposits.
3. Material may be physically removed from a local center (e.g. by the wind), leaving the surrounding ground higher.

The four maps for this chapter (all are in the back of the book) do not represent all of the above possibilities. Figure 7–1 is an area of depressions, the origin of which is debated. The other three maps all show limestone areas. However, in Figure 7–2 we use the depressions to determine the relative ages of two gently dipping formations, one of which is insoluble. In Figure 7–3 we use depressions to determine the relative ages of *three* such formations, two of which are insoluble. In Figure 7–4 we use the depressions to map a fault, of which only a short segment had been previously identified, presumably in the field.

You will also find depressions on several other interesting and provocative maps, all in the back of this book:

- ☐ Figure 3–2, Hyannis, Massachusetts (kettles)
- ☐ Figure 3–3, Greenfield, Massachusetts (kettles)
- ☐ Figure 6–1, Manly Peak and Trona, California (playa depression)
- ☐ Figure 11–3, Tulelake, California (graben)
- ☐ Figure 12–2, Rawlins, Wyoming (deflation?)
- ☐ Figure 13–2, Chattanooga, Tennessee (solution)

It is important to determine, if possible, the processes or agents responsible for the depressions that appear on our maps. But it is equally important to use our knowledge of these depressions and their origins to solve *other* geologic and/or geomorphic problems which we have identified on the maps.

Anton, Texas

(Figure 7–1 in back of book, C.I. 10 feet)

This is a small fragment of the extensive Llano Estacado, or Staked Plains, of northwest Texas. The extreme youth of the surface is evidenced by the fact that it has been cut by only a single stream. Elsewhere the area consists of a gently undulating plain, pocked by shallow, undrained depressions. Obviously this is a karst area.

At least it *looks* like an area of widespread solutional activity. However, the origin of these depressions has long been the subject of debate and confusion. It has also been proposed that they are (1) buffalo wallows, (2) deflation basins or blowouts, or (3) the product of differential compaction of the underlying sediments.

The lesson to be learned here is of great importance to the map interpreter, the photogeologist, or the interpreter of any other remotely sensed imagery: Several different processes can result in the same topographic appearance.

Upon encountering topography such as this, if we lack prior knowledge of the area and its problems, we should propose karst as a *possible* explanation. Most geologists, upon seeing this map for the first time, might identify these depressions as the products of solution, and let it go at that. After all, in most areas, solution produces this kind of topography. But it depends in part on where you are from. Someone from Canada or the north-central United States might assume it to be a typical *morainal* area.

Mammoth Cave, Kentucky

(Figure 7–2 in back of book, C.I. 20 feet)

The map name removes any doubt as to the role played by solution in this area. The numerous depressions may be identified as sinkholes and similar landforms.

However, this area displays a second distinct terrain type. A dissected upland rises 200 feet or more above the limestone lowland. It displays such nonsolution characteristics as well-defined valley networks and stream channels, and large expanses devoid of undrained depressions.

This map area lies within the Interior Low Plateaus geomorphic province, a region of structural simplicity and nearly horizontal sedimentary strata. Glaciation did not extend this far south, so the topography must be explained in terms of differences in rock type. We may think of this, then, as an area dominated by two principal lithologies.

One of the lithologies is carbonate, most likely limestone. (Dolomite, though somewhat soluble, is considerably more resistant and would not be expected to underlie such low-lying areas.) The southern one-third of the map area is thus readily identified as a limestone area.

More-resistant noncarbonate rocks, the other lithology, dominate the northern two-thirds of the area. These may be tentatively identified as resistant clastics such as sandstone, conglomeratic sandstone, or even conglomerate.

Note the isolated hills, such as **I** in the east, completely surrounded by limestone. Also note the solution depressions, such as **B** in the northeast, completely surrounded by clastic rock. This attests to the fact that these two rock types, carbonate and clastic, do not lie next to each other (neither steeply dipping nor in fault contact). Rather, they must lie one atop the other.

The resistant clastic must overlie the limestone. Erosion has separated the caprock of hill **I** from the main clastic plateau to the north. Hill **I** is thus an outlier of the plateau, just as the clastic rock which caps this hill is an outlier. Smaller hill **K** is also an outlier, a landform which in this region is called a *knob*.

On the other hand, solution of the limestone beneath the resistant caprock, accompanied and followed by the collapse of the caprock into the cavities, explains the numerous depressions (such as **B** and **A**) within the clastic plateau area. The limestone cropping out in these depressions is the older rock. Hence these outcrops are inliers, the counterparts of the clastic outliers to the south. The presence of both inliers and outliers over such a large area is convincing evidence of extremely gentle dip.

So we have large-scale solution, in an area where a resistant noncarbonate unit caps a limestone. This is ideal for creating the breathtaking underground scenery for which Mammoth Cave is world-renowned.

In this area of nearly horizontal strata, we would expect a well-defined dendritic drainage pattern, and a random distribution and orientation of topographic elements such as hills, ridges, valleys, lowlands, and depressions. Such randomness does prevail, but there are some notable exceptions.

Note that a definitely linear lowland/valley has developed along line **JC** in the east-central part. Such a linear topographic element is unlikely to be the product of haphazard activity. It is far more likely to be produced by selective solution and erosion along a fracture or fracture zone (joint or fault). There are several smaller linear features, both valleys and divides, that are just a bit too linear. Note east-west-trending valley **GH,** for example. And observe the arcuate line of depressions along **DEF** in the west.

Monticello, Kentucky
(Figure 7–3 in back of book, C.I. 20 feet)

Having studied Figure 7–2 (Mammoth Cave, Kentucky, in back of book), you should be able to see at a glance the numerous depressions and the several knobs (e.g. **I** and **J**) depicted on this map. Note that there are no knobs to the northwest of hills **I** and **J**. And note that most of the higher ground is in the southeastern 40% of the map area. These distributions are shown in Figure 7–3a, a highly generalized sketch of the area viewed from the southwest.

It is interesting, and significant, that in southeastern area **SV** the closed depressions are limited to valley bottoms. In the northeast-southwest-trending

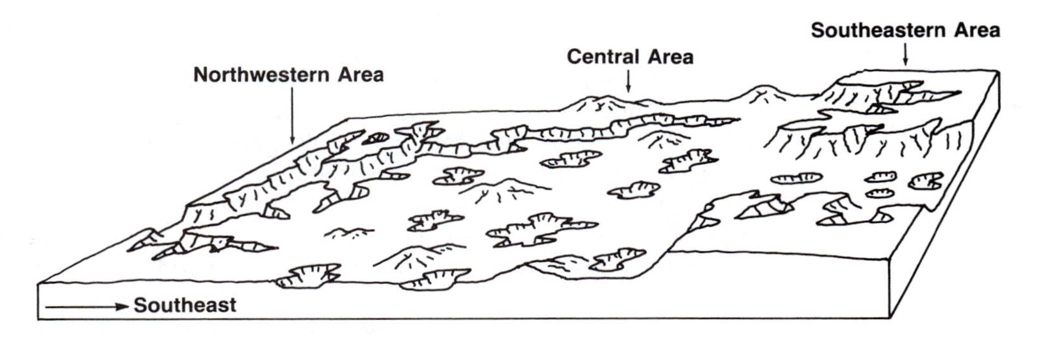

FIGURE 7–3a
Block diagram of the area depicted in Figure 7–3, viewed from the southwest.

central belt **QH,** which extends across the entire map area, closed depressions are in the lowland, above which the knobs rise. In northwest area **AD,** closed depressions are extremely small and few, occurring only on the top of the upland. In this northwestern area, all of the stream valleys, main and tributary, are cut below these upland depressions.

In the central belt, the depressions and shallow valley bottoms are at the same general elevation. In the south, the numerous tributary valleys are cut into the flanks of the highlands that rise well above the level of the depressions.

Does this make geological sense? Perhaps at first it seems not to. But let us concentrate on one area at a time.

Bedrock Geology. In the south, the limestone (it must be limestone to be so soluble) must lie stratigraphically beneath the sequence of resistant noncarbonate rocks that make up the highlands. Dip must be very gentle, since the solution depressions (hence the limestone outcrops) are distributed far back into the highland (plateau) area, probably beyond the southern edge of the map area.

Another indication of gentle dip: outliers **J, M, O,** and **I** of the resistant layers well to the north of the plateau front. The limestone in the valley floors of the south is the same as that underlying the central lowland that is overlooked by the outliers.

The bedrock in the northwest is not the same rock that forms the plateau in the southeast. How do we know this? And if these rocks are not the same, what are their respective ages? The plateau-forming resistant strata in the southeast are younger than the limestone. In the northwest, the solutional features are restricted to the divides and upper surfaces, below which the stream network is incised. Thus, the noncarbonates in that area must lie beneath the limestones.

Hence, there are three mappable stratigraphic units in the map area, the oldest outcrops in the northwest, and the youngest in the southeast. The regional dip in the area, which we deduced to be extremely gentle, is apparently toward the southeast.

The rocks that underlie and overlie the limestone sequence are themselves noncarbonate. May we further speculate that they are similar? They may be similar in that they are both noncarbonates. But, they do not display identical topographic characteristics.

The older rock unit in the northwest is finely dissected (high drainage density). The younger unit in the southeast is considerably more coarsely dissected (much lower drainage density). This contrast suggests that the younger formation is more coarse-grained, possibly a sandstone or a sandstone-conglomerate complex. And it suggests that the older formation consists of finer-grained strata, such as a sequence of siltstones, mudstones, and shales, with some sandstone interbeds. We cannot state that the older unit definitely includes all these rock types. But we can interpret the rocks as probably being finer-grained than the plateau-forming younger clastics.

That is as far as we can go on the bedrock geology, its structure, and lithology. Let us now turn to specific geomorphic features.

Geomorphic Features and Streams. Consider stream **F** in the north. It follows a remarkably linear (controlled?) southwesterly course, almost two miles in length. Then it turns to the northwest at **E,** to join stream **B** at point **C.** In this area of nearly horizontal strata, we would expect the dominant drainage pattern to be dendritic, a system of randomly oriented bending and branching insequent streams and tributaries. Thus linear stream **F** stands out as distinctly anomalous. It would be reasonable to designate it as subsequent along a fracture or fracture zone, be it joint or fault.

Once attention is called to this linear feature, and we become aware of the possible influence of fractures on the topography, we start to see other examples. For instance, note the three-mile-long line, trending northeast-southwest, defined by three linear stream segments—northeast-flowing tributary **K**, southwest-flowing segment **L**, and southwest-flowing tributary **G**.

Now return to streams **F** and **B**, particularly to their confluence at **C**. Stream **B** occupies a 200-foot-deep valley, although its course is clearly that of a meandering stream. The valley-side slopes around the outside of the meander bends tend to be noticeably steeper than those around the inside of the bends. This is an ingrown stream, an incised stream that has continued to cut laterally as it cut downward.

Of particular interest is the divide between stream **B** and its tributary **F**. The divide has been almost completely destroyed at **E**, due to lateral corrasion by stream **B**. Tributary **F** is about to be captured by its own master stream! When that capture occurs, the water of stream **F** will be diverted to join stream **B** at **E**, leaving the downstream segment of its present channel (from **E** to **C**) abandoned and dry. A byproduct of this activity is the creation of the isolated elongate hill between **C** and **E**, a feature which once had been merely the divide between streams **F** and **B**.

Hill **N**, another small, isolated elongate hill in the valley of stream **B**, lies a couple of miles to the southeast of **E**. There is no evidence that this feature was also produced by the capture of a tributary by the master stream. There is no tributary in the vicinity. Lateral corrasion did play its part. But, read on.

A few miles further to the southeast at **P**, note the extremely tight ingrown meander, the neck of which is being impinged upon from the south. This entire neck is in danger of being severed and isolated. Should this occur, all that will remain of the neck will be an isolated elongate hill standing above the valley floor. Perhaps a better explanation for hill **N** is that it, too, is all that remains of a former ingrown meander neck.

Finally, let us look at the larger streams that have cut down into the southeastern plateau. So long as they were eroding the upper noncarbonate sequence, they flowed in well-defined channels and valleys, such as streams **R** and **U**. However, once such streams penetrated the younger beds and encountered the limestones, solution depressions proliferated. You can see one on the floor of valley **W**, near the southeast corner of the map. In valley **T**, they have taken over, destroying the streams and changing the valley floor into a local karst lowland.

We suggest you complete Figure 7–3a by adding the two stratigraphic contacts to the front of the block.

Trenton, Kentucky

(Figure 7–4 in back of book, C.I. 10 feet)

Surface bedrock throughout the entire map area is the essentially horizontal Mississippian St. Genevieve Limestone, an extremely soluble rock (*Geologic Quadrangle Map of the Hammacksville Quadrangle, Kentucky-Tennessee*). The only structural detail shown on this segment of the geologic quadrangle is the east-west-trending fault **BD** in the northwestern part of this figure.

We consider this map of great value to the student of topographic-map interpretation. It is important because the mapped fault **BD** is so clearly reflected in the topography (although the up/down on the fault is not apparent). The western and central thirds of **BD** are expressed by linear, deep, steep-sided sinks. The eastern third is expressed by a linear subsequent segment (**CD**) of a stream, which is one of the two major streams that flow north-south across the map area.

A second reason this is an important map is that the topography reveals numerous unmapped structural features. Most are probably faults, but some may only be fractures. All of these features either were missed by the geologists who mapped the area in 1964, or else they were considered too insignificant to map. Whatever the reason, we are now free to map them.

Are they faults, or are they fractures along which no slippage has occurred? The topography does not provide the answer. But it does tell us that something has controlled the development of this linear stream segment, or that linear sink, or this alignment of sinks, or that linear slope, or alignment of a variety of topographic elements.

We can extend fault **BD** to the western border of the map (**A**). We can follow it eastward to point **E**. After we skip over to **F**, we can continue to **G** along what must be a subsequent linear segment of the easterly main stream.

A zone of variable width extends across the southern part of the map area. It is a complex of linear slopes, sinks, and stream segments that can be traced from point **J** to point **H**. A somewhat-arcuate similar band crosses from point **I** on the west border, through point **K** on the westerly stream, and on eastward to point **H**.

There are several other topography-revealed fracture lines that you may wish to delineate. A convenient way to map them without permanently marking the page is to place a transparent sheet, such as frosted Mylar, over the page. Trace the fracture traces onto it, together with the map corners and the main drainage lines. Indicate which line segments represent which type of feature by color-coding them. For example, trace subsequent stream lengths in blue, linear sinks or lines of sinks in orange, and linear slopes or scarps in brown.

Note the several places where stream-valley asymmetry—undercut versus slip-off slope—indicates flow to the south.

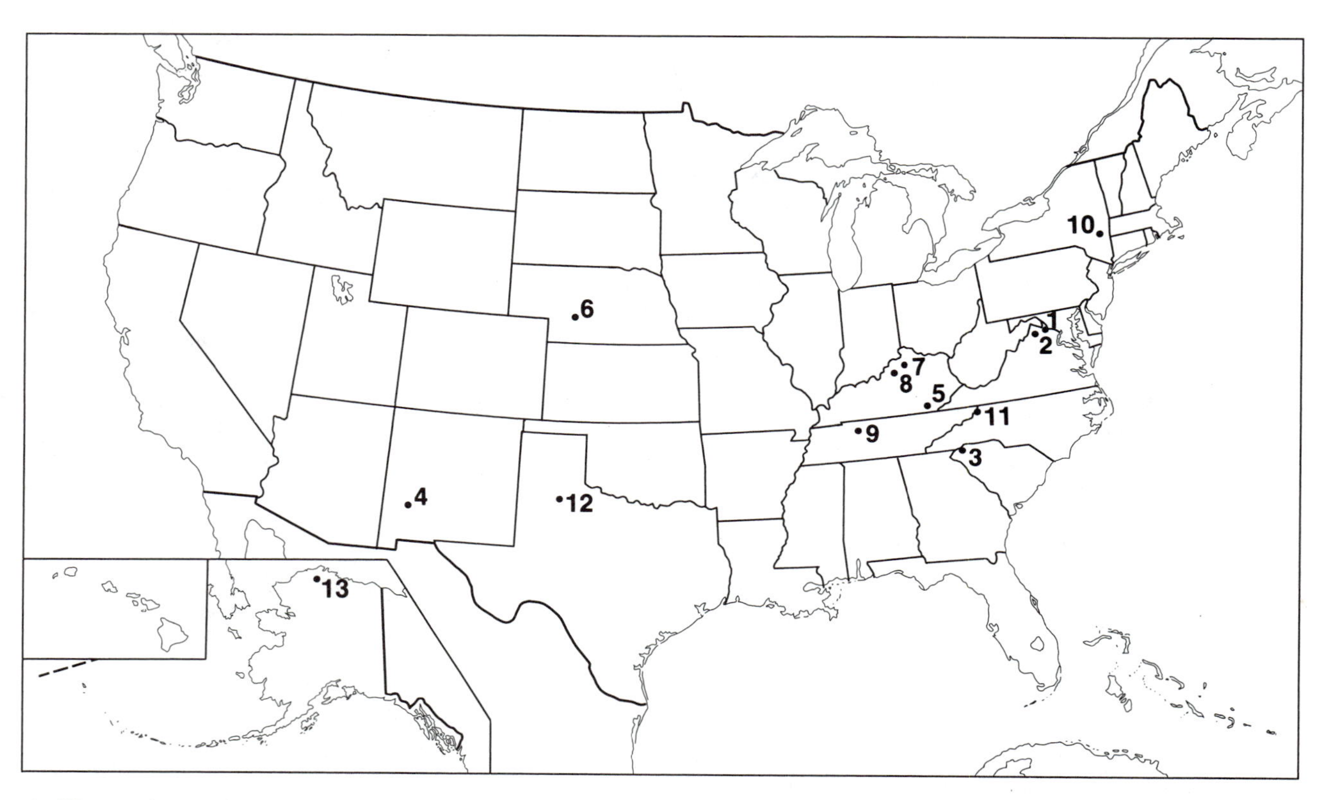

1. Figure 8–1. Bluemont, Virginia.
2. Figure 8–2. Orlean, Virginia.
3. Figure 8–3. Eastatoe Gap, North Carolina/South Carolina.
4. Figure 8–4. Santa Rita, New Mexico.
5. Figure 8–5. Kayjay, Kentucky.
6. Figure 8–6. Paxton, Nebraska.
7. Figure 8–7. Frankfort East and Frankfort West, Kentucky.
8. Figure 8–8. Gratz, Kentucky.
9. Figure 8–9. Ashland City, Tennessee.
10. Figure 8–10. Kaaterskill, New York.
11. Figure 8–11. Zionville, North Carolina/Tennessee.
12. Figure 8–12. Lubbock, Texas.
13. Figure 8–13. Lookout Ridge, Alaska.

8

Streams

INTRODUCTION

The study of streams demands that we consider more than just individual watercourses. We must also examine their valleys, their relation to surrounding drainage lines and terrain, their interaction with the underlying geology, and the role played by time.

You will find a generous variety of topography and stream types depicted in the 13 maps that we have assembled for this chapter. For example, Figures 8–7 and 8–8 illustrate recent changes in the valleys of ingrown meanders. In Figure 8–10 we concentrate on stream capture, while Figures 8–12 and 8–13 focus upon the results of large-scale divide migrations.

In contrast, Figure 8–1 shows the ways in which drainage patterns and stream-cut terrain reflect (or fail to reflect) variations and contrasts in bedrock types. In Figures 8–4 and 8–5, stream-cut topographic characteristics reveal the presence of large faults.

In previous chapters we examined areas in which streams play important roles. Excellent examples are the streams that contributed to the cutting of pediments and the building of alluvial slopes (Chapter 6). In the chapters which follow, we will examine area after area in which stream-study is essential to deciphering the geology and geomorphology.

Streams are probably the single most important agent in creating subaerial topographic features throughout the world.

Bluemont, Virginia
(Figure 8–1 in back of book, C.I. 10 feet)

This area of low topographic relief lies in the Piedmont[1] of northern Virginia, between Washington, DC and the narrow northern extension of the Blue Ridge. It is an area of complex Precambrian metamorphic rocks, the Ocoee Series. This series includes a considerable variety of lithologies, such as slate, quartzite, sandstone, conglomeratic sandstone, and others designated as "metamorphic rocks" on the *AAPG Mid-Atlantic Region Geological Highway Map*.

[1]A *piedmont* is a gently sloping surface developed at the base of a mountain front. The "Piedmont," on the other hand, is one of the major geomorphic provinces of the eastern United States. It is the broad, east- and southeast-sloping regional erosional lowland that lies between the Blue Ridge Mountains and the Coastal Plain.

The 1963 *Geologic Map of Virginia* identifies the rocks in this area as flows, schist, gneiss, arkose, conglomerate, and phyllite.

The homogeneity of the topography depicted on this map belies the significant lithologic differences and contrasts that characterize this area's complex structural assemblage. The drainage forms (essentially) a dendritic pattern, which generally indicates lithologic homogeneity or structural simplicity (e.g. horizontal strata). But neither is the case in this area, and it would be easy to deduce incorrectly from the map that the geology is very different from what it actually is.

This map, unexciting as it may appear, carries an important message: topographic variations and contrasts among various parts of a map area are usually reliable indicators of differences among the geological elements. However, the *absence* of such topographic variations in an area must not be taken to mean that the geology is *uniform*!

Note that we used the word *essentially* in reference to this area's dendritic drainage pattern. Superimposed upon this basic pattern is a here-and-there north-south topographic grain. Its elements range in direction from about N25°W (lines **AD** and **BC**) to about N15°E (line **EF**). There are other elements, both valleys and divides, that appear too linear to have resulted from erosion by insequent streams. (A good practical exercise is for you to map as many of these linear elements as you can.)

Although the area as a whole displays topographic homogeneity and dendritic drainage, the power of erosion to betray what lies beneath the surface has not been defeated completely.

Orlean, Virginia

(Figure 8–2 in back of book, C.I. 20 feet)

In contrast to the nearly uniform terrain of the area shown in Figure 8–1 (Bluemont, Virginia, in back of book), the terrain represented here may be divided into three distinct types:

1. The north-northeast-trending central lowland.
2. The large oval hill in the northwest.
3. The rolling, dissected intermediate area in the east.

The 1963 *Geologic Map of Virginia* shows isolated hill **2** to lie on the eastern edge of a mapped Cambrian-Precambrian unit. It consists of lava, schist, gneiss, arkose, conglomerate, and phyllite. There is no way we can determine from the map which one or more of these lithologies may constitute the bedrock in this hill. The geologic map also designates low area **1** and intermediate area **3** as part of a Precambrian formation. It includes granite, gneiss, and quartz monzonite.

What can we ascertain about the geology of the area by studying the topography as it is portrayed on this map? Like the area in Figure 8–1, this is also part of the Piedmont geomorphic province. It has been subjected to subaerial weathering and erosion for vast intervals of geologic time. It is an area in which differences in bedrock resistance might be expressed by perceptible, if not sharply defined, differences in topography. In attempting to interpret the geology of this area, we make the following observations.

The central lowland (area 1) is underlain by less-resistant rocks than those of the higher areas (**2** and **3**). We may think of this as a subsequent lowland, since it is occupied by subsequent stream **AB** and its tributaries. The overall topographic homogeneity of this lowland is misleading. Though it suggests that the lowland is underlain either by flat-lying strata or by a single nonresistant lithology, the geologic map indicates that it is almost certainly underlain by neither.

Large hill **2** consists of much-more-resistant rock than that underlying either lowland **1** or area **3**. It resembles, as a topographic feature, many of the so-called *monadnocks* of New England. (We say "so-called" out of deference to the ongoing argument over whether peneplains, and hence monadnocks, actually exist.) Whatever you choose to call them, there are many other such isolated hills and mountains standing along the western fringe of the Piedmont, as outliers of the frayed front of the Blue Ridge Mountains.

According to the geologic map, this large hill is underlain by the same suite of rocks that crop out throughout the entire area that is depicted in Figure 8–1. However, the contrast between this hill and the terrain shown in Figure 8–1 is startling.

Eastern area **3** is definitely more reflective of bedrock control than areas **1** and **2**. The topographic grain is well developed. The dominant factor in its formation is the selective etching of parallel northeast-trending planes or belts of nonresistance (probably fractures), and possibly less-resistant, steeply dipping rock units.

If you add drainage lines to the entire map (on a frosted Mylar or similar overlay), you will reveal an eye-catching array of subsequent streams in area **3**. In contrast, you will see many apparently insequent tributaries in areas **1** and **2**. Area **3** seems to be drained (overall) more by a parallel drainage pattern, or some variety of trellis pattern, than by a dendritic pattern.

Remember, both areas **1** and **3** are said to be underlain by the same Precambrian formation, a unit made up of several different rock types. But it appears that the two topographic units (areas **1** and **3**) are different because the specific rock types in one are significantly *different* from those in the other. They are different in their resistance and bedding, and/or other characteristics.

Eastatoe Gap, North Carolina/South Carolina
(Figure 8–3 in back of book, C.I. 40 feet)

The *Geologic Map of the United States* shows this Blue Ridge Mountains area to be underlain by schist, granite, gabbro, and gneiss. In such mountainous country, maturely dissected, with a topographic relief of over 1000 feet, it is impossible to extract a detailed geological interpretation from a topographic map. However, the map does reveal a few geological elements.

At first glance, the dominant drainage type seems to be the subsequent stream. Many such streams apparently are adjusted along a set of parallel, northeast-trending lines of weakness or nonresistance. There are numerous such streams, but they do not dominate throughout the entire map area. In fact, we can divide the map into two distinctly different topographic sections, separated by line **AB**.

To the northwest of **AB,** the topographic grain is displayed by uplands and divides, and by the larger drainage lines. The grain is oriented along the northeast-southwest trend.

To the southeast of **AB,** several linear northeast-southwest-oriented streams are prominent. Yet, the drainage as a whole branches out in all directions, to form more of a dendritic pattern than a parallel or trellis pattern. We might refer to the drainage pattern here as a "dendritic pattern, modified by an auxiliary or secondary parallel pattern."

If we were to field-check this area, we would investigate the possibility that **AB** divides two rock types or assemblages that are sufficiently different to have caused the development of the two dissimilar terrains.

Santa Rita, New Mexico

(Figure 8–4 in back of book, C.I. 25 feet and
Figure 8–4b in back of book, *Geology of the Santa Rita Quadrangle*)

Here is another area of considerable relief—up to 1000 feet—from which we can extract only limited geologic information based on topography alone. One thing that we can do, as we did in Figure 8–3 (Eastatoe Gap, North and South Carolina, in back of book), is to divide the area into two sections. The dividing line extends northwest-southeast across the map from **B** to **S**. Each section has its own distinct topographic characteristics.

The contrast between the two sections is shown in Figure 8–4a, a sketch of the area viewed from the east. Note how the uplands in southwestern section **1** and **A** rise several hundred feet above the highest ground in northeastern section **2**. Also, note that the drainage pattern in section **1** is dendritic. The drainage density in this section also is much lower than in section **2**. (Another way of stating this is to say that the topographic texture in section **1** is considerably "coarser" than in section **2**.)

The upper surface of section **1** consists of gentle, broad, rounded slopes. This kind of upland terrain suggests gently dipping-to-horizontal layers of rock. Such layers could be either a sequence of sedimentary rocks, or a sequence of pyroclastics and/or lavas.

The low bench, represented by relatively flat areas **N, O,** and **Q,** appears to be an erosional platform. It developed on a lower resistant unit from which the overlying beds have been stripped away.

USGS geologic map GQ-306, *The Geology of the Santa Rita Quadrangle, New Mexico,* shows this western section to be an area of Paleozoic strata, ranging in age from Silurian to Pennsylvanian. Generally gently dipping, the strata form a low plateau which rises above the lowland (section **2**).

Before feeling too confident over our possible explanation of section **1,** look carefully at Figure 8–4b, in back of book (taken from GQ-306). Note the many structural features that we cannot even recognize on the topographic map, let alone interpret. In the area we have designated as section **1,** 32 faults have been mapped. Section **1** includes local dips of 10°, 15°, 20° (e.g. at **A**), and

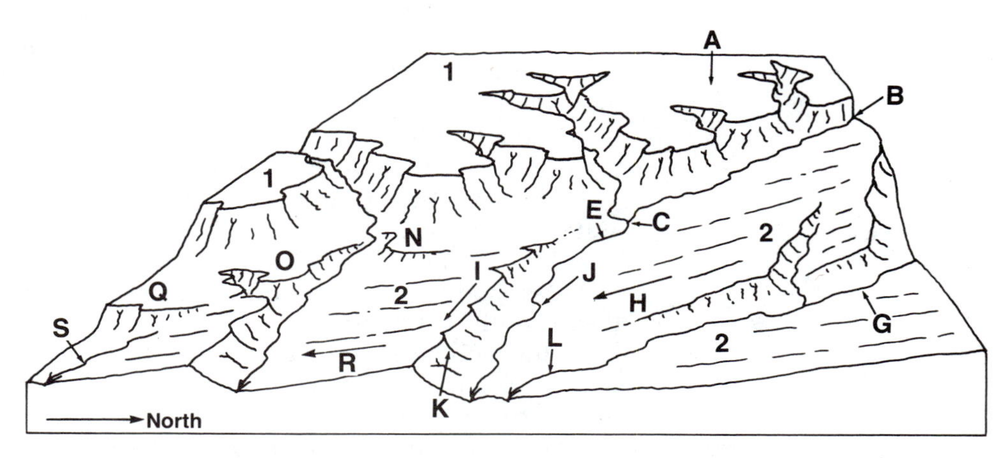

FIGURE 8–4a
Block diagram of part of the area depicted in Figure 8–4, viewed from the east.

even 25°. Most of the faults trend northeast-southwest, roughly normal to our dividing line **BS**.

We have consulted a detailed geologic map. We have learned that section **1** is an area rather severely cut by faults, and an area of variable dip magnitudes (and directions, for the beds dip in all directions). We must settle for the fact that most of these geologic complexities are not reflected in the topography (although a subsequent stream *does* seem to have developed along fault **Z** in Figure 8–4b).

However, the overall topography does suggest that this is what it is, a plateau area.

Section 2. Now, what about section **2** and line **BS**? We have already noted that section **2** is an area of higher drainage density (finer drainage texture) and lower topographic relief than section **1**. In addition, section **2**'s drainage pattern is assuredly not dendritic. However, we are not quite sure *what* it is!

Topographic asymmetry is one of this area's most outstanding characteristics. This asymmetry is well displayed along divide **IM**. North slopes along **IM** (slopes **K**) are far steeper and more intricately dissected than the divide's south slopes **R,** which are also much longer. Such asymmetry is very typical of the divide asymmetry (cuesta-like) that develops in a sequence of gently dipping strata. It is so typical that we suggest adding south-dip symbols to the map near **I** and **M**. However, it would be wise to indicate that these dips are questionable, since the asymmetry might have some other explanation.

Asymmetric divides are accompanied by similarly asymmetric valleys. Such asymmetry is almost flamboyantly displayed in the contrast between short, steep north slopes **K,** and across the valley from them, long, gentle south slopes **H**.

Section **2,** then, is truly different from section **1**. It may be lower because its rocks are less resistant, or because it has dropped down to a lower position. Or it may simply be a low area of accumulation that has not filled to the level of the adjoining upland to the west. The map cannot tell us which of these is the correct answer, if any. (There may be yet another explanation.) It does tell us, however, that the rocks in section **2** are different. Perhaps they are more fine-grained, or more poorly cemented, or of different permeability. Thus they would be subject to more-intricate dissection.

Whatever the structure of section **2,** it is a structure on which a well-defined topographic asymmetry has developed. Gentle south dip would certainly account for this. So would some kind of block faulting, each block tilted down toward the south.

Note that several streams in section **2** are not only quite linear, but are parallel to each other and to line **BS**. In fact, several are developed along line **BS**. The largest of these apparently subsequent streams and stretches are segments **BC** and **EJ** of the main northerly stream, stream **GL,** smaller tributary segment **DF,** and the two small valleys in the southeast that constitute line **PS**. All of these should be mapped as possible fault traces.

It should come as no surprise that GQ-306 shows a major fault, upthrown on the southwest, extending across the entire map from **B** to **S** (Figure 8–4b). This map indicates section **2** as occupied by "semiconsolidated gravel deposits" of Upper Miocene (?) and Pliocene age. It does not show a single fault trace or a single dip symbol in section **2**.

But we have seen the pronounced topographic asymmetry, and the anomalously linear drainage lines. We will stick to the maxim that the prisoner is guilty until proven innocent. We stand by our postulated dip and our fault control in this section, until they are conclusively disproven in the field.

Strong and widespread asymmetry and linear streams and valleys (particularly parallel linear) are simply not developed in undisturbed horizontal sediments!

Kayjay, Kentucky
(Figure 8–5 in back of book, C.I. 20 feet)

The topographic relief throughout this area is relatively constant, approximately 1000 feet. Only in that respect, however, can we consider the area to be a single geomorphic or topographic unit. We can divide it into three well-defined sections. Linear, northeast-flowing streams **CA** and **KI** serve as boundary lines that separate northwest section **1,** central section **2,** and southeast section **3**.

The drainage patterns in geomorphically homogeneous sections **1** and **3** are dendritic. The absence of oriented drainage or divides in these sections suggests a lack of local structural control such as faults, joints, or steep dips. The *AAPG Geological Highway Map of the Mid-Atlantic Region* shows that this entire map area is underlain by Pennsylvanian clastic strata, including thick, coarse-grained units. (The only exception is a narrow strip of Mississippian carbonates in section **2**.) Dendritic drainage patterns in such rocks are typically developed in areas of horizontal-to-nearly horizontal strata.

Linear, northeast-trending section **2**, actually an asymmetric ridge, contrasts sharply with sections **1** and **3**. Its linearity obviously sets it apart from these areas, in which the *absence* of linearity is a prime topographic element (see Figure 8–5a).

Section **2** is asymmetric in more than one way. Its basic asymmetry is that its north flank is steeper than its south flank. Also, its steeper north flank is quite concave in profile, its upper part being very steep, almost cliff-like. The overall simplicity of the north flank, its "smoothness," in no way resembles the complexly dissected, irregular south flank. To study this asymmetry, we suggest you draw a northwest-southeast profile through point **L,** from stream **CA** to stream **KI**.

This ridge displays the classic profile of a hogback, a homoclinal ridge developed on dipping beds of rock. Some beds (the ridge-formers) are more resistant than others (the ones that underlie the adjacent valleys). The steeper

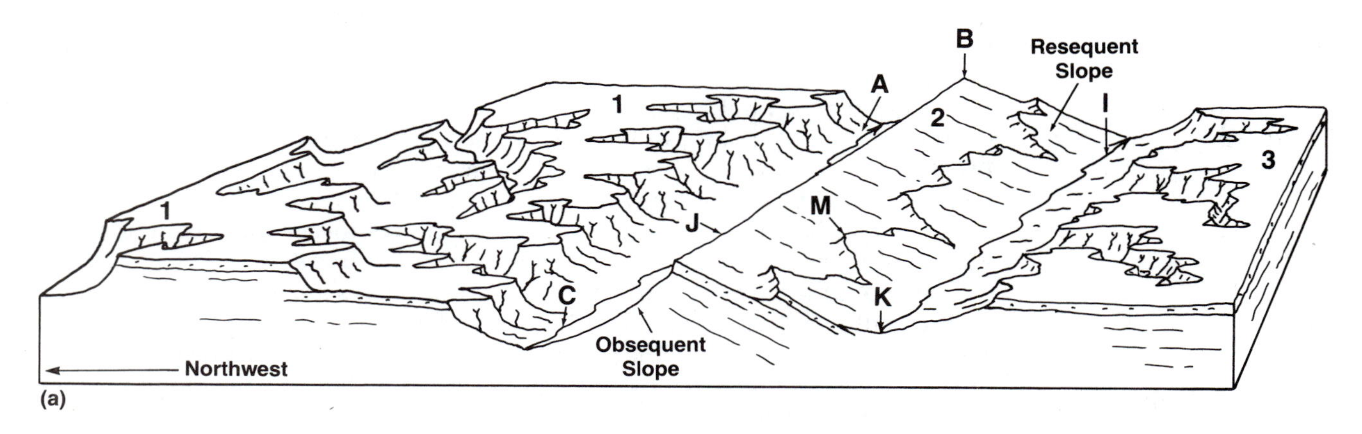

FIGURE 8–5a
Block diagram of part of the area depicted in Figure 8–5, viewed from the southwest.

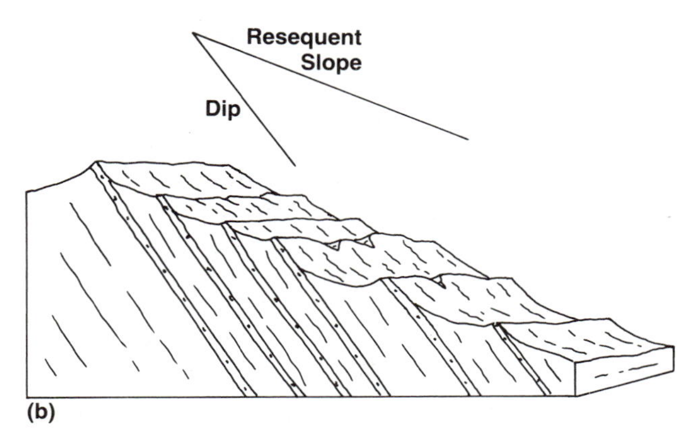

FIGURE 8–5b
Block diagram showing the relation between dip and resequent slope, where dip exceeds slope.

north slope, particularly its steepest upper part, displays typical obsequent-slope characteristics. (The obsequent slope is sometimes called the scarp slope.)

If this ridge resulted from recent tilting, its present south slope would be the dissected original slope. The streams draining it to the southeast would be consequent streams, and the slope itself would be a consequent slope or flank. However, this is an area in the Cumberland Plateau of Kentucky. Little major tectonism has taken place here since the end of the Paleozoic Era. The preservation of an original slope is extremely unlikely. So, the south slope is a resequent slope, or back slope (although not a dip slope—we will explain later).

Resequent Slope. Let us now examine the topographic details displayed on the resequent slope of this long ridge. Many of these details are extremely important to understanding the geologic characteristics of the strata underlying that slope, particularly the stratigraphic and structural characteristics.

Look at southeasterly sloping south flank **JBIK**. It is bordered on the south by stream **KI,** and on the north by ridge crest **JB**. Note the numerous minor topographic features, such as divides and valleys, that trend northeasterly, parallel to the ridge. We have indicated a few of these features: divide **L,** divide **D,** tributary **F,** and isolated small, elongate hilltop **G**. Also note the upside-down topographic V's, such as those traced by broken lines **M, E,** and **H**.[2] All point to the fact that this slope, though resequent, is by no means a dip slope.

Instead, it is a slope developed on the upturned edges of numerous, relatively thin, resistant and nonresistant beds. They combine to make a great, tilted stratigraphic sandwich of sorts. The beds all dip more steeply than the overall slope of the south flank. The resistant beds are apparently sandstones and somewhat finer-grained clastics, and the nonresistant interbeds are probably more shaly. The resequent slope, viewed from the southwest in Figure 8–5a, is perhaps too clean-cut and dip-slope-appearing to convey properly the true relation that exists between the dipping strata and the slope itself. For that reason we include in Figure 8–5b a more-detailed view of a section of the resequent slope. Note that the dip is much steeper than the overall slope.

The strata that underlie the main ridge-forming bed do not form scarp-slope, stair-step irregularities along the obsequent flank of the main ridge. This

[2]For a discussion of topographic V's, refer to Figure 2–3 and its accompanying text.

suggests that the north-facing slope lacks significant lithologic and resistance variations, in contrast to the south slope.

Faults. Strike faults are usually much more difficult to detect on topographic maps than faults that cut obliquely across the strike of the bedrock itself. For example, is line **CA** the trace of a steeply dipping fault plane, with its strike parallel to that of the upturned beds to the south? It could be, since we have established that **CA** separates nearly horizontal beds from dipping beds. There is nothing in the topography immediately to the northwest of **CA** to suggest that the beds in that area are dipping to the southeast. Instead, they seem to maintain their horizontality right up to line **CA**.

Line **KI,** on the other hand, appears to mark a zone of transition (perhaps a flexure) between the dip along the resequent flank of the ridge and the flat-lying strata of section **3**. Associated with this flexure may be some strike faults. These could account for the linear stretches of stream **KI**. In addition, there might be enough dip along this transition belt to expose some upturned nonresistant beds. This would have contributed to the development of subsequent valley **KI** along this general trend.

Line **CA** is the trace of the famous Pine Mountain thrust fault. The flat-lying rocks of section **3** have been thrust to the northwest along an undulating fault plane. It bends sharply upward along its front, rising to the surface as an extremely steeply dipping plane. Along the trace of the fault plane, stream **CA** has developed.

As mentioned, virtually the entire map area is underlain by Pennsylvanian clastics. Beneath these strata are Mississippian carbonates. They have been mapped in a narrow strip running along the base of the obsequent flank of the homoclinal ridge, immediately southeast of fault trace (subsequent stream) **CA**. You may wish to add this information to the profile you drew through point **L**.

Paxton, Nebraska
(Figure 8–6 in back of book, C.I. 20 feet)

As you can see at a glance, numerous sinks are the outstanding topographic features in the uplands that border the floodplains. Having established that obvious fact, we consult a copy of Raisz, *Landforms of the United States,* and find that he describes this area in such terms as "loess" and "Sand Hill Region." Suddenly the sinks are transformed into *depressions between and among eolian hills and hummocks*.

Beware the obvious!

However, our interest does not lie in the uplands. Our main concern is the two large streams that flow across the area from west to east. These are the North Platte River (**A**) and South Platte River (**B**), which join to the east.

The North Platte drains the entire southeastern quarter of Wyoming, plus part of the panhandle of Nebraska. The South Platte, on the other hand, drains a much-smaller area of northeastern Colorado, including the Colorado Piedmont and part of the Front Range.

The fascinating thing about these streams is that, in this area, they are so very different in character. The North Platte is a single wide river, its surface broken here and there by streamlined sandbars, all quite small (except one). Contour lines (C.I. 20 feet) cross the North Platte about every four miles. The gradient of this river is thus approximately five feet per mile.

The South Platte, on the other hand, consists of an intertwining network of small channels. This is a more obviously braided stream. As you can imagine, much of the water is moving downstream through permeable and porous sediments.

The gradient of the South Platte is greater than that of the North Platte. Contours cross the South Platte about every 2.5 miles, revealing a gradient of about 8 feet per mile (nearly *twice* that of the North Platte).

This map does not show us why the stream dynamics of the South Platte are so much different from those of the North Platte. No map of this small area could conceivably do that. But it clearly tells us that these are profoundly different streams—different in appearance, different in what they are doing, and different in how they are doing it.

Frankfort East and Frankfort West, Kentucky
(Figure 8–7 in back of book, C.I. 10 feet)

On this map we have an outstanding example of the elusive, sharing the spotlight with the obvious. First, let us label a few topographic features and relations. The two most eye-catching are the large oval features, which at first glance appear to be floodplain meander scars. Outside bluffs **CM** of the easterly meander and outside bluffs **GQ** of the westerly meander are both clean-cut. They could hardly have been carved by agents other than impinging stream bends.

However, in the center of oval **CM** stands 200-foot-high hill **D**. And in the core of oval area **GQ** stands a similar high feature, albeit of different outline, ridge **P**.

These are cutoff or abandoned meander bends. But they could not have been cut off as the streams snaked their way laterally across broad floodplains. Meander **CM** had been incised more than 150 feet below the upland surface prior to its being cut off at the neck, at or near point **E**. Meander **GQ** had been incised more than 250 feet below its nearby upland before it too was cut off, somewhere to the west of **O**.

Note the three gentle-slope arrows **H**. This is a *slip-off slope*. For one thing, it tells us that the stream that cut the oval valley segment, the meander, flows to the north (arrow **F**). As the ingrown meander cut down, it also cut outward in response to centrifugal force. (An *entrenched* meander would have equal side slopes, produced by the stream's downcutting while maintaining its horizontal or planimetric position.)

The highly elongated contour-V's at point **N** indicate that the easterly stream also flows to the north.

Note that the width of stream **F** "fits" the radius of curvature of its cutoff meander **GQ**. We can imagine a stream of this size describing just such a meander. Meander bend **CM** has about the same radius of curvature. So, we may assume that the stream that once flowed around that meander must have been about the same size as stream **F**.

Consider, however, the narrow width of stream **N**. And, to the south, note the radius of curvature (**S**) of one of stream **N**'s present meanders. Radius **S** is a reasonable dimension for a meander of a narrow stream such as **N**. But radius **S** is a far cry from the radius of meander scar **CM**!

How can we explain the size of meander scar **CM**? The map does not provide the complete answer, but it helps. It tells us that something has happened to the river since the cutoff occurred. Whatever happened, it apparently decreased both the size of the river itself, and the size of its meanders. What the map does *not* tell us is *what* happened, or *why*.

Abandoned channel **L** of meander **CM** is about 150 feet higher than abandoned channel **I** of meander **GQ**. Yet, these two empty channels are only about one mile apart. At the time meander **CM** was cut off, channel **L** was shifting laterally toward the west-southwest, toward channel **I**. And at the time meander **GQ** was cut off, channel **I** was migrating laterally toward channel **L**.

As it happened, both meanders were cut off, almost certainly at different times. But let us give free rein to our imaginations and picture what would have happened, had their necks not been severed. Eventually the two ingrown streams would have met, not head-on, but side-to-side, perhaps near what is now upland point **K**.

If that had happened, stream **N**, flowing from near **N** to **E** and then bending to the west and southwest, would have made its way around to point **K**. At that point it would not actually have encountered stream **F**, but would have found itself looking *down* upon stream **F** from a height of some 150 feet! In other words, stream **N** would have been captured by stream **F**.

Downstream from the point of capture (**K**), the entire channel of stream **N** would have been left abandoned, except for the waters delivered to it by tributaries. Any stream that might later have formed along the floor of that large abandoned valley, in the channel of the former large stream, would have been much too small for its new valley. It would have been an *underfit* stream.

Perhaps something like this series of events happened somewhere upstream (to the south) from this map area.

Why are tributary valleys **A**, **B**, and **J** so linear?

Gratz, Kentucky

(Figure 8–8 in back of book, C.I. 20 feet)

Given enough time, ingrown streams produce a suite of topographic features and relationships. These contrast sharply with features created by either actively downcutting streams or streams that meander across floodplains but do not downcut. The sharp contrast is particularly notable for ingrown streams that cut laterally at an appreciable rate, while downcutting much more slowly.

Look at Figures 8–8 and 8–8a. What sequence of events must have led to this topography? Let us begin by pinpointing the features we will need for interpretation and reconstruction.

Either stream **L** is a major tributary of stream **A**, or **A** and **L** are segments of the same river. The latter is more likely. Both have the same width, and the curvatures suggest that **L** leaves the map at **V**, swings to the west and north,

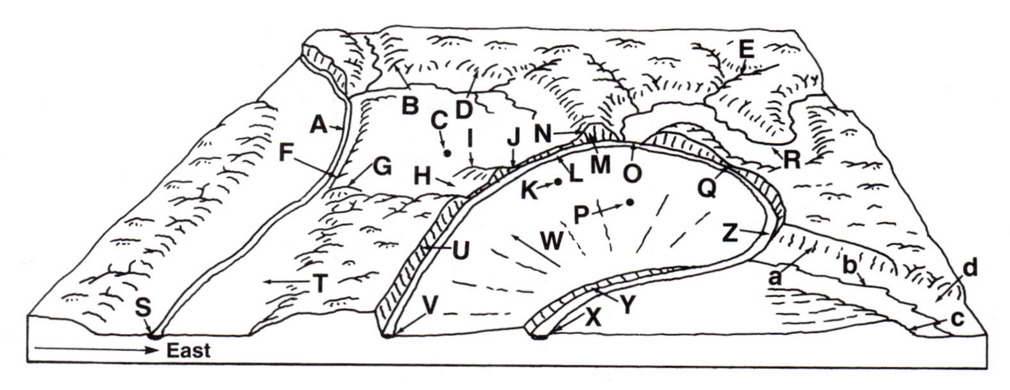

FIGURE 8–8a
Block diagram of the area depicted in Figure 8–8, viewed from the south.

and reenters the map at **S** to continue as **A**. Nevertheless, we shall discuss these two segments virtually as distinct entities. So, let us retain their separate designations.

Undercut slope **U** and gentler slip-off slope **T** reinforce our assumption that stream **L,** after leaving the map, loops back as stream **A**. A similar asymmetry is displayed between undercut slope **Y** and slip-off slope **W**. Segment **L,** entering the map area from the south, is deflected to the northeast near **Y**. This accounts for the undercutting along that stretch. Were the river to flow out of the map area at **X,** it would probably be slipping away from, rather than undercutting, at **Y**.

The undercutting at **Q** is most interesting. Very little additional cutting will be required before the divide (Figure 8–8a) between stream **L** and its tributary, stream **R,** is completely severed. When that cut-through is accomplished, stream **L** will continue to migrate laterally until it encounters stream **R**. At that time, all of stream **R** upstream from point **R** will be diverted (captured) into stream **L**. Such a capture, or beheading, of a tributary by its own master stream is more common than you might imagine.

When stream **R** is diverted, all of its downstream segment **RO** will be deprived of virtually all of its discharge. Valley segment **RO** will become a valley without a proper stream to occupy it. Tributary **E** and its smaller companions will continue to discharge into valley **RO,** so a very small stream will be maintained along this valley section, to remove the incoming waters of these minor tributaries. However, this new stream, which we may call stream "ro", will be a miniature vestige of its former self. Because stream "ro" will be much too small for its valley, it will be an excellent example of the type of misfit stream called an *underfit stream*. (Refer to stream **N** in Figure 8–7 in back of book.)

Note, in the southeast, tributary **b** and its valley **d**. The valley appears to have been cut by a much-larger stream than **b**. The map does not include enough area to the southeast for us to examine this valley further. However, this is an extremely interesting feature. It is beautifully displayed on the 1/24,000 Gratz, Kentucky quadrangle and its companion to the east, the 1/24,000 Monterey, Kentucky quadrangle.

The large meander of stream **L,** from **Y** to **Q** to past **U,** is a present-day feature. So are the undercut slopes downstream from point **Z**. What must be a former undercut slope, comparable but gentler, can be traced as follows: clockwise from point **B** (in the northwest) to **D,** to **N,** to **I,** to **G**. There is an interesting feature between **N** and **I**. Here the northwest-facing undercut slope needed to complete bend **BDNIG** is missing. Not only is it missing, but it is replaced by southeast-facing undercut slope **MJ**.

Discontinuous scarp or slope **BDNIG** originally must have been as steep and sharply defined as undercut scarp **ZQ**. But now, its entire length is considerably gentler and much-more dissected. We may say that these gentler dissected slopes are older-looking than scarp **ZQ**. Incidentally, the side slopes or bluffs overlooking valley **d** (slopes **a** and **c**) are also somewhat gentler than undercut slope **ZQ,** though they are not quite as gentle as slope complex **BDNIG**.

The infacing slope **BDNIG** is also broken between **I** and **G,** by low area **H**. **H** appears to be all that remains of the valley of a stream that once flowed into lowland **C** from the southeast.

Stream **L** is undercutting—hence, it is laterally migrating to the west at both **V** and **X**. Since the scarp along **V** is more than twice as high as that along **X,** we may assume that the rate of lateral shifting is greater at **X** than at **V**. Hence, meander bend **X** eventually will overtake **V,** cutting off the entire meander **L**. At that time, all further undercutting around scarp **ZQU** will abruptly cease. Thereafter these scarps will be subject only to normal downwasting processes. These processes will transform them into more-dissected and gentler slopes, similar to those of present slopes **BDNIG**.

GINDB and Area C. We are able to witness what is happening at **X** and **V**. This allows us to visualize what area **YZQJU** will look like after the cutoff at **V**. And this background prepares us to work backwards, in area **BDNIG,** from the present to the past. . . .

We can imagine stream **A** as it looped around in a great bend to the east, along the inside of then-scarp **GINDB**. We see it as it was actively undercutting the surrounding scarp, and slipping away from the much-lower central area **C**. As we move forward in time, we see its meander neck being cut off, probably just west of **A**. Note the high scarp at **F,** 50 feet or higher. This scarp resulted from the downcutting, and slight eastward lateral shifting, of stream **A** following the cutoff. If we may equate downcutting with time, we can say that the cutoff of the meander neck at **A** took place some 50 feet ago.

At the time of that cutoff, in-facing scarp **GINDB** was continuous. A divide existed between that meander scarp and the one to the southeast, which was being undercut by meander **L**. The divide must have extended northeast-southwest through point **K**. There is no way of knowing how wide the divide was at that time, but it most assuredly existed. Stream **L** may have flowed near point **P,** the center of the present meander loop. Later, well after the abandonment of meander lowland **C** by **A,** meander **L** worked its way laterally the rest of the distance through the divide, until it reached its present position along **MJ**. It not only destroyed the divide, but even consumed a lens-shaped slice of lowland **C** as well.

In conclusion, how do we know that lowland **C** was not invaded laterally by **L,** before the cutoff at **A** took place? Whether stream segment **L** is a tributary to stream segment **A,** or is merely an upstream part of **A** (which it is), the elevation of **A** must be below that of **L,** since it is downstream. It follows that, had **A** and **L** met by sweeping toward each other, then **A** (the lower) would have diverted **L**. Thus, from that time on, **L** would have flowed northwesterly into **A,** via "gap" **NI**.

Since it does not do this, we must conclude that the two meanders never met. Stream segment **L** may have been approaching **NI** from the southeast at the time that loop **A** was cut off, at or near point **A**. But it arrived too late. The high bluff along **MJ,** over 80 feet, shows that, by the time **L** arrived along that stretch, it (L) had cut down an additional 80-or-so feet below where it had been at the time of **A**'s cutoff.

Ashland City, Tennessee
(Figure 8–9 in back of book, C.I. 20 feet)

Situated in northwestern Tennessee, this is an area of extremely low regional dip to the northwest. Dip is away from the Nashville Dome (Nashville topographic basin), which occupies central Tennessee. The main stream, a tributary to the Cumberland River, is ingrown. This is indicated by the well-defined valley cross-section asymmetry (e.g. undercut slopes at **T, Q,** and **O,** and gentle slip-off slopes at **M, P,** and **N**.

The topography south and southeast of the river consists of steeper slopes, and has a higher drainage density than most of the area to the north and northwest. This contrast suggests two different lithologies in the area.

Since northwest regional dip prevails here, could the observed contrast be due to the disappearance of the older rocks (which crop out in the southeast) beneath the younger strata (which are preserved in the northwest)?

In areas of such gentle dip, we would not expect to find streams that are resequent, obsequent, and subsequent (along strike). Rather, the most-likely genetic stream type should be the insequent stream. Hundreds of these should combine to form a dendritic pattern.

There *are* many tributaries in this area that tend to bend and branch out haphazardly in all directions. Yet, closer examination of the map reveals that a surprising number of streams, either alone or in concert, define what could be linear and gently bending bands or belts of nonresistance, such as the traces of steeply dipping joints or faults.

The most apparent contrast is that between the meandering main stream, and the several long and short tributaries in the north (stream segments **A, B, D, E,** and **F**). These define what is apparently a belt of nonresistance (subsequent stream adjustment). This long band extends across the entire map, and is believed to be the trace of a fault zone. This zone is made up of several steeply dipping en echelon and parallel fractures and faults.

Whereas belt **AF** is the only one that crosses the map area from border to border, there are others which, though shorter, are equally well delineated. They are oriented in several different directions, suggesting that this area is broken by a *system* of several intersecting fracture sets. Some of these are clearly identified by subsequent streams and stream segments, such as streams **G, H, I, C,** and **L**. Others parallel these linear streams. In the southeast, east-west line **RU** is well defined by a segment of one stream, plus one of its small tributaries.

We wonder whether some of the "linear" segments of the main stream, such as north-south segment **KR,** are structurally controlled, and whether segment **KR** is merely a part of a much longer control belt that extends north-south from **C** to **S**.

It is not easy to decide. Since the scarp near **K** is steepened by undercutting, and since segment **KR** must therefore be corrading laterally to the west, its present position must be temporary. In fact, tributary **J** (the head of which is at **C**) is about to be captured, near point **J,** by the westward-cutting master stream.

Note the similarity between this pending lateral capture of a tributary by its master stream, and the capture that is in the future of tributary **R** in Figure 8–8 (Gratz, Kentucky, in back of book).

Kaaterskill, New York

(Figure 8–10 in back of book, C.I. 20 feet)

This is a portion of a quadrangle that was originally mapped nearly a century ago (1892), at a scale of 1/62,500. The mapping quality is remarkably good. The area shown is part of the eastern front of the Catskill Mountains. It is a glaciated area of gently westerly dipping clastic strata, many of which are resistant to weathering and erosion. Our principal interest is in the streams—what they have done, are doing, and will do to each other.

The key is at point **L** on Figure 8–10. This unexciting low divide lies between northwest-flowing stream **K,** which heads up at **M,** and canyon-cutting east-flowing stream **P**. The gradient of stream **K** is extremely gentle, in sharp contrast to the steep gradient of stream **P**. The rate of downcutting along stream **K** is obviously very slow. This is evidenced by the fact that, along most of its length, it flows more or less on the floor of its broad valley.

Stream **P,** on the other hand, must be actively downcutting. It is scouring out the head of its canyon, cutting deeper and deeper into the plateau front. At point **L,** then, we can imagine the headward growth of stream **P**. In the not-too-distant future, **P** will have worked its way westward to the channel of stream **K,** and will capture or behead it. When that happens, segment **ML** of stream **K** will be transformed into a minor headward tributary of stream **P**. It will be a *barbed tributary* resembling tributary **O**.

As we turn back the clock, we can witness the reverse of what we have just been imagining. The head of stream **P** will draw back to, and eventually beyond, point **N**. When the head of stream **P** was at (or just a short distance southeast of) point **N,** into which stream did tributary **O** flow? It must at that time have continued to flow westward. It was, in fact, the headwaters part of *stream* **K**. Point **N,** then, is an *elbow of capture,* just as point **L** will become an elbow of capture once stream segment **ML** has been diverted into the stream **P** drainage system.

The beheading of stream **K** by stream **P** is impressive. But it cannot compare to what has happened farther north (see Figure 8–10a, a sketch of the northern capture viewed from the south). The two lakes at **C** drain westward along stream **CG,** which virtually drops into the head of a ravine at **H**. There must be a reason why this stream's gradient so abruptly changes from gentle to steep. There must be a reason why this west-flowing stream is tributary to an east-flowing stream (**GJ**). Finally, there must be a reason why stream segment **EG** also abruptly increases its gradient at point **F,** which is 200 feet lower than **H**. (We may consider segment **EG** to be either the headwater stretch of stream **EGJ,** or a tributary to **GJ**.)

After analyzing the topographic features and relations regarding stream capture in the south, we can readily decipher what has transpired here in the north. At some previous time, lakes **C** drained westward along a continuous stream which passed through points **H, E,** and **D**. The extremely gentle gradient prevailing along this entire stream precluded active downcutting. In con-

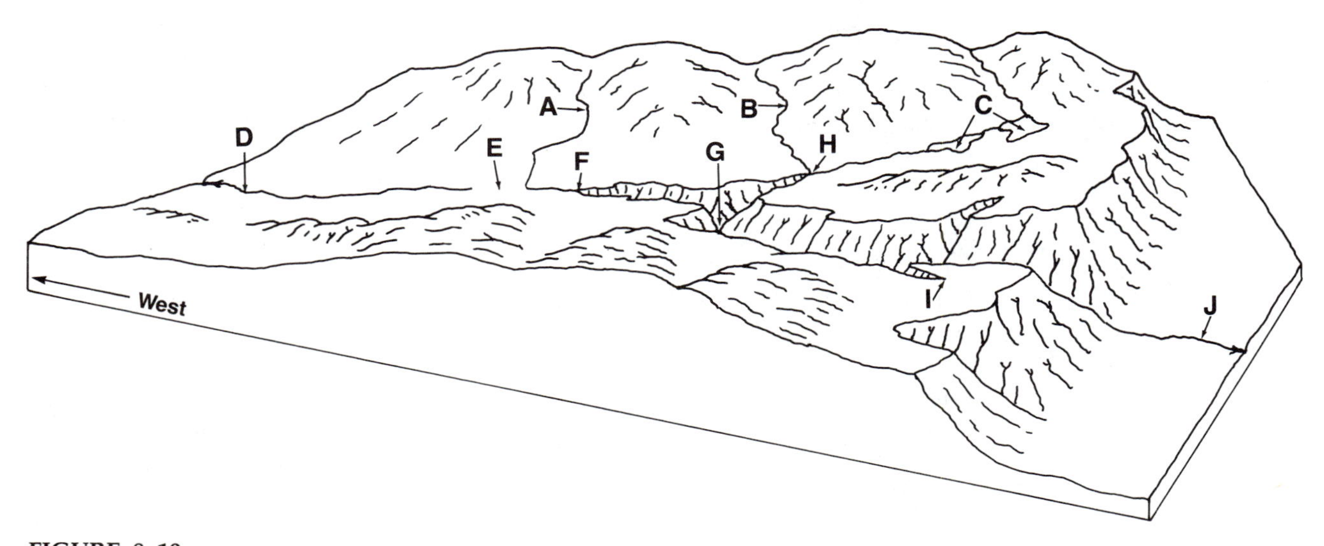

FIGURE 8–10a
Block diagram of part of the area depicted in Figure 8–10, viewed from the south. Stream **A** has "just" been captured near **E**.

trast was stream **GJ,** the head of which may have reached only to point **G**. At that time there was still a divide between the valley of stream **CD** and the head of stream **GJ**. The gradient of **GJ** was then much steeper than that of stream **CD,** and its bed was hundreds of feet below **CD**.

Stream **GJ,** with its steep gradient and its superior downcutting capability, eroded headward past point **G**. At some point on line **HF,** north of point **G,** it broke through the divide that had separated it from stream **CD**. That part of **CD** upstream from the point of capture was immediately diverted into the deep valley of the captor stream. Segment **GD** of the west-flowing stream, being deprived of most of its discharge, must have virtually ceased to exist as a downcutting stream. It was left high and dry.

The rapid downcutting and headward erosion that led to the capture did not cease once capture was accomplished. These processes progressed upstream from **G** toward **C**. To date the canyonization has reached point **H,** which is an excellent example of a *nickpoint* (shown in Figure 8–10a). Similar erosion progressed westward, down the valley of the beheaded stream. This canyon has reached point **F,** though the divide has migrated even farther, to **E**.

The capture of stream **CD** near **G** has of course incorporated the drainage network of stream **B** into the **EJ** system, **B** having all along been tributary to the old stream **CD**. Stream **A** has also "just" been captured about midway between nickpoint **F** and divide **E**.

Point **I** is the head of a northwest-flowing tributary to stream **GJ**. It is easy to visualize stream **I** as a tributary to a stream that once flowed farther to the northwest, to point **G**. Here it joined the other branch entering from the east (stream **CG**) to form the then-main trunk stream **GD**. Thus we see, in the barbed-tributary relation between stream **I** and stream **GJ,** the relic of a capture which occurred long before the more-vividly preserved capture of stream **CD** near **G**.

Zionville, North Carolina/Tennessee

(Figure 8–11 in back of book, C.I. 40 feet)

This area has a total relief of over 2000 feet. It lies in the Blue Ridge Mountains, just east of the North Carolina-Tennessee state line. According to the *AAPG Geological Highway Map of the Mid-Atlantic Region,* the bedrock throughout the map area consists of a complex of Precambrian metasediments, all or most derived from clastic strata.

This area has been subjected to subaerial attack since the close of the Paleozoic. Presumably, it has suffered little tectonic violence since that time. One would think that in this area the topography would have long since become adjusted to the structure.

But does this kind of geomorphic "stability" require a state of rigidity? Or can an area be stable, and at the same time undergo changes? Rather than attempt to answer these questions now, let us first examine the topography. Perhaps it will tell us something about the geology, the geomorphic history, and the geomorphic future of the area.

Essentially, the topography consists of relatively linear valley **JC,** flanked by two asymmetric divides **IB** and **KD**. We suggest that you very carefully trace these two divides and stream **JC** onto a frosted Mylar or similar overlay. Rather than thinking of these features as a valley flanked by two divides, envision them combined, as a broad linear upland which is drained by an axial valley (Figure 8–11a). Regardless of how we describe this topography, it almost certainly reflects some sort of structural and lithologic control.

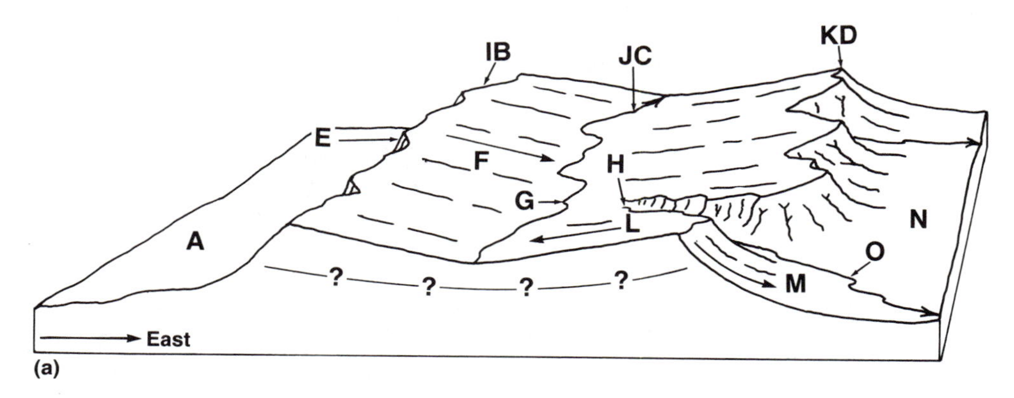

FIGURE 8–11a
Block diagram of part of the area depicted in Figure 8–11, viewed from the south.

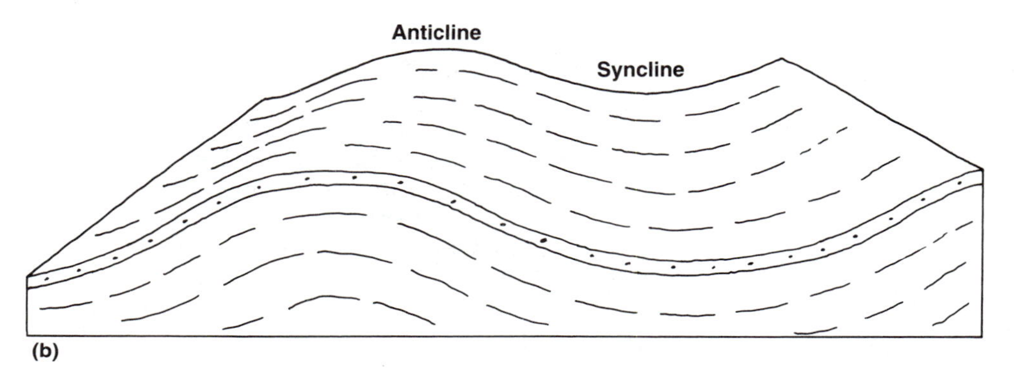

FIGURE 8–11b
Block diagram of anticlinal mountain and synclinal valley.

Note that the valley's slopes (e.g. **F** and **L**) are much gentler than those (e.g. **E** and **M**) that face away from the central upland belt. Divides **IB** and **KD** are clearly asymmetric. Such combined asymmetry often develops along homoclinal ridges that flank a synclinal valley (Figure 8–11c). But we do not know the geological details of this area, nor the geologic structure, which must be complex. So it would be unwise to designate this as a north-northeast-trending syncline.

On the other hand, we must think of it as some sort of possible "synclinal" feature, as depicted in Figure 8–11a, a sketch of the area viewed from the south. If you are one of the many who have difficulty with synclinal ridges, mountains or uplands, please refer to Figures 8–11b, 8–11c and 8–11d. These illustrate schematically the erosional transformation of an anticlinal mountain and synclinal valley into an anticlinal lowland and synclinal mountain or ridge.

Slopes **F** and **L,** then (Figures 8–11 and 8–11a), may be resequent slopes. Slopes **E** and **M** may be obsequent slopes. If this interpretation is correct, it follows that the rock which "holds up" the synclinal upland is more resistant than the rock beneath it. The rock beneath it is older rock which also underlies the topographically lower flanking areas, **A** to the west and **N** to the east.

Whatever may be the correct lithologic and structural explanation for the upland belt, one thing is certain: divides **IB** and **KD** are migrating. **IB** is pro-

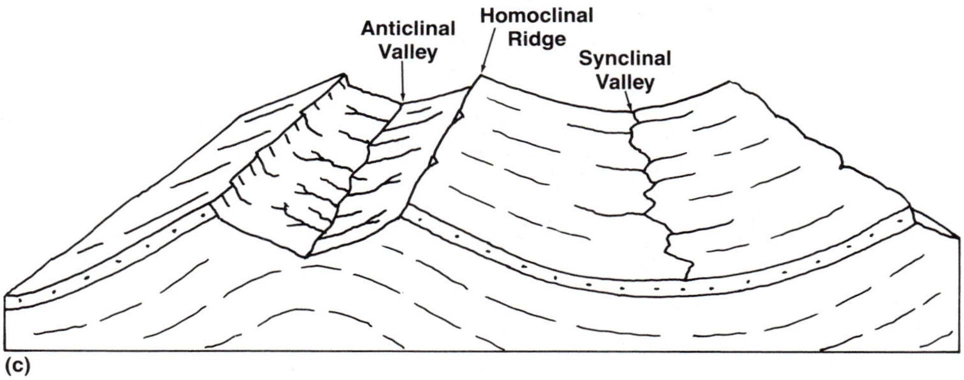

Anticlinal Valley
Homoclinal Ridge
Synclinal Valley
(c)

FIGURE 8–11c
Block diagram of anticlinal valley, synclinal valley, and homoclinal ridges (Figure 8–11b following erosion).

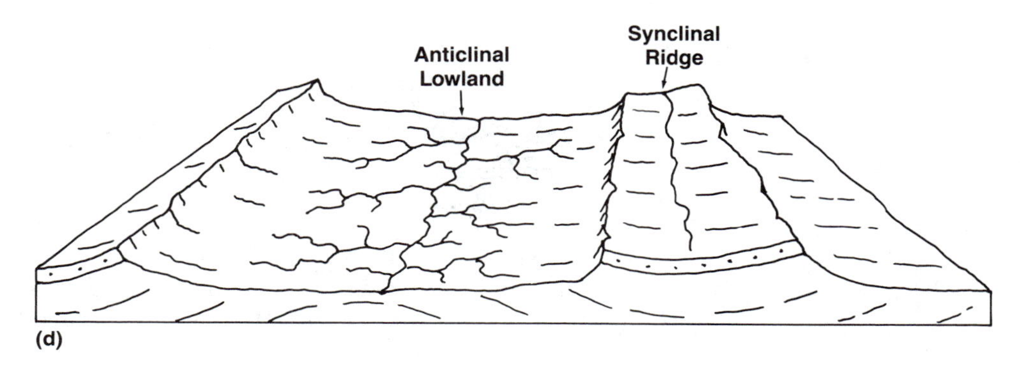

FIGURE 8–11d
Block diagram of anticlinal lowland and synclinal ridge (Figure 8–11b following further erosion).

gressing toward the southeast, and **KD** toward the northwest. Stream **JC** and its entire upland valley are doomed to complete destruction. Note, for instance, that the head **H** of stream **HO** is working its way toward point **G** on stream **JC**. When it reaches **G,** all of the watershed of stream **JC** upstream from this point (i.e. segment **JG**) will be diverted into the channel of stream **HO**.

During the time interval when capture is in progress, the drainage basin of stream **JC** is shrinking. This is because of divide migrations inward, toward **JC,** from both sides. Thus stream capture and overall divide migration are working hand-in-hand. Their goal is the destruction of stream **JC,** its valley and, ultimately, the upland itself.

Can such things happen in an area where the topography is adjusted to structure? The only way to prevent it would be to provide the area with a structure so uniform, and a lithology so homogeneous, that such processes as differential erosion, divide migration, and stream piracy would be precluded. It would be an area devoid of geomorphic features such as resequent slopes, obsequent slopes, and strike ridges.

The most significant thing about this map is that it permits us to visualize what is happening. At least we can see stream piracy and divide migration taking place. However, we can do no more than speculate on *why* they are taking place.

Lubbock, Texas
(Figure 8–12 in back of book, C.I. 100 feet)

We may divide this area into three distinct sections, each with its own topographic characteristics, and each with distinctly different events in progress. Particularly interesting things are happening along the borders that separate the sections. The sections are:

1. This plateau surface (in the northwest) is part of the Llano Estacado or Staked Plains, another example of which is depicted at a much-larger scale in Figure 7–1 (Anton, Texas, in back of book). In this region the Llano Estacado is a virtually undrained plateau. Its dissection is restricted to the bordering fringe belt (e.g. at **C**). It is along this belt that the persistent destruction of the upland is taking place.
2. Occupying over half of the map, this broad, gently sloping area lies south and east of Section **1**. It is bordered on the east by line **PJ,** and extends northeastward as the similarly undulating expanse to the north of line **JB**. All of the headwater streams in this section are attacking the marginal scarp that borders Section **1**. All of these streams are tributary to the low-gradient streams (e.g. **M, N, and A**) that traverse section **2**.
3. This is an area of steeper slopes and stream gradients, an area of intricate dissection (higher drainage density). It lies east of line **PJ** and south of line **JB**.

Note slope lines **D** and **G**. These slopes extend away from the base of scarp belt **C**. Surface **DG** is in fact the northwestern marginal portion of Section **2**. As the outer fringe of Section **1** is being destroyed, and as its border scarp retreats, its upland and scarp topographies are being replaced by such lower slopes as **DG**. We may identify these as pediment slopes.

When we first look at this map, we realize that the upland of Section **1** is being destroyed, and is being replaced by the lower terrain of Section **2**. We must conclude that the upland is the older surface, and that the undulating expanses of Section **2** constitute a younger topography. This is essentially correct, but we must not forget that, at one time, area **FI** lay at the foot of the scarp, which extended roughly along line **EH**. At that time, area **DG** was still part of the then-more-extensive Llano Estacado upland.

The lowland topography of area **FI** has been part of Section **2** much longer than area **DG** has been. We must therefore envision Section **2** as a surface that is growing headward, upslope, toward the retreating border fringe of Section **1**.

However, the upslope growth of Section **2** is not the complete story. Let us return to line **PJB** and examine what is happening along this border zone. Note the extremely gentle slopes at and to the west of **K,** and the low gradients of streams **M** and **N**. They are essentially toward the southeast, as indicated by long arrow **K**. A short distance to the east of **K,** line **LO** defines the headwaters of many closely spaced, steeper-gradient tributaries of stream **Q**. All flow to the east. In fact, all the drainage immediately east of the entire line **PJ** flows to the east. It is extremely important that the gradients of these tributaries are much steeper than the ubiquitous gentle southeasterly slopes and gradients (arrow **K**) of Section **2**. In fact, stream **Q** (at **Q**) is 200 feet lower than stream **N** (at **N**).

Just as border **C** of Section **1** is retreating under the onslaught of numerous high-gradient headwater streams, so border **PJ** is retreating westward under the attack of the many steep tributaries to main stream **Q**. Similar attack and destruction, though less-dramatically illustrated, is also taking place along line **JB**.

Thus it is that Section **2,** as presently delineated, is a transitional entity. While it is growing along its border with higher Section **1,** it is being destroyed along its border with lower, more-highly dissected Section **3**.

There are many additional questions which must be asked about this area, though we will not attempt to answer them here. They include:

1. Why is the drainage density of Section **3** so much higher than that in Section **2**?
2. Is stream **N** destined to be captured by the tributaries of stream **Q**? If so, at what point within the map area will this capture occur? How would such a capture affect the continuing destruction of Section **2**?
3. Of Section **1**?

Lookout Ridge, Alaska

(Figure 8–13 in back of book, C.I. 100 feet)

This area is situated on the north flank of the Brooks Range in northwest Alaska. We can divide it into three sections.

Section 1, in the north, is a low-lying area of gentle slopes, low drainage density, and an apparently dendritic drainage pattern. Its southern border with Section **2** is line **AD**.

Section 2 is an area of intermediate elevation and relief. Within it, certain linear divides (e.g. **B** and **C**) and tributaries (e.g. **J**) suggest structural and lithologic influences, such as steeply dipping, east-striking, resistant and nonresistant formations. Dip direction is not revealed (no apparent asymmetry). Thus, it is not possible to determine whether these beds are homoclinally dipping, tightly folded, or complicated by strike faults. The various obliquely oriented linear (subsequent) streams, such as **E** (in the west), suggest steeply dipping oblique faults.

The absence of such apparent fault control and strike ridges and valleys in Section **1** may reflect the absence of steeply dipping structural planes in that northern area. In Section **1,** then, we speculate that (1) steep-dip structures, if they exist, are effectively masked by overlying formations, or (2) north of line **AD** the structure is much-less complex.

Section 3 is the area south of line **LR**. For the most part line **LR** is the major divide between the southerly drainage of Section **3** and the northerly/northeasterly drainage of Section **2**. This section is topographically higher than Section **2**. Its dissection appears to be more intricate, as shown by the pronounced crenulations in its contours. Strike ridges such as **M** indicate steep-dip structure in this section, similar to that in Section **2**. But the higher drainage density (finer topographic texture) of the southern area suggests that the nonresistant units of Section **3** may be finer-grained than those in Section **2**.

Fine-grained clastics, however, are not particularly resistant. It is difficult to explain Section **3**'s greater elevation in terms of overall greater resistance. The key may lie with stream **P**. It is the only north-flowing tributary that has eroded headward past divide **LR**. This divide seems to be the strike ridge of a steeply dipping resistant formation which, in effect, protects the southern upland. The numerous steep-but-small headwater streams (e.g. **G**) that are attacking this barrier from the north call to mind a force of armed invaders trying unsuccessfully to scale the walls of a medieval fortress.

Why, then, is Section **3** higher than Section **2**? We cannot determine the answer solely by studying this map. The base level of the streams that drain Section **3** may be higher. Possibly, if the base level of all the streams in the map area is the same (sea level), the south-flowing streams of Section **3** are more distant from the mouth of their master stream (or streams) than are the streams of either Section **1** or Section **2**.

The important thing is that the topography of Section **3** must have developed independently, beyond any influence from the other sections. By the same token, the more complex topographic history of Section **2** was apparently not influenced by events in Section **3**.

Changes are taking place. Stream **P** is reaching into the upland terrain of Section **3**, and locally replacing it with deeper valleys and steeper slopes. The asymmetric divide between stream **P** and stream **Q** is migrating to the east. The divide between stream **P** and stream **N** is migrating to the west. In fact, stream **N** is about to be captured and its head destroyed. The resistant bed that forms the divide (two dots) between the heads of streams **P** and **O** will temporarily deter divide migration to the south. But in time, stream **P** will extend its head to and beyond point **O**.

The north-south-trending divide **HI** (in Section **2**) is fundamentally different from divide **LR**. It is migrating to the east. Its migration is indicated by the fact that the streams draining its western flank have steeper gradients and a lower local base level (stream **F**), than do the streams which drain its eastern flank. Unlike divide **LR**, which is maintained by an upturned resistant formation, this divide trends at right angles to bedrock strike. It thus merely separates two unlike drainage regimens. It affords no natural protection.

In his analysis of the geomorphic history of this area, Kalaswad (1983) proposes that, at one time, Section **2** was drained by stream **K**. According to his interpretation, the upstream (western) portion of that drainage network (e.g. the area west of present divide **HI**) was beheaded by stream **F**. Stream **F** then encroached headward into this area, much as stream **P** is currently encroaching into Section **3**.

The limited area shown on this map cannot provide sufficient data with which to confirm or refute his hypothesis. Yet, it has merit. Capture of stream **K**'s headwaters by stream **F** would explain the pronounced disequilibrium between the opposite flanks of divide **HI**. This divide's existence and asymmetry cannot be explained as responses to either geologic structure or differences in bedrock resistance. As pointed out, divide **HI** trends north-south, whereas the geologic structures strike east-west.

1. Figure 9–1. Cumberland, Maryland.
2. Figure 9–2. Yakima East, Washington.
3. Figure 9–3. Bluemont, Virginia and Round Hill, West Virginia.
4. Figure 9–4. Delaware Water Gap, Pennsylvania/New Jersey.
5. Figure 9–5. Independence Rock, Wyoming.
6. Figure 9–6. Lake Killarney, Missouri.

9

Water Gaps and Wind Gaps

INTRODUCTION

Surely the most-important word in the map interpreter's vocabulary is *why*. Why is this slope long, and gentle, and smooth, while that slope is short, and steep, and intricately dissected? Why is this stretch of shoreline so irregular, with its many embayments and headlands, while that stretch is nearly linear? Why is this area so mountainous, while that area is dominantly low-lying? Why? Why? Why?

Water gaps also have their whys, but their whys have something special about them. When we ask why a particular water gap is situated where it is, we tack on an additional note of puzzlement: "Why *here,* of all places?" There is something almost unreasonable about the very presence of most water gaps. Why did this stream cut this gap here? It would have been so much easier for it to have continued around the ridge, hill, or mountain through which it cuts!

Wind gaps, true wind gaps which began as water gaps, add yet another "why" dimension. Not only must we ask why the stream elected to cut the gap there, of all places. We must add, "and why was that stream later diverted away from its course through that gap?"

We use the expression "true wind gap" advisedly, because there are many saddles and sags in ridge and mountain crests that are not the product of stream downcutting. It is not always easy—sometimes impossible—to tell which is which. But we must try.

Cumberland, Maryland

(Figure 9–1 in back of book, C.I. 20 feet)

This area lies in what is variously called the Folded Appalachians, the Newer Appalachians, and the Ridge and Valley (or Valley and Ridge) geomorphic province. The map, though not "typical" of the region, displays topographic features and relations common to such a geologic setting. This is an area of folded Paleozoic strata. They include such sedimentary types as limestone, shale, sandstone, and conglomerate. Differential erosion has been in progress since late in the Paleozoic.

Two remarkable gaps attract us: large water gap **BK** in the north, and less-obvious wind gap **N** in the south. Figure 9–1a is a sketch of these gaps viewed from the southeast.

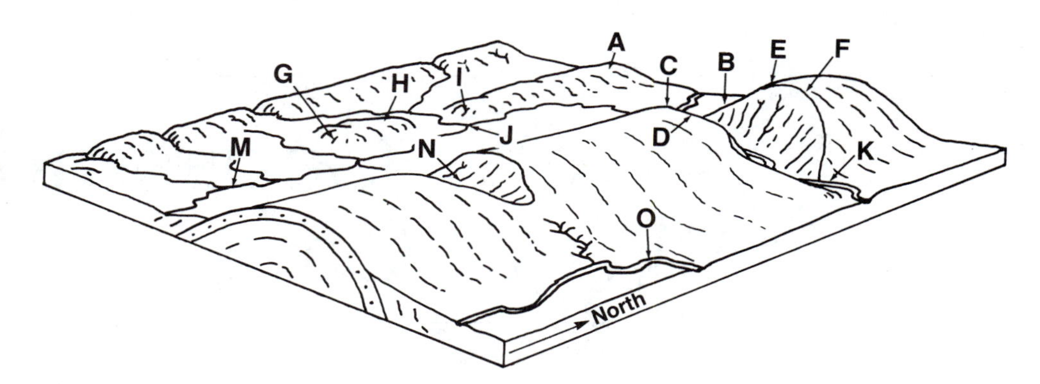

FIGURE 9–1a
Block diagram of part of the area depicted in Figure 9–1, viewed from the southeast.

Sharp contour bends along lines **EB** and **CB** (dashed-line V **EBC**) define the outcrop V[1] of a steeply west-dipping resistant bed (say it is a sandstone). A similar outcrop V, though of an east-dipping bed, is defined by similar sharp bends along lines **DK** and **FK** (dashed-line V **DKF**). These opposing dips disclose the anticlinal structure of the large mountain ridge through which the river has cut the gap. Since the strata were folded during the late Paleozoic, it would be pointless to speculate as to whether this is an antecedent stream.

There is a possibility that superposition is not the answer either. The offset of the ridge-forming sandstone at **K** suggests displacement along a cross fault, which could have facilitated the development of the gap.

It is also possible that the stream has been superposed upon the fold. Regional superposition has long been one of the hypotheses advanced to explain such stream-gap-ridge relations throughout this region. In this particular instance, however, there is a third, though less-dramatic possibility. It involves what we (the authors) call *conformity superposition,* as opposed to traditional superposition through an unconformity (*unconformity superposition*).

Consider the following *possible* scenario.

Dip at **B** is to the west. At **K** it is to the east. The strata that occupy the lowlands immediately west of **B** and east of **K** must be nonresistant. (At least they are much-less resistant than the sandstone that supports the ridge.) In addition, both must be the *same* nonresistant formation, since both are directly underlain by the sandstone. Let us assume that this is a shale sequence.

It follows that, at one time, this nonresistant shale unit must have extended over and across the anticline, as depicted in Figure 9–1b. At that time the resistant sandstone was entirely underground. It was unable to exert any influence on the topography, which was then developed exclusively on the shale. The stream that now flows west-east from **B** to **K** could thus have established this course across that shale lowland (stream segment **bk,** Figure 9–1b).

As the stream later cut down, it encountered the top of the sandstone fold, into which it began to entrench itself. Its progression downward, then, would have been from anticlinally folded, nonresistant younger shale, into conformable, anticlinally folded, resistant older sandstone. This is a superposition of sorts: not superposition through an unconformity, but through a normal stratigraphic contact instead.

---

[1]You may wish to refer back to Figures 1–7, 1–8a, and 1–8b, which illustrate the ways in which contours depict outcrop V's and dip.

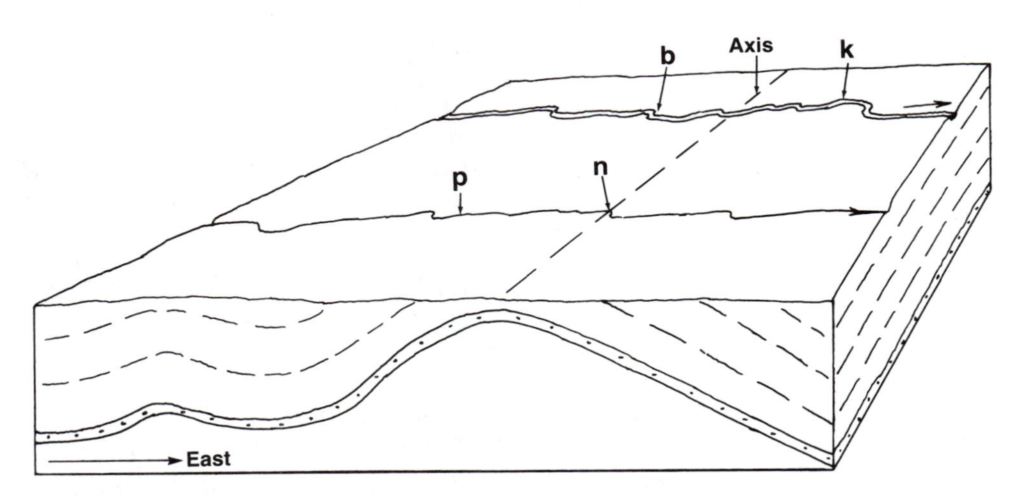

FIGURE 9–1b
Block diagram of part of the area depicted in Figure 9–1, as it may once have appeared.

Wind Gap N. This gap does not at first appear very impressive, especially compared to gap **BK**. It is, however, from 150 to 200 feet deep and about 1000 feet wide. It must have been cut by a stream of appreciable size. Stream **MB** flows, as a *subsequent* stream, along the shale belt. This stream could not have cut and then abandoned gap **N,** though obviously some stream must have. Surely river **O** could not have once flowed east-west through it.

Let us once again turn back the clock, to when the sandstone was underground and the area between **M** and **O** was occupied by the overlying shale. A stream similar to **bk** could have flowed west-east across this shale lowland (stream **n,** Figure 9–1b). Later, it too could have cut down until it encountered the sandstone, into which it continued to cut. As time passed, the shale above the uparched sandstone would have been stripped away. The sandstone "ridge" would have come into being as a low, gentle topographic arch. As this arch grew, as more of the sandstone was uncovered by the stripping away of the shale from its flanks, this water gap deepened and widened, controlled by through-flowing stream **p**.

It is impossible to determine the precise upstream course of this stream. But it must have been situated along a line such as that currently followed by stream segment **LM**. At that time, of course, this stream did not continue on to **B**. Instead it must have been part of the drainage system passing through gap **N**. The head of present stream **MB** may then have been somewhere in the vicinity of point **J**.

Stream **JB,** flowing in a shale belt, had a distinct advantage over stream **LMNO** (we assume such to have been a reasonable course for the stream that cut gap **N**). The latter stream was cutting with great difficulty through the sandstone at **N**. But stream **JB** was free to downcut with ease and to erode headward toward the flank of stream **LMNO**.

Stream capture!

When did it happen? We cannot say in millions of years, but we can state it in terms of feet. The capture happened about *300 feet ago*! Segment **LN** of the east-flowing stream must have had a gradient toward the east. The point of capture, at or near point **M,** must have been some 300 feet above the present elevation of **M**. In other words, valley **MB** has been lowered about 300 feet since the time of the capture.

Before we leave this map, study ridges **GH** and **IA**. Could these be landforms developed on a second resistant formation that overlies the shale? Possibly, though its absence south of **G** makes such a possibility unlikely. It might, on the other hand, be a small anticlinal ridge developed on an upfolding (south plunge) of the same sandstone that underlies the main mountain.

Such a structure is suggested by the gentle symmetry of ridge **IA**. If this were a homoclinal ridge developed on a younger, resistant unit, it would probably be a much sharper ridge (similar to that in Figure 1–7). It would probably extend south, from point **G** to the southern border of the map. South plunge readily explains the absence of this ridge trend to the south of **G**.

Yakima East, Washington

(Figure 9–2 in back of book, C.I. 20 feet)

On many of the foregoing maps, considerable geomorphic and geologic information is provided by such things as contour spacing, crenulations, linearities, variations, similarities, and contrasts. The topography depicted on this map, on the other hand, reveals little. The river flows through an almost surgically cut water gap in a ridge which rises between 700 and 800 feet above the flanking flatlands. And that is about all we can venture, knowing nothing about the regional geologic structure, stratigraphy, and history. We can, however, make a few observations on which to base some highly tentative suggestions.

What is the significance of the apparently flat floor of the gap, which ranges from a quarter-mile to a half-mile in width? It is possible that the river has repeatedly swung laterally and cut this flat surface. It is also possible that, after it cut a deep gap through the mountain, the river underwent a change in regimen, and was transformed from a downcutting stream to an aggrading stream. Partial filling could thus account for the flatness of the gap, just as a more-widespread aggradation could explain the extensive flatlands both north and south of the ridge and gap. The braided character of the river, particularly in the north, would fit such a recent history.

Geologists and geographers familiar with the geomorphology of the Valley and Ridge, the Rockies, and the Arkansas Valley, know that differential erosion is the key to understanding ridges, valleys, and gaps. Professionals of such background might rush to the conclusion that this is a ridge because it is underlain by resistant rocks, and similarly that the lowlands are low because they are underlain by nonresistant rocks. In the areas mentioned, a ridge like this might be homoclinal, anticlinal, synclinal, or fault-bordered. It could be composed of any rock type, sedimentary or crystalline, so long as the rock was resistant.

The asymmetry of the ridge at several places (e.g. profiles **AC** and **BD**) suggests south dip (homocline). However, such a profile could equally reflect an asymmetric anticline, or an anticlinal fold complicated by strike faulting (particularly along the north side).

Since we have no idea of this area's geologic structure and history, we can only describe and ponder. There is no way to determine whether the gap is (1) the product of stream adjustment to a zone of nonresistance (e.g. fracture or fault zone) that cuts across the ridge-forming rocks, (2) the result of antecedence, (3) the result of traditional superposition (unconformity superposition), or (4) the result of the type superposition depicted in Figure 9–1b (conformity superposition).

Now, let us add some geologic information. According to the *Geologic Map of the United States,* this entire area is underlain by Miocene and Pliocene volcanics (plateau basalts). This single fact contributes greatly to our understanding of the geomorphic features and relations. It tells us that, since deformation occurred in late Tertiary time or later, we are apparently looking at topography that directly reflects structure. The ridge is *structurally* as well as topographically high.

The gap?

The gap must have been cut as the ridge rose across the course of the *antecedent* stream. It is hard to imagine basalts as young as these having been buried and later exhumed, which would have been required for superposition to have occurred.

Is this map interpretation? Not in the strictest sense. Instead, it is the combining of what we can see on the map (ridge, transverse stream, and gap) with what we have learned elsewhere (the age of the rock—and hence of deformation—from the geologic map).

We have made a geological *deduction,* a deduction based on information obtained in part from the topographic map.

Bluemont, Virginia, and Round Hill, West Virginia
(Figure 9–3 in back of book, C.I. 10 and 20 feet)

The Blue Ridge of northern Virginia is scored by several extremely large and famous wind gaps. Snickers Gap (**B**) is one of them. Prior to its capture by the Shenandoah River (**A**), a stream of considerable size flowed west-east through this gap. (The Shenandoah eroded headward to the southwest along a belt of nonresistant rocks.)

At that time the low areas to the west and east of the Blue Ridge must have been at or above the level of point **B**. After the capture, the less-resistant rocks of the lowlands were further lowered, at a far greater rate than the more-resistant rocks of the ridge. Whereas the ridge now stands from 1100 to 1250 feet above the low areas, at the time of capture it must have been a much-less imposing feature (300 to 600 feet high).

Most of this history of the gap and surrounding areas comes from analyses conducted by others throughout the entire region. All this map tells us is that an otherwise even-crested broad ridge is cut by a large gap which is not currently occupied by a through-flowing stream. It also tells us that, if capture indeed occurred here, it must have taken place 600-plus feet ago.

One final point. If we had no knowledge of the geomorphic history of this general area, we would be ill-advised to state dogmatically that this wind gap is the result of capture. Since the ridge itself is the product of differential erosion, there is no reason why the gap could not have the same explanation. For example, suppose an east-west trending fault zone cuts across this map area. Would not normal erosional processes etch out the fault zone to form such a gap?

Delaware Water Gap, Pennsylvania/New Jersey
(Figure 9–4 in back of book, C.I. 20 feet)

Here is a classic example of a stream that is discordant to the geology across which it flows. As is well shown by the overall grain of the topography, the stratigraphic units in this area strike northeasterly. In contrast, the Delaware River flows to the southeast. At **F** it has cut its famous gap through the main ridge, the unlikely name of which is Kittatinny Mountain. The main features of the map area are depicted in Figure 9–4a, a sketch of the area viewed from the south.

As we have seen on previous maps, water gaps such as this may result from antecedence, as in the Yakima area (Figure 9–2 in back of book), or of some variety of superposition, as in the Cumberland area (Figure 9–1 in back of book). We must always consider the possibility that such gaps may also be structure-controlled (e.g. cross faults). When confronted by a gap, how do we choose? Perhaps this map will give an answer.

What can we deduce about the structure of the main ridge? Its pronounced asymmetry, consistent throughout its entire length, is the asymmetry of a homoclinal ridge: dip to the northwest, steep obsequent scarp to the southeast. On the other hand, this ridge could be a recently tilted fault block, its border fault trace extending along the base of its southeast flank. The contours can only tell us what the ridge's structure *might* be. If we cannot be more certain of the structure and its age, we are in no position to select antecedence over superposition, or vice versa, or structural control, to explain the water gap.

Before leaving the area near gap **F**, note ridge **BC** to the north. In contrast to the main asymmetric ridge, ridge **BC** is not only quite symmetric, but it is smoothly rounded in profile, its crest gradually descending to the southwest toward the river. There is no trace of this ridge to the west of the river. We cannot interpret this ridge as a homoclinal ridge, because its profile does not have the right sharpness, and it is not asymmetric. And, the crest of a homoclinal ridge does not ordinarily descend as this one does.

Could ridge **BC** be a gently southwest-plunging anticline? And may not adjacent valley **AD**, which separates the two ridges, be a southwest-plunging syncline? Note the similarity between ridge **BC** and ridge **AI** of Figure 9–1. We see both as plunging anticlines, and for the same reasons.

A fairly well-defined, if not very steep, south-facing "scarp" extends along line **GE**. This scarp is matched by north-facing "scarp" **KI**. Together, these scarps overlook the linear valley of tributary **H**. What is the significance of these two in-facing scarps? They do not display the arcuate irregularities so frequently encountered in bluffs produced by the meandering of a stream across its floodplain. These scarps are not only remarkably linear, but they also parallel the grain of the entire map area.

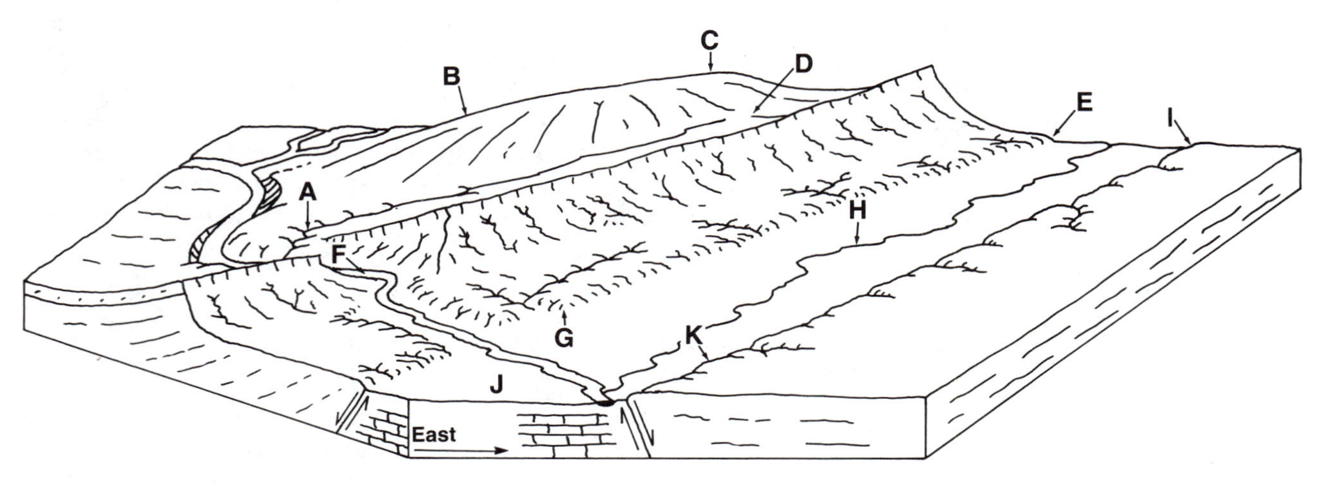

FIGURE 9–4a
Block diagram of part of the area depicted in Figure 9–4, viewed from the south.

So far we have noted several features, and made a few hesitant suggestions. But we have not come forth with a clear-cut interpretation of this assemblage of topographic and hydrologic features. What will happen if we now consult the geologic map of the area, which is conveniently printed on the back of the Delaware Water Gap quadrangle?

We find that all rocks in this area are Paleozoic sedimentary rocks. The oldest unit is a carbonate. It occupies the belt bordered on the north by scarp **GE** and on the south by scarp **KI**. This limestone, which continues across the river to **J** on the map's south border, is Cambro-Ordovician in age.

Stratigraphically overlying the limestone is a thick sequence of Ordovician shales. These shales occupy the entire area southeast of scarp **KI** (area **KIL**), as well as the belt between scarp **GE** and the base of the main ridge. Traces of steeply dipping faults extend along the bases of both scarps **GE** and **KI**.

Block **GEIK** must be a horst, since between the faults, the older limestone is faulted (up) against the younger flanking shales. The valley of stream **H,** therefore, must be the result of differential erosion. Border scarps **GE** and **KI** are necessarily obsequent fault-line scarps.

Atop the Ordovician shales lies a resistant Silurian conglomerate. It in turn is overlain by a series of nonresistant, finer-grained beds. It is the conglomerate that forms both the main ridge and ridge **BC**. The northwestern lowland is floored by the younger finer-grained clastics. A finger of these extends up valley **AD** toward, but not quite to, point **D** itself.

The ages of the stratigraphic units permit us to confirm the homoclinal structure of the major ridge, the synclinal structure of valley **AD** (southwest plunge), and the anticlinal structure of ridge **BC** (southwest plunge).

The structures of this area, like the strata themselves, are extremely old, formed during the Paleozoic. This is all we need to reject antecedence as an explanation for the water gap. It does not, however, constitute *proof* of either superposition or structural control.

Independence Rock, Wyoming

(Figure 9–5 in back of book, C.I. 20 feet)

There is a remarkable similarity between the geomorphology and history of the water gap depicted here, and the geomorphology and history of the gaps in Figure 9–6 (Lake Killarney, Missouri, in back of book). This despite the dissimilarities in climatic setting, lithologies of some of the rocks, and the time and conditions under which the sedimentary units were deposited.

Gap **B,** in the middle of this map area, is the famous Devil's Gate of central Wyoming (see Figure 9–5a, a sketch of part of the map area viewed from the south). This feature must have been viewed with mild curiosity, if not with searing awe, by countless thousands of covered-wagon pioneers as they passed en route to Utah, Oregon, and California. We know they passed by here, because a few miles to the east is Independence Rock, a prominent marker along the trail.

The lowlands to the east, northeast, and southwest of **B,** and between the main hills in the west and hills **D** in the southeast, are all underlain by Tertiary sedimentary rocks. This fact is provided by the *Geologic Map of Wyoming* (we cannot interpret such things from a few contour lines).

The ridge in Figure 9–2 (Yakima East, Washington, in back of book) is either an anticline, homocline, or fault block formed in relatively recent geologic time. The ridge in Figure 9–1 (Cumberland, Maryland, in back of book)

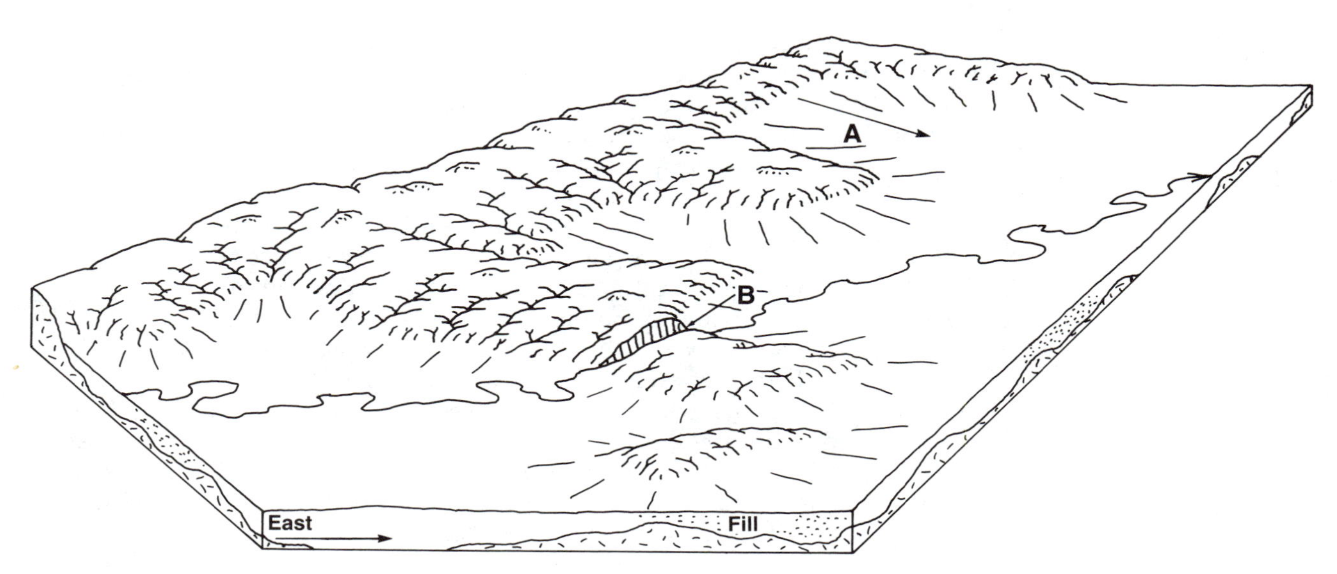

FIGURE 9–5a
Block diagram of part of the area depicted in Figure 9–5, viewed from the south.

is the product of differential erosion of a much-older anticlinal fold. But the amoeba-like outline of the hills in this area is not a form that we can readily attribute to either block faulting or folding. However, it might be:

- □ The product of accumulation (e.g. volcanism), followed by erosion and partial destruction.
- □ An irregular mass of resistant intrusive rock, from the flanks of which the less-resistant surrounding rock is being removed.
- □ An irregular mass of older rock, around which the Tertiary rocks have been deposited, partially burying it.

According to the geologic map, these hills are carved in Precambrian igneous rocks collectively designated "granite." Since the rocks being stripped from them are Tertiary in age, these must be Tertiary hills (or mountain tops) which were partly or even completely buried. If the latter, they are currently being exhumed.

Of particular interest is the wasp-waist restriction of the main stream (at **B**) as it cuts through the granite gap. In this single, short stretch, the stream must devote all of its energy to downcutting into the resistant granite. Both upstream and downstream, where it encounters only the less-resistant Tertiary clastics, it is free to meander.

What we have here is a beautiful small-scale example of superposition. The only way the stream could have cut gap **B** is for it to have cut down to the granite from above. This means that the Precambrian rock (at least) in the **B** area must have been completely buried by the Tertiary strata. These strata have since eroded away, down to the level of the present lowland surface.

Piedmont Slopes—Fans or Pediments? Look at several small features, which are all of a kind. If we can identify and explain one of them, we probably will have identified and explained the others. These features are piedmont slopes, examples of which are **A** and **C**. (A *piedmont slope* is simply a slope lying or formed at the foot of a mountain. Do not confuse it with a *pediment*, which as we saw in Chapter 6, is a gently inclined erosional surface, similarly situated.)

The obvious alternatives for such slopes are alluvial fans and pediments. Some geographers and geologists have claimed the ability to distinguish between them, based on slope, slope variations, and contour crenulations. But we lean toward searching the map for the most-reasonable answer.

As we see it, the solution lies in the geometry—not of slopes **A** and **C**, but of gap **B**. In profile, this gap is virtually V-shaped. It has no appreciable "floor." We may assume from this that, along the stretch where it is restricted to the gap, the river is still downcutting. At the very least, it is not aggrading. In other words, the gap may be thought of as a very short, actively erosional valley segment.

The contours tell us that, along its entire length, the stream's gradient is extremely gentle, a few feet per mile at most. Within the gap, the stream is not aggrading, and has not aggraded (unless it has just begun to do so, which we cannot detect). We may thus assume that, both upstream and downstream from the gap, it is not and has not aggraded.

The slopes, divides, valleys, and other geomorphic features of the dissected hills are clearly the products of erosion and dissection. They constitute the topography that developed as a consequence of removing the Tertiary cover. The overall geometry of these high areas may be inherited from the topography that prevailed at the time of burial. But the smaller details are probably of more recent (postexhumation) vintage.

What about slopes **A** and **C**? Are they more probably fans or pediments? With no evident history of aggradation, the main stream may be thought of as the base level to which all the tributary streams and marginal slopes such as **A** and **C** must be graded. Surfaces **A** and **C**, and numerous similar slopes, may be strewn with relatively thin veneers of detrital material from the nearby uplands, but this alluvial cover is more incidental than significant.

Like the hills, these lowlands must be areas of erosion. They are a complex of moderate, gentle, and extremely gentle slopes, each adjusted to its orientation and distance from the major base level, the main stream.

It is important to distinguish between the *sediments* that accumulated around and over the granite hills, and the *topography* which has developed on these sediments. Though the materials in the lowlands are accumulated, the topographic features in the lowlands are the products of erosion. Thus many, perhaps all, of these lowland fringe slopes must be *pediments*.

Lake Killarney, Missouri

(Figure 9–6 in back of book, C.I. 20 feet)

Two obvious and outstanding gaps in this area are water gaps **A** and **C**. We selected this map because of them. However, there is another feature in the area, one to which we paid no attention when choosing the maps for this book. Yet, it is more tantalizing and perhaps more worthy of study than gaps **A** and **C**. It is feature **E**, a saddle so well-defined that we may designate it a wind gap. This is provided that we define a wind gap as a gap of any origin, not occupied by a through-passing stream. Figure 9–6a is a sketch of the area viewed from the west.

The question we must ask is whether gap **E** was originally occupied by a stream, i.e. was it a water gap? It has all the geometric characteristics of an abandoned water gap. It is about 200 feet deep and has a symmetrical

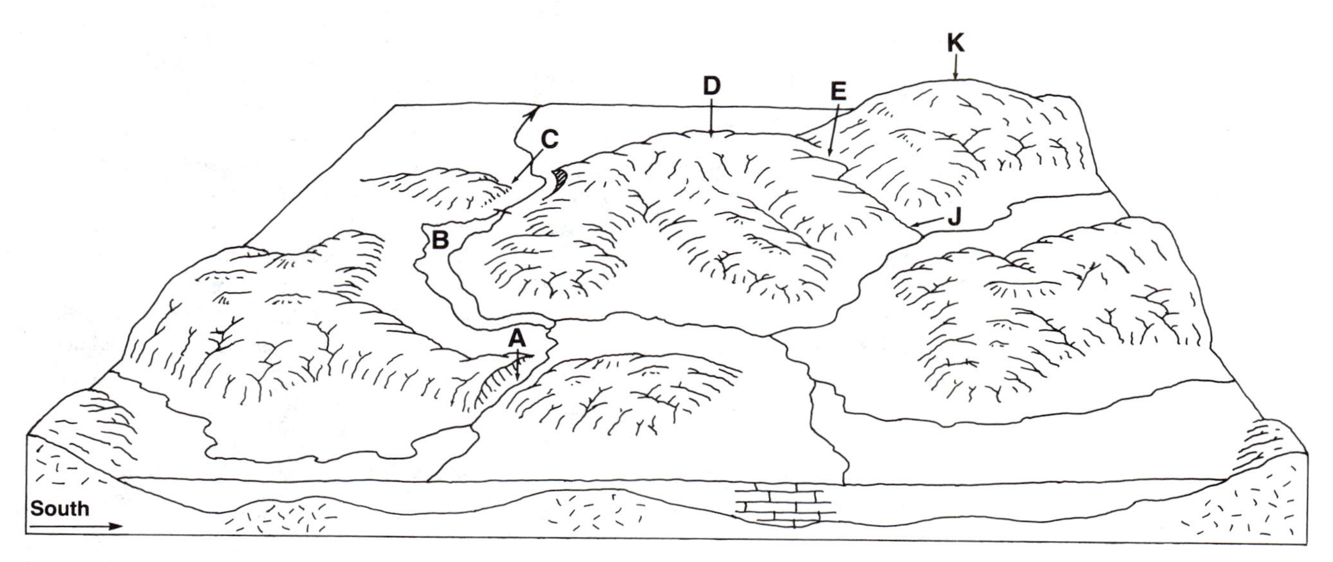

FIGURE 9–6a
Block diagram of the area depicted in Figure 9–6, viewed from the west.

V-shaped profile. If it was a water gap, the stream has long since been diverted by capture. If that capture took place, it happened more than 350 feet ago (at a height over 350 feet above the present level of point **G** on the main stream in the area). There is no way we can now determine the course of that former stream, nor the point of capture, nor the reasons for it.

Must gap **E** have been a water gap? Must *all* such passes or saddles be abandoned water gaps? Surely not! Could this gap have formed in some other way? It may be significant that the heads of streams **J** and **F** oppose each other at **E,** though these are small, otherwise insignificant streams.

As we have seen throughout this book, topographic map interpretation is frequently enabled or strengthened by a knowledge, however generalized, of the area's geology. Such knowledge is most often obtained from published geologic maps, and occasionally from reports and other documents.

The large and small hills, such as **I** and **H,** and those through which gaps **A** and **C** are cut, are a complex of Precambrian crystalline rocks, known at times as "granite." (Source: *AAPG Geological Highway Map of the Mid-Continent Region.*) The lowlands, including such small, isolated ones as that occupied by lake **B,** are underlain by Cambrian carbonates. These rest upon a basal conglomerate.

This is yet another area where we see demonstrated the importance of relative resistance to erosion in the development of topography. By tracing on a frosted Mylar overlay the lines that separate the hills from the low areas, we can actually delineate the contacts between the granite and the carbonates—in other words, make a geologic sketch map of the area.

The geologic/geomorphic history of this area is similar to that of Figure 9–5 (Independence Rock, Wyoming, in back of book). There is a difference: in that area, the mountains and hills were buried by continental sediments during the early Tertiary. In this area, the hills and mountains were entombed in lower Paleozoic marine sediments. In both cases, the hills and mountains that were buried had been developed in Precambrian crystalline rocks ("granites"). Rejuvenation of the drainage in these areas resulted in extensive downcutting and regional downwasting. These led to the uncovering (exhumation, resurrection) of the tops of these high features. In places, it also led to superposition of the streams, and the creation of such water gaps as the Devil's Gate in Figure 9–5 and gaps **A** and **C** in this area.

In our analysis of wind gap **E,** the important thing is that the topography prior to burial (i.e. the topography that was buried) was undoubtedly a typical, dissected, crystalline-rock highland. It consisted of hills, valleys, knobs, saddles, and passes—saddles and passes such as we now observe as gap **E**.

So gap **E** need not be a latter-day landform resulting from geologically recent dissection of a Precambrian mass. In all probability, it is a Cambrian saddle or wind gap that was buried. Much later, it was uncovered by removal of the much-less-resistant overlying and encasing Paleozoic strata.

Visualize this scene from the Cambrian: these hills stood as islands in a sea. The sea had risen to the level of the present lowland surface. At that time, hilltop **D** would have loomed 400 to 450 feet above sea level. Hilltop **K** was 50 feet higher. The floor of gap **E** stood a mere 225 to 275 feet above the water's surface.

Water gaps **A** and **C** must be explained as the result of superposition. But gaps such as **E** need no such interpretation. It is improbable (though possible), that superposition and later capture did produce gap **E**. However, in the absence of strong evidence for this sequence of events, we are on more solid ground in viewing this feature as an exhumed saddle.

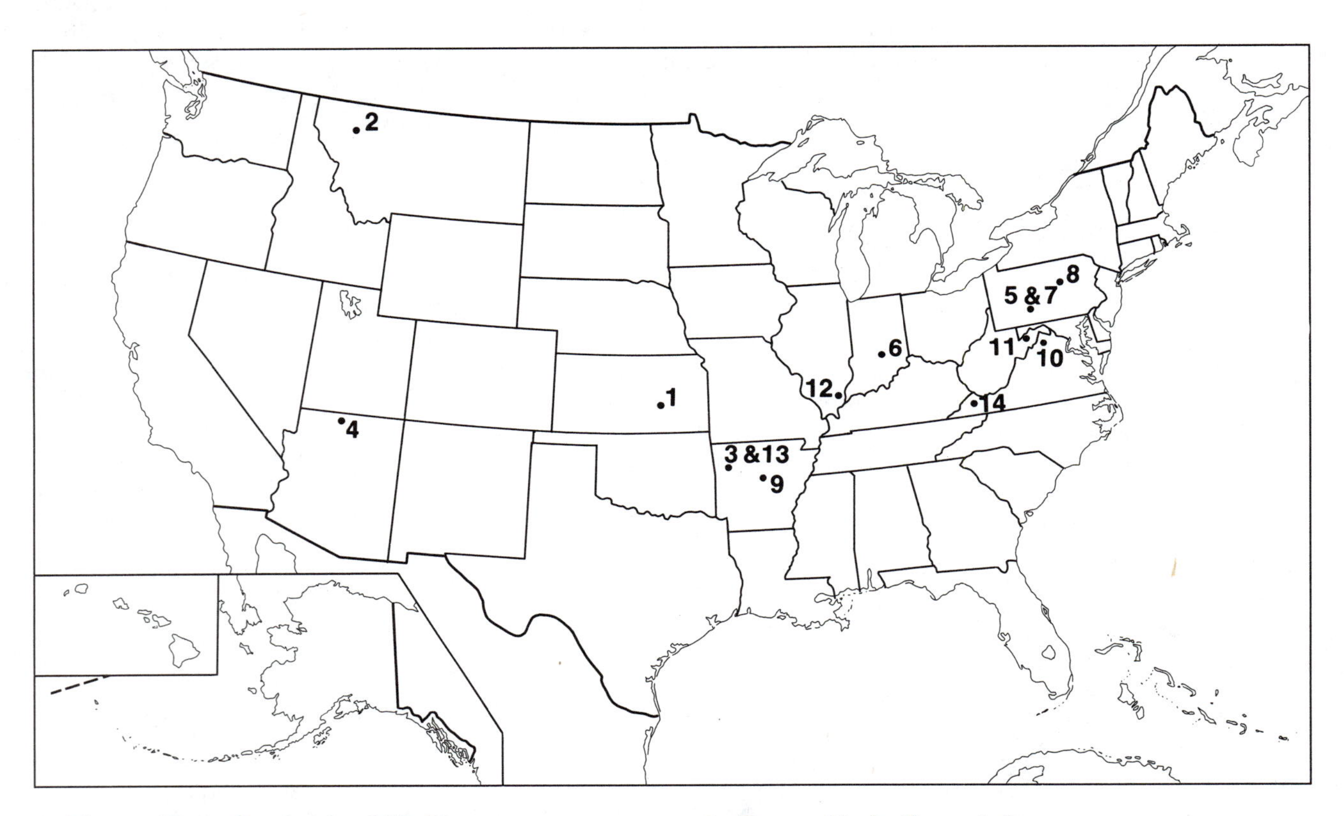

1. Figure 10–1. Cambridge NE, Kansas.
2. Figure 10–2. Marmot Mountain, Montana.
3. Figure 10–3. Booneville, Arkansas.
4. Figure 10–4. Jacob Lake, Arizona.
5. Figure 10–5. New Enterprise, Pennsylvania.
6. Figure 10–6. Nashville, Indiana.
7. Figure 10–7. Everett West, Pennsylvania.
8. Figure 10–8. Millheim, Pennsylvania.
9. Figure 10–9. Cato, Arkansas.
10. Figure 10–10. Strasburg, Virginia.
11. Figure 10–11. Greenland Gap, West Virginia.
12. Figure 10–12. Karbers Ridge and Herod, Illinois.
13. Figure 10–13. Booneville, Arkansas.
14. Figure 10–14. Mendota, Virginia.

10

Dips and Folds

INTRODUCTION

If there were no such things as weathering and erosion, areas of folded strata would retain their pristine appearance. They would resemble the terrain sketched in Figure 8–11b. All anticlines would form ridges, and all synclines would form valleys. However, there *are* such things as weathering and erosion.

If there were no such thing as variable rock resistance to erosion, all rocks would weather and erode at the same rate (given uniform climatic conditions). However, there *is* variable rock resistance to destruction. Rocks do not erode at the same rate. As a result, erosional attack on a sequence of layered rocks usually produces a sequence of landforms resembling those in Figures 8–11c and 8–11d.

This is what happens when differential erosion is given free rein in such a suite of folded rocks. Anticlinal mountains may be eroded into anticlinal valleys, and synclinal valleys may be transformed into synclinal mountains.

Add variations in dip, and we encounter varied homoclinal landforms such as:

- ☐ *cuestas* (low dip)
- ☐ *hogbacks* (steep dip)
- ☐ *ridge-and-valley asymmetry* (low-to-medium dip, as in Figure 8–4 in back of book)
- ☐ *ridge-and-valley symmetry* (steep dip).

In addition, the dip (plunge) of fold axes in areas that have undergone differential erosion leads to development of anticlinal and synclinal noses.

In areas such as that shown in Figure 8–11d, the rule of thumb is: the *older* rock units are exposed in the core of a breached anticline, and the *younger* rocks are protected in the axial part of a syncline. Many beginning students attempt to remember these rules verbally, and fail to visualize the eroded structure. We promise that, if you remember what Figure 8–11d looks like, and the principles it depicts, you will never confuse the relations among oldest/youngest rocks, and anticlines and synclines.

Streams flowing across the trend of dipping and folded rocks must cross these various homoclinal, anticlinal, and synclinal ridges. The only way they can do so is to cut water gaps. In such gaps, the intersections between dipping beds and valley and gap slopes create the familiar outcrop V's. (We introduced these in Figures 1–7, 1–8a, and 1–8b, and they are nicely depicted in Figure 9–1.)

It is absolutely astounding that such an endless variety of landforms has been produced by the combination of varying strata and varying streams! Strata vary in folding (gentle, tight, symmetric, asymmetric) and bedding (thin, thick, and interbedded). Streams vary in size, and in their adjustment to the geology (well adjusted, more or less adjusted, and defiantly unadjusted!). Because each combination of structure, lithologic sequence, and drainage system is a special case, interpretation of topographic maps of such areas is always a delight and a never-ending challenge.

Cambridge NE, Kansas

(Figure 10–1 in back of book, C.I. 10 feet)

As pointed out in Chapter 1, asymmetry in divides and valleys is one of the most-reliable topographic indications of dip. Such asymmetry is restricted to where the dip is gentle enough to influence the inclination of the slope that coincides with dip direction. Since true dip slopes are extremely rare, we are talking about the more-common resequent slope.

On the other hand, ridges upheld by steeply dipping resistant beds are generally symmetric in profile. Two such symmetric ridges, small hogbacks, are ridge **L** in Figure 10–3 and ridge **L** in Figure 10–13 (both Booneville, Arkansas, in back of book). As you will see when we study these figures, we can determine the dip direction along each of these ridges, without the aid of asymmetry. All that the contours tell us about each hogback is that the dip is steep. These are strike ridges.

In Figure 10–1, northeast-trending divide **FA** separates two very different topographies. The area to the east consists of a stair-step terrain, in which slopes range from gentle to steep. Topography to the west is similar, in that it displays a series of benches. However, it is different in that the steeper slopes are much less so. Overall, the western area is gentler and more subdued.

Divide **FA** appears to be the crest of a cuesta which has developed, through differential erosion, on a sequence of alternating, more- and less-resistant beds. If this is a cuesta, the gentle regional dip is toward the west. The steep slopes to the east are the obsequent flank, and the gentle descent to the west is the resequent flank.

The stair-step characteristic of this terrain may be ascribed to resistance differences in a series of extremely gently dipping layers. But does it follow that, because the topography is cuesta-like, the structure is that of a cuesta? Is it not possible that the asymmetry has some other explanation? In fact, there is an alternative answer.

Let us assume that these beds are perfectly horizontal. Let us further assume that the streams draining both the west flank and the east flank have the same base level. Finally, let us assume that the west-flowing streams are twice as far from their base level as the east-flowing streams. The average gradient of the western streams would have to be one-half that of those to the east. (In fact, it is.) Longitudinal valley profiles and divide slopes would tend to be much gentler in the west than in the east. (In fact, they are.)

Could it be, then, that this is not a cuesta? And that the beds do not dip to the west? Consider secondary divide **C** (long arrow **C**), which descends eastward from divide **FA**. Along arrow **C** we can count four steeper slopes and four gentler slopes. Could we say that eight stratigraphic units, four of which are more-resistant and four of which are less-resistant, crop out along this line? Obviously secondary divide **C** cannot be a dip slope. Composite slope **C** cuts across eight units. These beds, if not horizontal, are dipping very gently. (They may, of course, be horizontal.) They may dip gently to the west. They may even dip gently to the east (if they do, their dip must be much gentler than the overall slope of secondary divide **C**).

Now consider slopes **B** and **D**. In what ways do they differ from slope **C**? For one, they are much gentler than slope **C**. For another—and this is critically important—they are *smooth* slopes, not stair-steps. In fact, they are virtually uniform, gentle, westerly slopes, apparently remnants of a uniform regional slope.

Such slopes are typical of uplands that have developed on top of a gently dipping resistant bed. They have developed following extensive erosion, which has stripped away any overlying strata. This is what a dip slope looks like (and what an "almost" dip slope looks like, too). This is a *resequent slope,* **FA** is a *cuesta,* and dip is to the *west.*

That gentle westerly dip does not necessarily explain why virtually all the western slopes are gentler than their counterparts in the east. The contrast between slope **C** and slopes **B** and **D** is the reasonable difference we would expect between nondip slope (**C**) and (virtual) dip slope (**B** or **D**). But the overall contrast in slope magnitude cannot be attributed to obsequence versus resequence.

Overall difference in gradient is frequently seen as the direct response to difference in distance from base level. Such a gradient difference appears to exist in this area. The western area is subject to much-slower downcutting than the eastern area. Consequently, in the western area, all slopes are gentler than their counterparts in the east. Even the "steeper" western slopes, those developed on the more-resistant units, are much gentler than the steeper slopes in the east.

The following exercises should help you visualize and appreciate the contrasts between the east and west slopes.

1. Draw a simple block diagram of the area, viewed from the south.
2. Draw topographic profiles along lines **C** and **D** (extend the latter to **E**).
3. Draw longitudinal profiles along streams **c** and **d**.

Marmot Mountain, Montana
(Figure 10–2 in back of book, C.I. 40 feet)

Another part of this quadrangle has already appeared as Figure 3–5 (in back of book) in the chapter on glaciation. In fact, the two maps overlap slightly. The northeastern corner of Figure 3–5 extends into the southwestern corner of this figure.

We used Figure 3–5 to illustrate topographic features produced by mountain glaciation. But the topography depicted in this area is apparently devoid of the impact of glaciation. Note that all the valleys are distinctly V-shaped; there are no scooped-out troughs in evidence.

We have designated two sets of points which extend across the map from south to north: zigzag line **RUOMIEB** and parallel zigzag line **QSNLHDA**. (We suggest you connect them with a pencil and straightedge.) To the west of the latter, the mountain terrain is massive. The drainage network is dendritic and coarse textured. This massiveness is typical of topography developed on a homogeneous body of resistant bedrock, such rocks as "granites," "metamorphics," and the like. Resistant "crystalline" rock, whatever its proper petrographic designation, would be a likely candidate.

Now, using a straightedge, draw a line connecting **E, M,** and **U**. Extend each end of the line to the edge of the map. Large triangular features (e.g. **MIE**) appear to be plastered onto the ends of several main east-northeast-trending ridges. These landforms virtually shout out, "dipping strata"! Line **QSNLHDA** not only separates the crystalline topography of the west from the dip topography of the triangles. It must also define a narrow band of relatively less-resistant bedrock. Note that several linear and apparently subsequent tributaries (e.g. **C, G,** and **J**) have developed courses along this outcrop belt.

A word of caution. These tributaries follow the outcrop band of a specific stratigraphic and lithologic unit—a nonresistant one—but they are not "strike" streams. The simple reason is that they do not flow horizontally. Strike is close to N27°W, whereas tributary **G,** for example, flows nearly due north. These streams are adjusted along component or apparent dip lines.

The map area includes a third division, the large lowland in the northeast. This is part of a broad valley floor across which major stream **F** now meanders. This area may be topographically low because it is occupied by nonresistant strata that overlie the more-resistant rocks forming the big triangular landforms. On the other hand, it may be nothing more than the valley of a large river, though the fact that it follows the strike of the upturned beds suggests the influence of low resistance in this area.

What, exactly, are the triangular features? They are not triangular facets (see Figure 2–4). Rather than cutting across structure, or developing where there is no structure to influence them, they are definitely dip- and strike-associated. In fact, such slopes as **K, P,** and **T** (arrows) are nearly, if not precisely, dip slopes.

We may think of these triangular features as short segments of a once-continuous hogback-like ridge. If connected, hilltops **R, O, I,** and **B** would form a strike ridge. If this were a hogback, it would be a very special kind. It is quite literally plastered up against a topographically homogeneous mountain mass.

Thus, these triangular features must be, and are, fine examples of *flatirons* (see Figure 2–5). According to the *AAPG Geological Highway Map of the Northern Rocky Mountain Region,* they consist of "Paleozoic" strata which lie upon a crystalline Precambrian basement.

Booneville, Arkansas
(Figure 10–3 in back of book, C.I. 20 feet)

Note the position of the north arrow. This map has been rotated to the northeast. The grid lines that extend northeast-southwest across the page are actually north-south lines. Therefore, the grid lines should be used for orientation—**not** the page margins.

The ridge-and-valley asymmetries displayed on this map are truly what one would term "textbook" quality. The three main ridges (**E, M,** and **S**) are clearly asymmetric, as is valley **B**. A bit more study is required to perceive the asymmetry of broad valley **R,** which separates ridges **M** and **S**. Part of the map area, viewed from a generally southerly direction, is depicted in Figure 10–3a.

All of the steep (obsequent) ridge flanks face to the north-northwest. All of the much-longer, gentle (resequent) slopes are south-southeast-facing. By labeling some slopes obsequent and others resequent, we indicate that dip is to the south-southeast. The extreme asymmetry reflects the gentleness of the dip. Along most of their lengths these are true strike ridges.

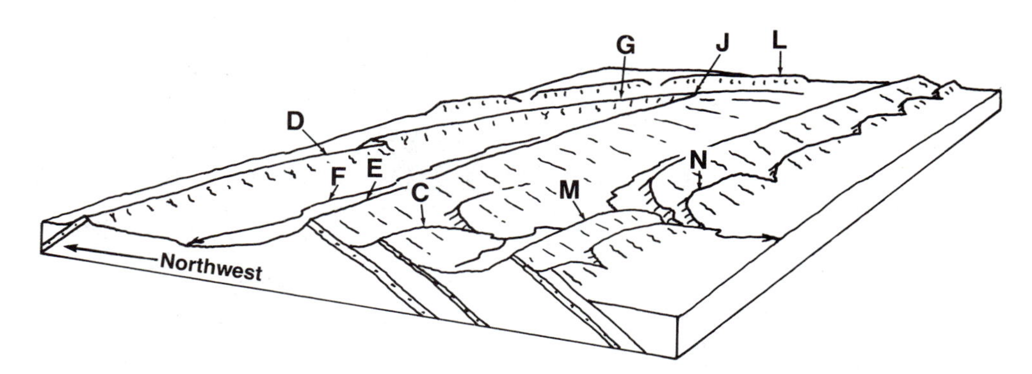

FIGURE 10–3a
Block diagram of part of the area depicted in Figure 10–3, viewed from the southwest.

According to the *AAPG Geological Highway Map of the Mid-Continent Region,* this area lies in the Arkansas Valley of central Arkansas. It is a structurally low area in which Pennsylvanian clastics, dominantly sandstones and shales, achieve thicknesses of several miles. In this area it is safe to think of the ridges as sandstone ridges, and of the intervening valleys as shale (subsequent) valleys.

There are more than three sandstone beds in the sequence. The three main ridges (**E, M,** and **S**) are developed on the three most-resistant sandstones—probably the three thickest. Ridge **C** is developed on another. Yet another forms secondary crest **N,** just *upsection*[1] from **M**. Near the middle of broad lowland belt **R,** in the thick shale section, is small sandstone ridge **Q** (thin, resistant unit). Elongate, rounded, low hill **T** in the south suggests an even-younger, thin sandstone.

To review, let us make a downsection traverse northward from ridge **T** to valley **F,** youngest to oldest. Please refer to Table 10–1, in which we will use lithologic units rather than topographic forms.

Note that we can only trace ridge **E** northeastward to point **J**, where it appears to run into the side of narrow symmetric ridge **L**. Whereas the dip along ridge **E** is at most moderately steep to the south-southeast, all we can say about the dip at **L** is that it is very steep. Also, the strike of sandstone **L** is east-west.

We cannot determine dip direction along symmetric ridge **L,** though farther to the west, ridge **D** is asymmetric. Its asymmetry is that associated with northward dip. If the dip at **D** is to the north, the sandstone that forms ridge **D** must directly overlie the shale of lowland **F**. In other words, sandstone **D** is the same as sandstone **E,** the resistant unit that forms ridge **EJ**. Lowland **F** is thus the axial valley of a breached asymmetric anticline.

We can trace this sandstone from **E** to **J**. Here there are minor structural complications and the ridge continues as ridge **GD**. Ridge **M** extends almost to the right side of the page and appears at the top of the page as Hickory Ridge. Note that subsequent valley **BH** is quite broad as a result of the gentle dip, but its continuation in the northeast, where dip is steep, is extremely narrow. (It is followed by small subsequent streams **K** and **I**.)

When we refer to ridge-and-valley asymmetry, we think in terms of contrast in slope magnitude. Such contrast is exactly what we see in this area. The

[1]A stratigraphic unit is *upsection* from the older unit that it overlies. The older unit is thus *downsection* from the younger unit which it underlies.

TABLE 10–1
Downsection Traverse Northward from Ridge **T** to Valley **F**.

Relative Age	Lithologic Unit	Map Position (Figure 10–3)
Youngest	sandstone **T**	Southernmost
↓	shale **U**	↓
	sandstone **S**	
	shale **V**	
	sandstone **Q**	
	shale **P**	
	sandstone **N**	
	shale **O**	
	sandstone **M**	
	shale **B**	
	sandstone **C**	
	shale **A**	
	sandstone **E**	
Oldest	shale **F**	Northernmost

northwest flank of ridge **M**, for example, is much steeper than the southeast flank. However, these opposing flanks display another contrast: they are also dissected quite differently. Undoubtedly the steep northwest flank is scored by many obsequent valleys too small to be represented at this map scale. But as mapped, it appears almost as a smooth steep slope or front. By contrast, the southeast flank is mapped as being well dissected. This kind of contrast is also clearly demonstrated by ridge **S**. However, in this case, both northwest and southeast flanks are shown to be dissected. The southeast flank presents a considerably higher drainage density.

A word of caution. In most cases you will find the obsequent slope to be the more-finely dissected of the two. This difference is usually brought about by exposure, in the obsequent slope, of such fine-grained rocks as shales and clays. These underlie the more-resistant ridge-forming beds, the latter being the rocks into which the coarser topography of the resequent slope is carved. An outstanding example of this type of dissection asymmetry is depicted in Figure 13–9 (Hilton, Virginia, in back of book).

Along some homoclinal ridges, the resequent flank is more-finely dissected. Along others, it is more-coarsely dissected. This makes a telling argument against attempting to use anything like "keys" in map interpretation.

Jacob Lake, Arizona

(Figure 10–4 in back of book, C.I. 50 feet)

This is a marvelous map on which to study structure/topography harmony. It is also an ideal map on which to practice your knowledge of stream classification. The only problem is that you cannot do these things until you know what you are looking at. And you cannot know what you are looking at until you learn how the topography came into being. Catch-22? Fortunately, no.

The *Geologic Map of Arizona* can tell us what we need to know in an instant. But let us see what we can discover on our own. First, what are the most obvious and easily identified (not interpreted!) features? Three major stair-step-like topographic surfaces (**K, N,** and **J**) stand out clearly (see Figure 10–4a, a sketch of the area viewed from the south). They look a bit like Lake Bonneville shorelines (wave-cut terraces), except these are of monstrous dimensions. Separating the three nearly horizontal surfaces are two north-trending belts of steeper ground, **E** and **H** (arrows). We shall return to these large-scale features.

On a smaller scale, there is the three-mile-long linear scarp **AB,** which trends obliquely to the major belts. Whatever the nature of those large belts, we are reasonably secure in classifying this scarp as structure controlled. It is too straight not to be. Most likely it is fault controlled. At this point, using only this map as our guide, we would be ill-advised to label this a fault scarp. Nor may we claim that it is a fault-line scarp; we simply cannot tell. There are several other anomalously linear features in the area, including trend **MG**.

A Traverse. Let us make a topographic traverse west-east, from upland **K** to lowland **J**. The general slope from **K** to **E** is convex, smoothly rounded rather than abrupt. In the central part of belt **E,** the slope is uniformly steep. To the east of letter **E,** the slope becomes rounded and concave, bringing us down to surface **N**. After crossing broad surface **N,** we begin the convex descent toward **J**. Steep slope **H** is somewhat cut up by streams. But we can still note that the steepening from **N** to **H,** like that from **K** to **E,** is a convex rolling-over, rather than a sharp break in slope.

However, the concavity between steep slope **H** and flat surface **J** is sharper and more angular than the concavity between slope **E** and surface **N**. This break in slope is more like the intersection of two planes, than the bending of a single plane.

This map area lies in the Colorado Plateau. It is a region teeming with vertical cliffs and scarps developed on resistant horizontal and gently dipping strata, primarily sandstones and limestones. This is the country of the Grand Canyon, Monument Valley, Zion National Park, and the Paria Plateau. Yet, in this particular map area, the only steep slopes that we might regard as scarps are a few canyon walls such as **O,** and linear scarp **AB**. Everything else is *slope,* be it steep or gentle. Steep slope zones **E** and **H** are not the scarps of a canyon, mesa, butte, or plateau. This is not a topography carved in a sequence of nearly horizontal, resistant and nonresistant formations.

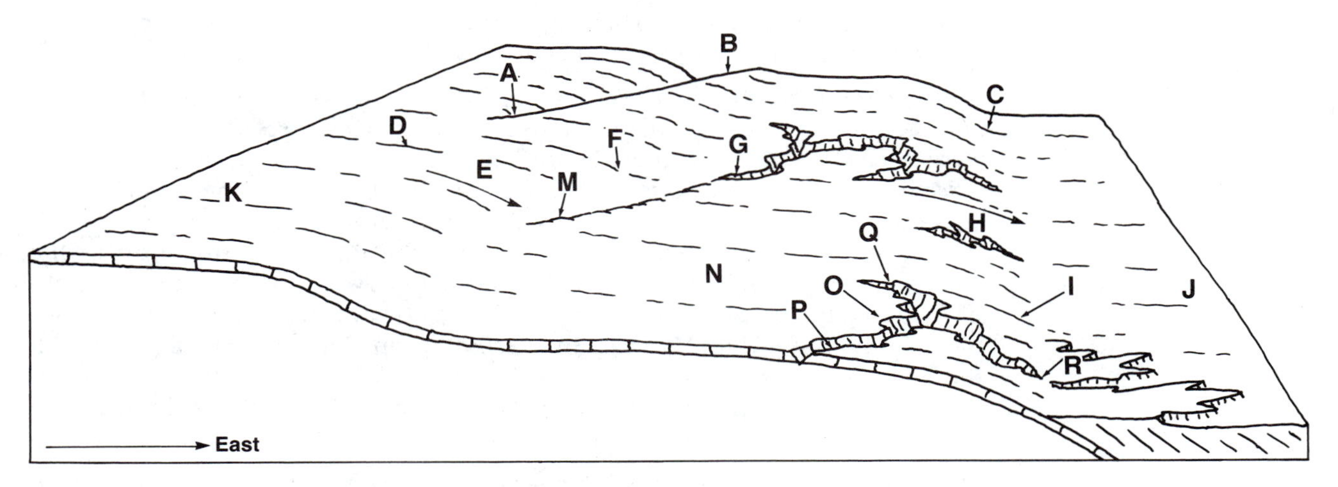

FIGURE 10–4a
Block diagram of the area depicted in Figure 10–4, viewed from the south.

We may think of scarp **AB** as the steep, linear, northwest flank of an asymmetric triangular ridge, **ABF**. The asymmetry of this ridge suggests east dip. (It does not *prove* east dip, since scarp **AB** is fault associated.) Since the easterly slope of surface **ABF** is the northern extension of slope **E,** the possibility of east dip in area **ABF** permits us to consider that slopes **E** and **H** could be reflections of east dip as well.

Geology. Now it is time to look at the *Geologic Map of Arizona*. Surface rock throughout the area west of line **CI** is the Permian Kaibab Limestone. This is the caprock of the Grand Canyon's north and south rims. Lowland **J** is underlain by the Triassic Moenkopi Formation (redbeds), which consists of fine-grained clastics.

In the southeast part of the map, a series of sharp contour bends defines dashed line **QR**. With some imagination you can "see" another contour-bend dashed line, **PR,** on the south side of the valley. Together, these two lines form a surprisingly well-defined outcrop V. East dip. That makes sense. The Permian limestone dips to the east beneath the nonresistant Triassic redbeds. Line **CI** is the contact. Slope **H** is virtually a dip slope.

Extremely non-linear scarp **QOP,** which borders the head of canyon **R,** stands in sharp contrast to outcrop-V **QRP**. This is the scarp of a horizontal or nearly horizontal resistant bed. Letter **O,** then, lies on nearly horizontal rocks. What we are looking at is a roll-over, a steepening of dip toward the east, a monocline. This explains the absence of major east-facing scarps extending north-south across the entire map area. Since the dip at letter **O** is so gentle, we may assume that such gentle dip extends westward across surface **N** to the foot of slope **E**. Surface **N** is developed on a structural terrace (Figure 10–4a).

The possibility of east dip in area **ABF** and along slope **E** is supported by the east dip which we have confirmed along slope **H**. We may now propose that the easterly slopes of belt **E,** like those of belt **H,** are dip slopes, or very nearly so (see Figure 10–4a).

Look at the top of belt **E,** where the upper slopes become more gentle as we look westward. We can visualize a corresponding decrease of dip until we reach **K,** where the exceedingly gentle dip appears to be to the north, toward **D**. In fact, the gentle regional slopes (arrow) west-northwest from **K** suggest correspondingly gentle west-northwest dip as well. Divide **KD** may thus be the crest of an extremely asymmetric north-plunging anticline.

Though there may be northwesterly dip from **K,** the surface to the south of **K** is virtually horizontal. This suggests that, in this southwest corner of the map area, the limestone may also be horizontal.

Structure/Topography in Harmony. What we are saying is that the topographic contours west of line **CI** may be read and interpreted as though they were *structure contours*. It must be understood that we are speaking about the *generalized* spacings and orientations of the contour lines. Minor crenulations are obviously topographic details resulting from dissection.

In review, from west to east, we have the following structures: in the southwest corner, horizontal rock at **K,** bordered by monocline **E** (to the north of **K** these structures are replaced by asymmetric anticline **KD**); structural terrace **N**; and monocline **H**. Since the base of steep slope belt **H** coincides with the limestone-redbeds contact, slope **J** must be an erosional slope that truncates the more steeply dipping redbeds. We depict this relationship in Figure 10–4a.

Streams. In the beginning of this section, we said that this is an ideal map on which to practice your knowledge of stream classification. Into what categories do the streams in this map area fall? The entire limestone area consists of a near-dip-slope expanse from which the nonresistant redbeds have been stripped. Obviously, none of the slopes in this area can be thought of as "original" slopes. Hence we rule out *consequent*.

The several linear streams that have developed by selective headward erosion along fault traces are of course *subsequent* streams. And the many streams that flow in the downdip direction (such as stream **L**) are all *resequent* streams.

Fan or Pediment? To the east of canyon-mouth **R** several crenulate contours depict what appears to be a dissected alluvial fan, extending out on top of erosional surface **J**. However, since this local feature might also be erosional (pediment remnant?), we may only suggest, rather than declare, that it is a fan.

New Enterprise, Pennsylvania
(Figure 10–5 in back of book, C.I. 20 feet)

When it comes to ridge asymmetry, differential erosion, topography/structure, topography/lithology correlations, and variations in fold geometry and scale, few areas can match the Valley and Ridge of Pennsylvania. This map is an excellent sample.

The outstanding topographic feature on this map is sinuous ridge **BFR** (see Figure 10–5a). This is a hogback, a ridge developed on a dipping, resistant stratigraphic unit by differential erosion. Its profile is asymmetric in various ways. In the southeast corner, for example, slope **RK** is a complex of gentle and steep segments. But slope **RQ** is smoothly concave. Point **S,** in fact, lies on a bench. This segmented bench can be traced along the entire ridge, to **D** in the north. Note that at **C** and **J** it is surmounted by small knobs.

At no point is the smooth, slightly concave southwest flank of ridge **BFR** interrupted by such a feature as bench **SD**. At point **F** (not incidentally its highest point) the ridge bends sharply. Slope **FL** is extremely uniform and gentle. In contrast, opposite slope **FG** is much shorter and steeper. It is easy to visualize slope **FL** as a dip slope. Slopes **BA** and **RQ** are therefore also dip slopes, or nearly so. Together, all the slopes on this side of ridge **BFR** constitute its resequent slope or flank.

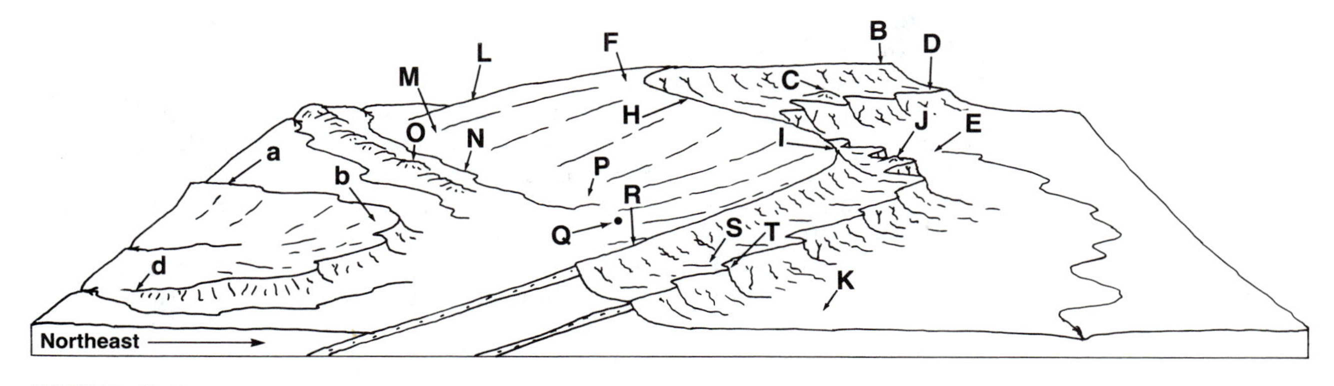

FIGURE 10–5a
Block diagram of area depicted in Figure 10–5, viewed from the southeast. The northwestern corner of the Figure 10–7 area is included here. (Horseshoe ridge at left does not appear on Figure 10–5. It is shown on Figure 10–7.)

How can we account for bench **SD,** on the opposite (the obsequent) flank of the ridge? A traverse from **Q** to **K** (updip[2], downsection) will tell us. (Study the cross section of the structure-lithology-topography relation in Figure 10–5a, a sketch of the area viewed from the southeast.)

In going from **Q** to **R** we are rising up the backslope, or the resequent slope, or the dip slope (whatever you choose to call it) of the hogback. The rock we are crossing is the top of the resistant unit that forms the backbone of the ridge. The upturned edge of this unit crops out at **R**.

As we start down the east flank, we progress downward in the section, into the older units that underlie unit **R**. Most of these beds, from **R** down to **S**, must be less resistant than the cap rock, since they are being eroded back rapidly. Their removal is undermining or sapping the cap rock. This is why the upper part of slope **RS** is so steep.

For some reason, however, the rock at **T** is not being eroded back so rapidly; it is resisting. This rock, which must be another more-resistant unit, stratigraphically underlies the nonresistant **RS** sequence. In turn, unit **T** is underlain by yet-older, less-resistant strata, since at point **T** the slopes are also quite steep, leading down to lowland **K**. It is the upturned edge of this older, resistant unit **T** that projects as knobs **C** and **J**.

Anticline and Syncline. At all points along ridge **BFR,** the strike of the beds coincides with the local trend of the crest. Thus at **B** the strike is about N20°E, with dip to the northwest. At **F** the dip is toward **L,** S30°W, whereas at nearby **H** dip is S10°E. These dips define a southwest-plunging anticline (note symbol on map) which is expressed by anticlinal nose **FL**. Note that here, where dip along the ridge is lowest, the ridge crest is highest. At **B,** where the dip along the ridge appears to be steepest, the ridge crest is lowest. The ridge is also narrowest along this segment.

East of **H,** at **I,** the dip is due south, whereas farther to the southeast, at **R,** dip is to the southwest. The dips at **H, I,** and **R** define a south-plunging syncline (note symbol on map), companion to anticline **FL**. This syncline can be traced back northeastward to **E** (dashed line).

Incidentally, the nose from **F** to and beyond **L** is an anticlinal mountain. The ridge northward from **F** to **B** is a homoclinal ridge. So is the ridge from **F** to **R**. Some earth scientists think of ridge segments such as **FIR** as "synclinal" ridges because they define a synclinal structure.

We must not forget, however, that segment **FI** is also part of the south flank of anticline **EFL**. Surely segment **FI** cannot be both a synclinal ridge and an anticlinal ridge! There are no synclinal ridges in this map area. The triangular lowland defined by points **C, G,** and **J,** however, is a small breached anticline or anticlinal lowland.

Contour trend **MP** coincides with strike, as does parallel low divide **O**. We may safely assume that along this ridge, the dip is also to the southeast. Ridge **O** is upheld by a minor resistant unit. It lies stratigraphically above the nonresistant beds, which in turn overlie the main ridge-former. The nonresistant beds are followed by subsequent stream segment **N**.

Note: horseshoe ridge **abd** is shown in Figure 10–5a (left side). But this ridge does not appear on the map for this section (Figure 10–5). Instead it is

---

[2]Dip is always measured down from the horizontal. *Down-dip* means in the direction of dip. *Up-dip* means in the direction opposite to dip direction. In the **aP** area (Figure 10–5a), dip is to the south. A traverse from **a** to **P** progresses in the opposite direction (up-dip). Such a traverse goes from younger to older strata (down-section).

shown on Figure 10–7 (Everett West, Pennsylvania, in back of book) as ridge **ADB**.

Nashville, Indiana
(Figure 10–6 in back of book, C.I. 10 feet)

This is one of the most-important maps in this book. It is also one of the most fascinating and misleading. It provides a valuable lesson in map interpreting.

Two characteristics stand out glaringly. One is the linearity of several main valleys and divides. The other is the pronounced asymmetry of many valleys and divides.

Indiana is a state not particularly renowned for its faults or measurable dips. Yet here is a south-central Indiana area in which the obvious explanation for the topographic linearities and curvilinearities is fault control. Streams **CB, EJ,** and **NG,** for example, exhibit classic characteristics of streams that either (1) have adjusted by headward growth along lines or zones of nonresistance, such as the traces of steeply dipping faults, or (2) have formed along valleys created directly by block faulting.

Similarly, consider such extremely asymmetric divides as **DA, FI,** and **PH**. All suggest measurable dip to the south and southeast. This dip is apparently also associated with tilting, which would accompany block faulting. In the southwest corner, ridge **MO,** which lies at right angles to the end of ridge **PH,** indicates probable dip to the southwest. Near the center of the map, the clear-cut asymmetry of minor divide **KL** indicates dip a bit to the south of west.

Can the steep flanks of these possible fault-block divides and ridges be classified as either fault scarps or fault-line scarps? We are not equipped to answer at this point. In any case, the various apparent dip directions and fault traces are convincingly indicated by the topography.

So far we have found nothing especially remarkable or noteworthy about the topographic features and relations, and the various structures that they appear to reflect. But there is a problem. This area lies in a region of extremely gentle regional dip. The dip is away from the Cincinnati Arch (to the east) and toward the Illinois Basin (to the west). Regional dip of 10 to 20 feet per mile is to the west.

Such dip clashes with the ridge-and-valley asymmetries displayed in this map area. For one thing, regional dip direction is almost at right angles to the dip ''direction'' indicated by most of the asymmetries. Secondly, even if the low regional dip direction were to the south, it could hardly bring about the pronounced and clear-cut asymmetry depicted on this map. The regional dip is much too gentle.

The truth is that the asymmetries in this area are *not* dip-associated. The lesson is that *asymmetry does not always indicate dip*.

This part of Indiana was subjected to periglacial climatic conditions during various advances by Pleistocene ice sheets. At such times, the microclimates on north- and east-facing slopes were significantly more severe than on south- and west-facing slopes. More-severe, more-frequent, and longer-lasting freezing on the north and east slopes deterred their being lowered by either erosion or ''lubricated gravity.''

The linearities and curvilinearities are another matter. For the present, we may postulate some kind of fracture control, despite the fact that the block-fault prospect seems to be without support.

Repeat: *Asymmetry does not always indicate dip*!

Everett West, Pennsylvania
(Figure 10–7 in back of book, C.I. 20 feet)

The undependability of asymmetry is strikingly displayed in Figure 10–6 (Nashville, Indiana, in back of book). But here, asymmetry is reliable to the fullest.

Let us first find some dip-denoting asymmetries. Please refer to Table 10–2.

If you add dip symbols for these points to the map, you will delineate a doubly in-plunging syncline, shown by the dashed axial line. What is dip direction at **B**? At **D**? How can you tell?

Figure 10–7a shows the area, viewed from the southeast. But the most-effective way to visualize the structure in this area of fragmented topography is to simplify it mentally by filling in the missing pieces (Figure 10–7b).

There are two main resistant units in the area. The older (outer oval) is represented by ridges **FABD, K, L,** and **U**. At first glance, you might think this unit continues southwestward and westward from **U** across the valley to **P**. However, from **P** it would then continue northward to **H,** paralleling itself in an odd spiral pattern. The fact is that ridges **P, H, J, S,** and **R** are the topographic expressions of a younger resistant unit that is nested within the downfolded older unit. Racetrack valley **OCT,** which separates the two, is underlain by nonresistant strata, probably shales and maybe limestones.

In Figure 10–5 (New Enterprise, Pennsylvania, in back of book) we observed an inverse relation between dip magnitude and ridge size (width and crest height). This relation is also beautifully represented by ridge assemblage **PHJSR**. Segments **J** and **S** are the broadest, highest, and most asymmetric. Here the dip is least steep. Segments **P** and **H** are narrower and lower, and they tend to be less asymmetric. Dip along these segments is steeper. At and near **R,** the sandstone is represented by a mere line of small hills. Here the dip is steepest.

There is a third ridge-forming unit, represented by ridge **G** and **I** and upland **Q**.

This map area lies immediately south of that depicted in Figure 10–5. Major homoclinal ridge **BFR,** which we studied in Figures 10–5 and 10–5a, cuts across the northeast corner of this map as ridge **E,** and then continues to the south as ridge **N**. Broad bench **DS,** clearly shown in Figure 10–5a as extending along the obsequent flank of ridge **BFR,** can be detected in only a few places along ridge **N**. This ridge segment is narrower, more nearly symmetric, and considerably lower than it is in the Figure 10–5 area. Along this southern segment the dip must be appreciably steeper, for the obsequent slope bench is absent along most of its length.

TABLE 10–2
Dip-Denoting Asymmetries.

Point	Dip Direction	Point	Dip Direction
F	SE	**J**	W
A	SE	**S**	W
K	SW	**R**	NW
L	NW	**Q**	E
U	NW	**G**	SE
P	NE	**I**	WNW
H	SE		

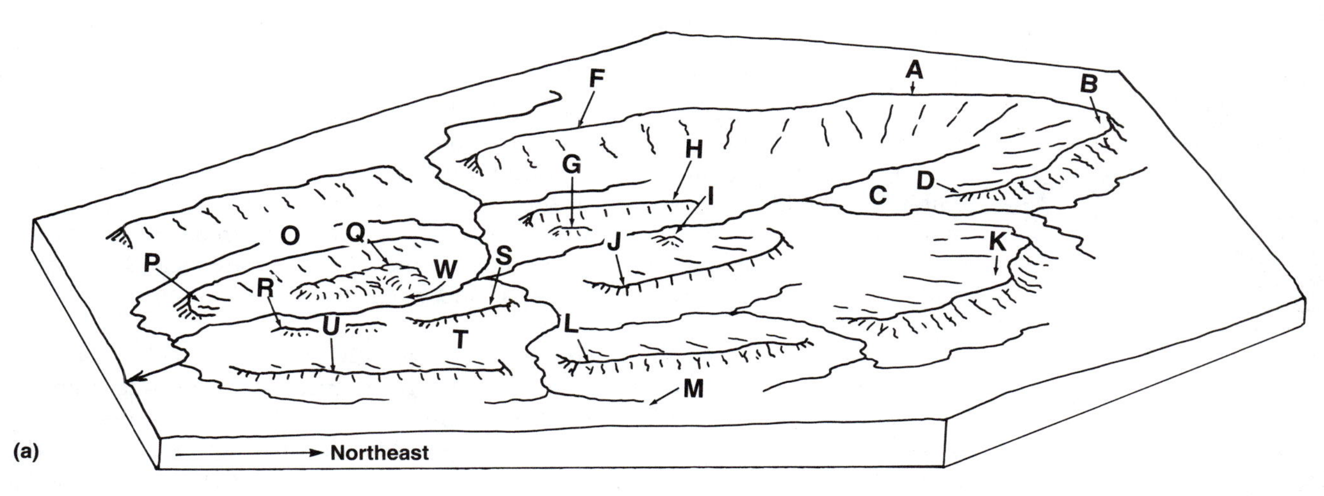

FIGURE 10–7a
Block diagram of part of the area depicted in Figure 10–7, viewed from the southeast.

Southeastern lowland **V** correlates with the **EK** lowland of Figure 10–5. These lowlands are underlain by the oldest stratigraphic units of the two map areas. The younger beds that overlie unit **V** are:

resistant unit **N**
nonresistant unit **M**
resistant unit **L**
nonresistant unit **T**
resistant unit **S**
nonresistant unit **W**
resistant unit **Q**.

By tracing the contacts between these several topographic/stratigraphic units, you can make a fairly accurate geologic map of the area.

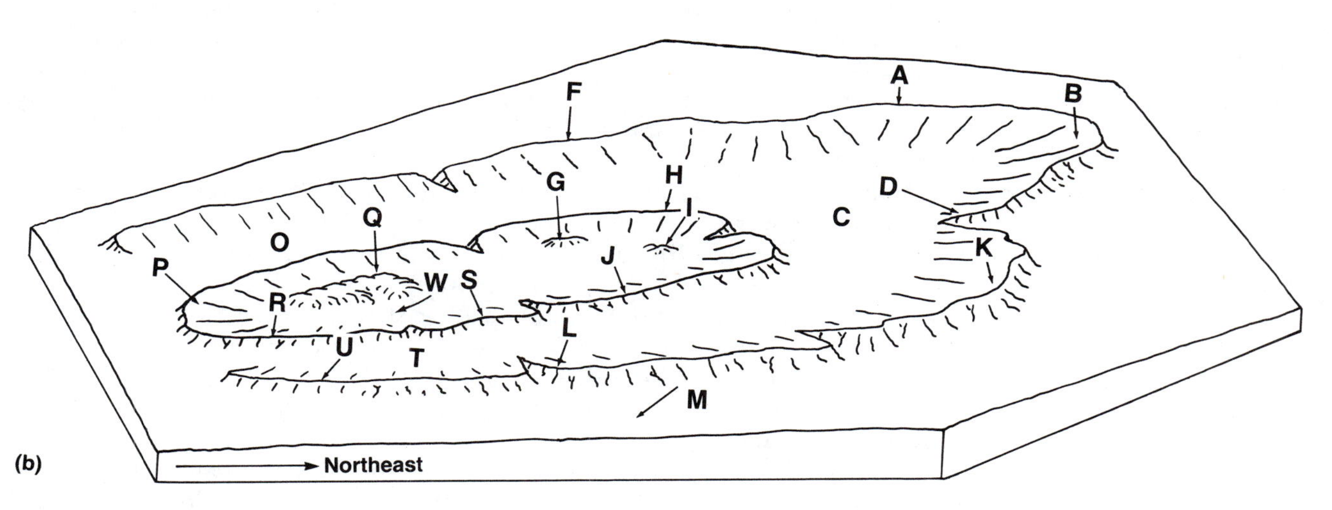

FIGURE 10–7b
Block diagram of part of the area depicted in Figure 10–7, viewed from the southeast (topography more generalized and simplified).

Millheim, Pennsylvania
(Figure 10–8 in back of book, C.I. 20 feet)

If a single quadrangle had to be selected to represent the Valley and Ridge province, it would be the Millheim 1/62,500-scale map. Not because it is typical of the province—no limited area is truly typical of such a varied, large expanse. Millheim would be chosen because, in this small area, there are so many outstanding examples of a wide range of fold/topography features. These include homoclinal, synclinal, and anticlinal mountains; anticlinal valleys and synclinal valleys; ridge asymmetry; water gaps; and subsequent, obsequent, and resequent streams.

This figure covers only about one-tenth of the Millheim quadrangle. Yet even in this limited area, there are two plunging anticlines, a plunging and a nonplunging syncline, and part of a possible second plunging syncline. The *Geologic Map of Pennsylvania* shows that there are (surprisingly, perhaps) only two resistant and two nonresistant formations exposed in the area we show. All we need to determine which is which, and where, is a single reliable dip. In other words, an asymmetric ridge.

Ridge **J** will serve. Its asymmetry clearly reflects south-southeast dip. Knowing this, we can state that at **N** the dip is to the east, and that at **P** it is to the north. Point **N** is a good example of a *synclinal nose*. Plunge is to the east, as indicated by the dashed axis and plunge arrow. Lowland **M** is developed in nonresistant rocks that underlie the ridge-forming unit of ridge **JNP**. Note the steep drop-off around the outside of the nose. In contrast, note the extremely gentle resequent slope (and probably dip slope) **NO**.

Within the curvature of ridge **JNP** is a second hairpin-shaped ridge, **LRS**. It also has steep outside slopes and a nearly flat top. This too is a synclinal nose, developed on a second (younger) resistant formation. The steep obsequent slopes that wrap around the end of this ridge, like the steep slopes bordering ridge **JNP,** are underlain by nonresistant strata (**Q**). These are eroding back rapidly, undercutting or sapping the caprock.

Recapping, from west to east along the synclinal axis, we have from oldest to youngest: nonresistant **M,** resistant **N,** nonresistant **Q,** and resistant **R**.

Let us assume that this area has not been complicated by strike faulting (oblique and dip fault offsets are not apparent). We may now extend our geologic interpretation, restricting it to the northern part of the map. If you wish, you may continue to the south as well. In fact, we strongly suggest that you obtain a copy of the Millheim quadrangle and interpret it.) (The Millheim map is an old 15-minute map, now out of print. You can obtain equivalent coverage by purchasing four 7.5-minute maps: Coburn, Millheim, Weikart, and Woodward.)

Valley **H** is merely an eastward extension of lowland **M**. It is apparent that it too is underlain by nonresistant formation **M**.

Segmented ridge **GIK,** actually a bench on the flank of ridge **A,** is similar to bench **DS** in Figures 10–5 (New Enterprise, Pennsylvania, in back of book) and 10–5a. Knobs **G, I,** and **K** in Figure 10–8 strongly resemble similar knobs on bench **DS—C** and **J**. Ridge **GIK** is also a resistant formation that lies stratigraphically below the main ridge-forming unit that composes ridge **A.** This lower resistant unit, which lies directly above the nonresistant beds of unit **M,** must correlate with resistant unit **N**.

Beyond **GIK** is ridge **A,** which must be upheld by upper resistant unit **R**. Steep slope **F,** which leads up to the caprock, must be underlain by nonresistant unit **Q**. Unit **Q** separates resistant units **N** (below) and **R** (above).

There are but four formations in this map area, two resistant and two nonresistant. Here are some interesting exercises:

- □ Propose a reasonable structure for ridge **A** and ridge **D**.
- □ Identify the formations that form lowland **B,** steep slope **C,** ridge **D,** and steep slope **E**.
- □ What is the *structure* of ridge **GIK**?
- □ What is the *structure* of valley **H**?
- □ What is the dip at **F**?

Cato, Arkansas

(Figure 10–9 in back of book, C.I. 10 feet)

Like the areas shown in Figures 10–3 and 10–13 (both Booneville, Arkansas, in back of book), this map area lies in the region of folded, resistant and nonresistant Pennsylvanian clastics known as the Arkansas Valley (from the *AAPG Geological Highway Map of the Mid-Continent Region*). This geomorphic division includes the part of the Arkansas River watershed that falls between the Boston Mountains plateau (to the north) and the complex imbricate structures of the Ouachita Mountains (to the south). This large area of folds is situated mostly in central Arkansas, though it extends into easternmost Oklahoma.

The asymmetry of the main sandstone ridge **ALM** is very well defined. It indicates that at **A** the dip is to the south, at **L** it is to the west, and at **M** it is to the north. As you can see, the axis of a west-plunging syncline extends from east of **L** westward to **B** on the map's western edge. You may wish to draw this axis (but if you do, wait until we have discussed the **DGFE** area). The topography and structure are shown in Figure 10–9a, a sketch of the area viewed from the southwest.

Most of the strata in this area are nonresistant. For convenience, we can think of them as shales, though they may include other nonresistant rocks such as siltstones, mudstones, and clays. Similarly, we assign sandstone to ridge **ALM,** though it may include other coarse-grained resistant lithologies.

Is ridge **ALM** a synclinal mountain? As we pointed out in our discussion of Figure 10–5 (New Enterprise, Pennsylvania, in back of book), a ridge may be designated a synclinal ridge or mountain only if its internal structure is that of a syncline, and if the mountain is exactly at the fold axis. A sinuous or bent

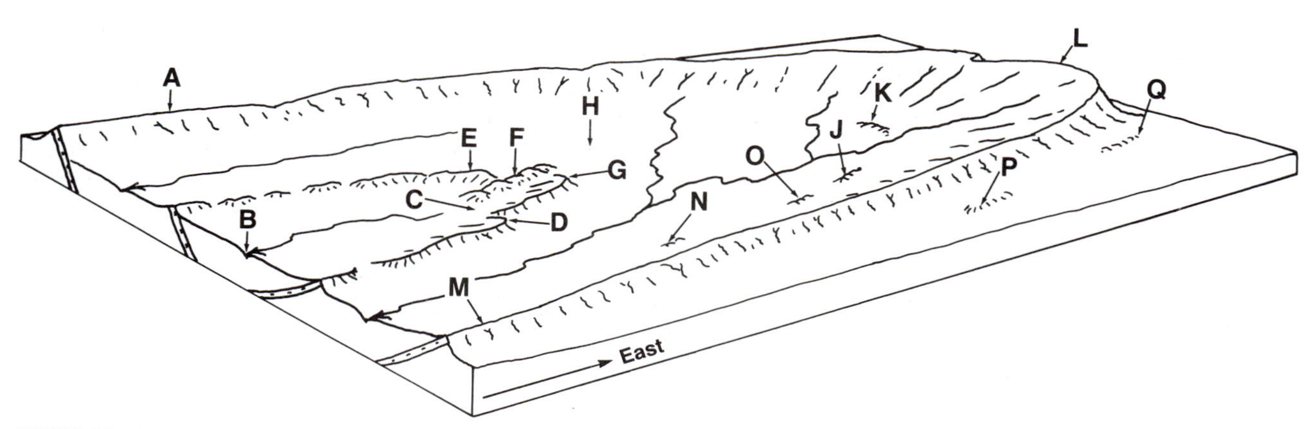

FIGURE 10–9a
Block diagram of the area depicted in Figure 10–9, viewed from the southwest.

homoclinal ridge is still a homoclinal ridge, whether its strike and dip variations define a syncline or an anticline. Thus ridge **ALM** is a homoclinal ridge throughout its entire length, though its dips define a syncline.

What is the dip at **P,** a short, low, linear hill? This hill lies on the flank of a large syncline. It is almost parallel to the nearby segment of the main homoclinal ridge that dips to the north-northwest. We may reasonably assume that the dip here is also north-northwesterly. The small hill **P** per se can tell us only what its strike is. However, hill **Q** is apparently developed on the same minor resistant unit that underlies hill **P**. The asymmetry of hill **Q** is that of a north-dipping bed. This supports the assumption that at **P** the dip is toward the synclinal axis.

Another minor resistant bed is expressed as several aligned knobs (e.g. **K** and **O**). Its presence is also indicated by the telltale configuration of the contours at **N,** where the contour bends are similar to those at **I**.

It is only by chance that **I** has closed contours (e.g. **J**) within it. The closed contours are there solely because of a fortunate combination of height, elevation, and contour interval. There is almost certainly a similar, though smaller, knob within bend **N**. But it does not happen to be "caught" by a contour plane. It might show up very nicely if the contour interval were decreased to five feet.

A relatively resistant, younger sandstone underlies cuesta-like feature **D**. **D**'s resequent (northwest) slope is extremely gentle. Such gentleness of slope suggests similarly gentle dip—much gentler, for example, than at **M**.

Note that the major homoclinal ridge is widest, highest, and most asymmetric in the vicinity of **L,** around the synclinal nose. As noted, the reason for this is that dip is lowest near the axis, and the lower the dip, the wider, higher, and more asymmetric the homoclinal ridge. The steepest dip along ridge **ALM** is in the northwest, at and near **A**. The extreme gentleness of the dip at **D** is explained by its proximity to the synclinal axis.

It is instructive to compare the smooth curvature of ridge **ALM** with the rather odd zigzag of the crest of ridge **DGFE**. The dip at **D** is surely to the northwest, toward the west-plunging synclinal axis. The asymmetry of the ridge at **E** indicates that the sandstone is dipping to the southwest. At **G** the dip is to the northwest. Should not the dip at **F** be to the west-southwest?

How can a west-southwest-dipping sandstone form such an east-northeast-trending ridge segment? Does not ridge segment **F** bear a remarkable resemblance to ridge segment **G**? Is something wrong with our interpretation?

There must be some structure here that has imparted a unique combination of topographic features and relations in this map area. We must consider this to be an anomalous area.

There are several possibilities. But the best one is that the simplicity of the synclinal folding has been complicated by some faulting. Along a line connecting points **C** and **H,** we would map a possible or a questioned fault, upthrown on the north. We would also indicate questioned north-northwest dip at **F**. We would add (dot-dash line) a questioned small, west-plunging synclinal axis, between the southwest dip at **E** and the northwest (?) dip at **F**.

These local structures would all be superimposed as minor complications on the overall **L-B** syncline, so well defined by major ridge **ALM** and by ridge **DGE**.

Strasburg, Virginia
(Figure 10–10 in back of book, C.I. 40 feet)

This confusing-looking area lies in the Great Valley of northern Virginia. According to the *AAPG Geological Highway Map of the Mid-Atlantic Region,* the bedrock, all sedimentary, ranges in age from Ordovician to Devonian. Let us assume that this is all we know as we begin our study of this map.

Obviously there are resistant and nonresistant formations in this area. How much structure can we decipher? Let us begin with the more obvious dips, as revealed by the better ridge asymmetries. Ridge segments **NO** and **Yb** are both asymmetric, their steep flanks facing southeastward. They also have sharp crests and northwest dip. These ridge segments are homoclinal. The resistant formation that forms ridge segment **NO** is underlain (lowland **Q**) and overlain (lowland **M**) by nonresistant strata.

Ridge **Yb** is developed on the same resistant unit that underlies ridge **NO**. Hence the strata at **Z** correlate with those at **Q**, whereas the younger strata at **W** correlate with those at **M**.

The broad valley floor immediately west of **M** is occupied by a large meandering stream, whereas immediately west of **W** there is the southwest end **V** of spear-shaped ridge **aV**, which projects southwestward as an appendage to ridge trend **ONbY**. Ridge **aV** narrows and becomes uniformly lower from **a** southwestward to **V**. It is essentially symmetric. Lowland **M** continues southward to **U**, just west of ridge **aV**.

In the west, asymmetric ridge **RB**, steep on its northwest flank, is another sharp-crested homoclinal ridge (southeast dip). Just to the east of **B**, at **E**, this ridge abruptly turns back to the south, to **C**. Then it continues easterly to **F**. As at point **a**, a branching ridge extends southwestward from **C** to **L**. Here it encounters water gap **K**. Then it continues as segments **J**, **H**, and **T**. From **C** to **L** this ridge is also spear-shaped, also narrows, and also declines in height. Though it is not dissected identically on both flanks, it is nevertheless relatively symmetrical, in contrast to the asymmetry of ridge **RB**.

Note that the sharp bends **E** and **c**, both of which are bordered on the outside by steep slopes, strongly resemble synclinal nose **N** of Figure 10–8 (Millheim, Pennsylvania, in back of book). As stated, southwestward from **B** the dip along ridge **BR** is to the southeast. At **B** we see the dip beginning to change a bit toward east of south. At **E** the dip is southwest, and at **D** it is toward the west. A southwest-plunging synclinal axis passes through **E**, which is a synclinal nose.

Ridge bend **Ybca** is a mirror image of ridge bend **RBEC**. Bend **c** is another synclinal nose. This syncline also plunges to the southwest. After all, the southwest dip at **c** is that fold's plunge, just as the southwest dip at **E** is the plunge of the western syncline.

In Figure 10–5 (New Enterprise, Pennsylvania, in back of book) is a rather broad, open anticline, the nose of which is **FL**. If it were a tighter fold, its nose would closely resemble ridges **aV** and **CL**, both of which are anticlinal ridges. The axis of the western anticline extends from **C** to **L**, continuing through **J**, **H**, and **T**. Its southwest plunge apparently ends at gap **G**. Here both ridge crest descent and ridge narrowing cease. In fact, farther to the southwest, the plunge is back to the northeast, to **G**. Anticline **aV** plunges southwest from **a** to, and probably beyond, the map border.

Syncline **E** must continue southwestward to and beyond **S**, just as syncline **c** must continue southwestward to and beyond **W**.

We studied water gaps in Chapter 9. There are five interesting gaps (**K**, **I**, **G**, **X**, and **P**) in this map area. Gaps **K**, **I**, **G**, and **X** all cut through anticlinal ridges. If we could determine how any one of them developed, we should solve the question of origin for all four. Gap **P** might have an entirely different explanation.

As discussed in Chapter 9, there are water gaps and there are wind gaps. There are gaps cut by antecedent streams. There are gaps produced by stream superposition, whether of the type we call "unconformity" or that which we refer to as "conformity superposition." There are gaps resulting from structural control.

Since this chapter focuses upon the criteria for recognizing dips and folds, we will not undertake an analysis of these gaps. How do you explain gaps **K, I, G,** and **X**? You might begin by asking whether there was a ridge along line **LJHT** at the time the main stream was flowing 700 or more feet above its present elevation. (It now flows northeasterly along lowland **UM**.)

Gap **P** is quite another matter. There might be one or more northeast-southwest-striking faults here that could promote selective erosion. There might be some other influence exerted on the drainage. All we can do, judging from what this map shows, is admit that we cannot confidently explain stream **UM**'s cutting through the ridge at this point.

Finally, note the atypical "meanders" of major streams **A** and **d**. It might be more appropriate to think of these bends as loops that connect successive parallel stream stretches. Surely the linearity and parallelism of these many stretches must be explained by some structural control. One possibility is that control is exerted by a set of tension joints, normal to regional strike and fold axes, and thus parallel to tectonic compression.

Greenland Gap, West Virginia
(Figure 10–11 in back of book, C.I. 50 feet)

Regardless of how reliable certain criteria may be for interpreting topographic maps, it is always good to exercise caution. This explains the selection of this map. At first, it may appear to be a duplication of areas we have already studied.

Ridge **QE** is just a rounded ridge of unknown structure, until we zero in on water gap **MO**. Here the unique bends in the contours clearly define the anticlinal reversal of a resistant caprock. Refer back to Figures 1–7, 1–8a, 9–1 (Cumberland, Maryland, in back of book), and 9–1a.

What we have are two classic examples of outcrop V's. One (**CON**) shows dip to the east, and the other (**CMN**) shows dip to the west. Slight asymmetries along this ridge, such as the one across **L** (steeper on east), disclose fold asymmetry rather than homoclinal dip direction.

Two other clearly defined outcrop V's (east dip) are those at water gaps **T** and **P**. Segmented ridge trend **SG** (broken by water gaps **T** and **P**) is thus revealed to be a homoclinal ridge. However, at and near points **S** and **G,** this ridge is also asymmetric. Its east flank is steeper than its west flank. This asymmetry is the asymmetry of a west-dipping resistant formation. Which should we believe, the V's or the asymmetry?

In this case we would choose the V's. Not only are they beautifully etched out, but the dip direction they indicate is in accord with the nearby dip along the east flank of anticlinal ridge **QE**.

Though the steep east flank of ridge trend **SG** may be in discord with the dip direction indicated by the V's, it is very much in accord with the steepness of the dip revealed by those V's. We believe the gentler slopes on the west flank of ridge trend **SG** (e.g. slope from **S** to **R**) are obsequent slopes. They lead down to several subsequent streams (e.g. **R** and **F**). These have developed along the nonresistant outcrop belt that separates anticlinal ridge **QE** from homoclinal ridge **SG**. These subsequent valleys are thus homoclinal valleys.

The geology/topography relation along segmented ridge trend **ID** (in the northwest) is not so readily disposed of. There is an upturned resistant formation here, but how can we determine which unit comprises it? If it is the same unit that forms ridge **SG,** it should have west dip. Yet, if it is another unit, such as the one that forms anticlinal ridge **QE,** it should have east dip.

We show a part of this ridge's divide as dashed line **IJKBA**. Segment **IJK** is the crest of a homoclinal west-dip ridge, if we can believe in the infallibility of ridge asymmetry. Unfortunately, we can also "see" an equally convincing outcrop V, in divide segment **KBA,** which looks very much like steep *east* dip! See Figure 1–8b.

What can we do about this quandary? Could the beds in segment **KBA** be *overturned* west-dip? Perhaps. Could the bend **IJK** simply reflect slight changes in strike? Not likely. It is not always possible to come up with the correct and convincing answer!

Finally, note wind gap **H**. Could capture have done this? If so, what captured what? When? And why?

Karbers Ridge and Herod, Illinois

(Figure 10–12 in back of book, C.I. 20 feet)

The topography of this area is at once extremely complex and remarkably simple. Its very high drainage density indicates that a considerable part of the bedrock is fine-grained. If you are interested in visualizing just how fine-textured this terrain is, draw in all the streams you possibly can, regardless of how short and closely spaced they may be. (Do this on an overlay, e.g. frosted Mylar.) This drainage map should prove most informative as a key to the geologic structure.

The tedious procedure of tracing many hundreds of streams is not required however, for you to see, despite the intricate topography, the few essential large elements that comprise this area.

The central oval pattern, with its concentric high and low rings, must have developed over either a structural dome or basin. In the midst of all this geomorphic confusion, is it possible to detect features and relations (such as asymmetries) that will permit us to choose between dome or basin?

Before we find out, consider one additional possibility. In speaking of a choice between basin and dome, we are thinking in terms of dip. This is because dip, and variations in dip direction, are apparently fundamentally important in this area. However, note that in addition to combining to form a fine-textured, trellis-like drainage system, some of the streams are anomalously linear. Linearity in a setting such as this almost certainly indicates fracture control, if not fault control. One outstanding linear valley is **P** (dashed line) in the southeast. It is reasonable to consider that faults complicate the otherwise simple fold structure of this area.

In the north, curvilinear, steep, south-facing hillside slope **B** contrasts sharply with the overall gentle decline to the north. This strongly suggests north dip (steep obsequent and gentle resequent slope). Slope **B** is but a small segment of an oval arrangement of steep slopes (e.g. **N** and **M**) that completely encircle and overlook a ring-shaped lowland (**GLO**). This lowland is drained by several apparently subsequent streams (**G, L,** and **O**). These streams combine to form a classic example of an *annular* stream pattern.

Some tributaries to these annular-pattern streams form a *radial* pattern in central highland **I** (e.g. tributary **H**). Others form a *centripetal* pattern, cut into infacing slopes **BMN** (e.g. tributary **F**). We cannot classify either the radial or centripetal streams as either resequent or obsequent, until we are sure whether this is a dome or a structural basin. However, the asymmetry at **B** suggests that it is a dome (north dip at **B**).

The divide between the centripetal streams (e.g. **F**) and a second, outer set of radial streams (e.g. stream **E**), lies much closer to annular lowland (letter **F**) than to surrounding lowland **D**. We indicate a portion of this divide by a sinuous dashed line. This entire divide, which encircles the annular lowland, is asymmetric. Its asymmetry, like that at **B**, also indicates outward dip around the flank of a structural dome.

Before we leave the structure of the dome, let us return briefly to central upland **I**. Note that the easterly slopes (e.g. at **K**) of this upland are much steeper than those that descend to the north, west, and south. If this upland were situated by itself, standing alone in a broad lowland, and not at the center of an annular lowland, its asymmetry might reflect a west-dipping homocline. Steep slope **K** would be its obsequent slope.

This steep slope is indeed anomalous. If, instead of being an obsequent slope, it is a resequent slope, would not its steepness reflect a steep dip to the east-southeast? Not really, since that would not fit the considerable width of the lowland near **L**. Steep local dip should require a narrowing, not a widening, of the adjacent lowland.

The reason we are wandering through all of this conjecture is that we are convinced there are faults in this map area. A fault along line **KJ,** at the base of this steep slope, might answer many questions. The possibility of such a fault should be kept in mind.

Finally, note that the finest-textured drainage is situated in the higher areas. In low areas, such as **D, A,** and **C,** the texture is much coarser, the interfluves are broader, and the slopes are longer and gentler. This is the opposite of what is found in many areas. Coarse clastics such as sandstone often underlie higher topography, and fine-grained rocks such as shale often underlie low areas. In such areas, we find coarse drainage texture in the uplands and fine texture in the lowlands.

An interbedded sequence of thin sandstones and shales might explain the combination of fine-grained texture and fairly high resistance displayed by the higher areas. There are many Pleistocene deposits reported in this general region. Perhaps they underlie the coarse-textured lowlands. Perhaps not.

There is much that you can learn about this area, should you decide to ask and answer more questions about it.

Booneville, Arkansas

(Figure 10–13 in back of book, C.I. 20 feet)

Note the position of the north arrow. This map has been rotated to the northeast. The grid lines that extend northeast-southwest across the page are actually north-south lines. Therefore, the grid lines should be used for orientation—**not** the page margins.

The structure in this map area is simple enough to be deciphered from the topography, and complex enough to be fascinating and instructive. This is part of the Arkansas Valley, an area of folded Pennsylvanian clastics that attain thicknesses of many thousands of feet. For our purposes, we may think of the ridges as sandstone ridges, and of the valleys as shale valleys.

Let us begin with twisting ridge **TJKCL.** It is also depicted in Figure 10–13a, a sketch of the area viewed from the south. Segment **TJ** is extremely asymmetric. Its gentle southwest flank apparently is a dip slope, or very nearly so. The ridge crest coincides with strike from **T** to **J.** But the ridge at **J** bends

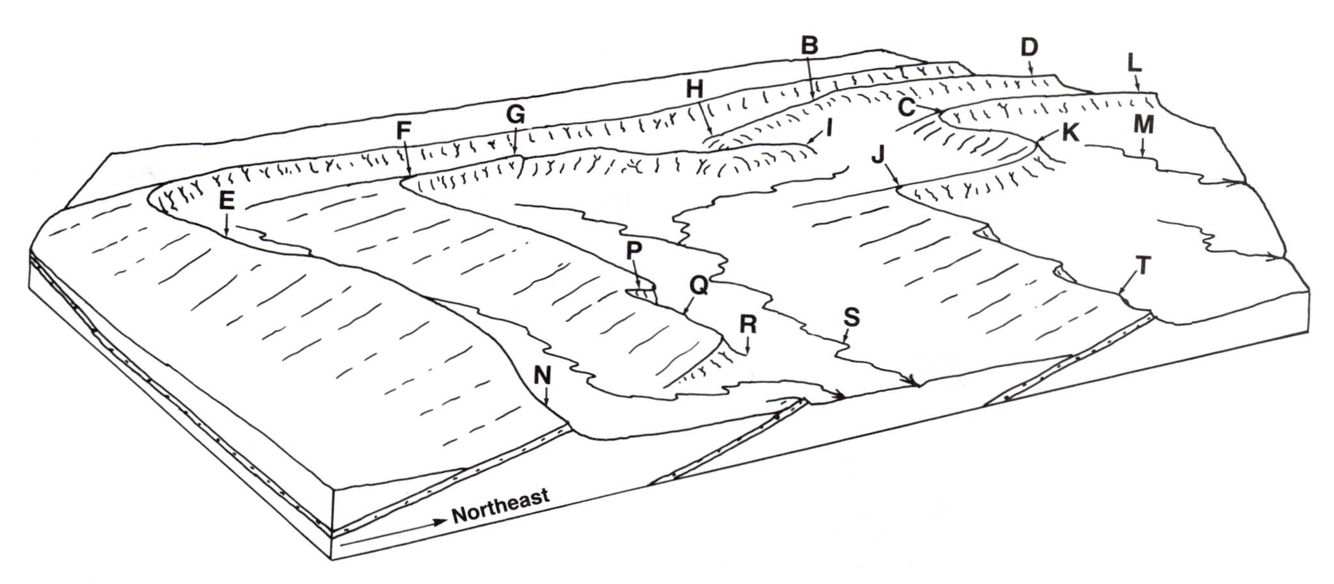

FIGURE 10–13a
Block diagram of the area depicted in Figure 10–13, viewed from the southeast.

to the north (west dip) and northeast (northwest dip). Then it swings back to the west at **K** (south dip). From **K** it bends generally to the west (south dip), and then around to the north again at **C** (west dip). Beyond **C** the ridge crest bends to the northeast (northwest dip), and follows this direction to and beyond point **L**.

These several dip and strike changes define two southwest-plunging anticlines, which we will call anticlines **J** and **C**, and a southwest-plunging syncline, syncline **K**.

Northwest-dipping limb **CL** and southwest-dipping limb **TJ** define a major anticline. Its axis may be extended (dashed axis) to the southeast from reversal **C**. Some might consider anticline **C** a minor wrinkle on the north flank of main anticline **J**. In that case, it would be axis **J**, and not axis **C**, that is extended to **M**.

One final note about ridge **TJKCL**: All segments of this ridge, from **T** to **C**, are asymmetric. Northeastward from **C** to **L**, however, it is symmetric. Nevertheless, we know dip direction along **CL**. Since the rocks in anticlinal lowland **M** are older than the ridge-forming unit, the dip must be northwestward, toward the younger units that lie to the north of **CL**, assuming it is not overturned.

Look at the second homoclinal ridge, **QFGIHBD**. Segment **QF** is extremely asymmetric (southwest dip), like segment **TJ**. An anticlinal nose at **F** indicates the westward extension (dot-dash line) of anticline **J**. The north limb of this fold is well expressed as ridge segment **FGI**. Steep dip along **GI** is reflected by that segment's symmetry.

When we jump to narrow symmetric ridge **BD** and work back toward **I**, we find that we can trace it only to **H**. Here it runs into the north side of ridge segment **GI**: *westerly* dip runs into *northwest* dip!

One would expect syncline **K** to project to, and west of, **I**. It may reach as far as **I**, but it apparently does not even exist to the west of that point.

Similarly, anticline **C** should project to, and west of, **H**. Though it may reach a point just north of **I**, it apparently does not exist to the west of that point.

How can we explain the intersection of ridge **BH** with ridge **GI**? If anticline **C** did continue westward, the westerly dip along ridge **BH** would constitute its northwest limb. Similarly, if syncline **K** continued westward, the northwest dip along ridge **GI** would be its southeast limb. Point **I** would be another synclinal nose. However, there is no indication of south dip in the **I-H** area that would constitute the southwest limb of the anticline, and the northwest limb of the syncline.

We believe that the line of dots separating **H** from ridge **GI** is the trace of a fault, upthrown on the northwest and downthrown on the southeast. This fault explains both the structural and topographic relations in this small area. Linear, narrow, low ridge **A** (to the west) may owe its orientation, location, and shape to its proximity to a westward extension of this fault (drag?).

We could devote many more pages to this map. But we will restrict our discussion to just two more features. The first is the en echelon offset of ridge **QF** at point **P**. There seems to be a small fault here, though it is difficult to establish its strike. If there is a fault, it does not extend very far to the west or southwest, since ridge **NE** is not offset.

Finally, the abrupt truncation of homoclinal ridges **QF** and **NE** along line **RO** virtually shouts, "fault"! However, upstream from letter **R,** stream **S** follows a subsequent course along the outcrop belt of nonresistant strata. (These strata overlie the sandstone of ridge **TJ,** and underlie the sandstone of ridge **QF**.) It may be that stream **S** at one time corraded laterally to the southwest, and merely eroded the broad, low expanse that now stretches from line **RO** to the base of ridge **TJ**'s southwest flank.

We cannot see enough of the surrounding area to decide between faulting and lateral corrasion. But what we can see demands that we consider fault offset as a possibility.

There is much more to be gleaned from this fascinating map, much geology you can add to our beginning.

Mendota, Virginia
(Figure 10–14 in back of book, C.I. 20 feet)

The *Geologic Map of Virginia* indicates but two major stratigraphic formations in this entire map area. One is a thick sequence of interbedded shales and sandstones. No single member of it is sufficiently thick or distinctive to leave much of an imprint on the topography. The sandstones render the formation relatively resistant, whereas the shales impart a low permeability and porosity. The resulting high runoff is responsible for the high drainage density in the areas underlain by this formation.

The other formation is composed of carbonates, largely limestones. These rocks obviously occupy lowlands **A** and **P**.

Extending northeasterly between areas **A** and **P** is high, maturely dissected upland belt **NC**. We may confidently assume it to be developed in the shale-sandstone formation. Note that uplands of this kind completely surround elongate lowland **P**. This lowland must be either (1) a structural basin, in which some of the younger carbonate unit is preserved, or (2) a breached dome, in which the older carbonate rocks are exposed. Whichever it is, its northeast-trending axis is shown as a dashed line.

If we can determine the relative ages of the two formations, we will automatically know the structure. Conversely, if we can ascertain the structure, the relative ages will be established.

For this to be a structural basin, the dips on both flanks, as well as the plunges at both ends, would have to be extremely steep. This is necessary to permit the limestone to drop low enough for some of it to be preserved, in a topographic basin whose walls rise over 600 feet above the basin floor.

In turn, such steep dip would produce linear-to-curvilinear stratigraphic contacts. These contacts would most likely be reflected in similarly linear slopes and bluffs and spur ends. However, none of these can be detected in this area. There are a few linear streams, such as **R** and **D**. But they probably have other origins which we will consider below.

For now, all we need to note is that linear stream **R** (subsequent stream **R**) flows northeasterly, but in a place where the strike should be northwest (whether this is a dome or a basin). If this is a basin, dip should be to the northeast. If a dome, dip should be to the southwest. In either case, strike would be northwest.

Not only is the clastic-carbonate contact neither linear nor curvilinear, but in many places it also seems to be extremely irregular in plan. For example, in the northeast, note the complex branching (dashed lines) of divides **H** and **J** (clastics), and their relation to depressions **I** (carbonates). As you look around the periphery of large carbonate lowland **P,** you will find numerous other interfingerings between carbonate depressions and odd-shaped clastic divides. Note too that there are no depressions whatsoever on the middle and upper slopes of the clastic terrain. There can be little doubt that the limestone is exposed only in those places from which the overlying clastics have been eroded.

Have we definitely established the relative ages of the two lithologic units, and hence deduced the structure? Let us first see if there is any direct evidence of dip and strike impressed on the topography, evidence independent of rock-unit age. As a matter of fact there is, though it does not reach out and bite us.

There is a most-suggestive asymmetry in the northeast. Gentle upland slope **G** (arrow), which seems to be a resequent slope if not a true dip slope, contrasts with steep slope **F** (arrow). North dip? Nearby, upland slope **L** and its companion, steep upper slope **K,** similarly suggest northeast dip.

Slope **Q** suggests southeast dip, as does slope **T**. Both have steep northwest slopes to complete their asymmetric profiles. Topographic asymmetries at **S** and **M** indicate possible southwest and west-northwest dip, respectively. It is such westerly dips as these that spotlight the anomalousness of linear stream **R**. Combined, possible dips **G, L, Q, T, S,** and **M** define a northeast-trending, elongate, structural dome or doubly plunging anticline.

In three places (**O, B,** and **E**), the topographic asymmetry appears to reflect dip back to the south-southeast. Here in the north, as around lowland **P,** the topography also suggests that the dip is away from older carbonate unit **A** and toward younger clastics unit **NC**.

Clastics trend **NC,** then, is apparently a synclinal *ridge,* the axis of which is shown as a dot-dash line. It seems that, throughout the map area, the process of *topographic inversion* is complete. The topography is the reverse of the structure. This is the kind of area that we would call *obsequent topography*.

Finally, we must consider the linear streams. They definitely appear to be subsequent streams, adjusted along the traces of steeply dipping fractures that parallel the structural axes and regional strike. These may be fault planes, though there is no topographic evidence of structural offset.

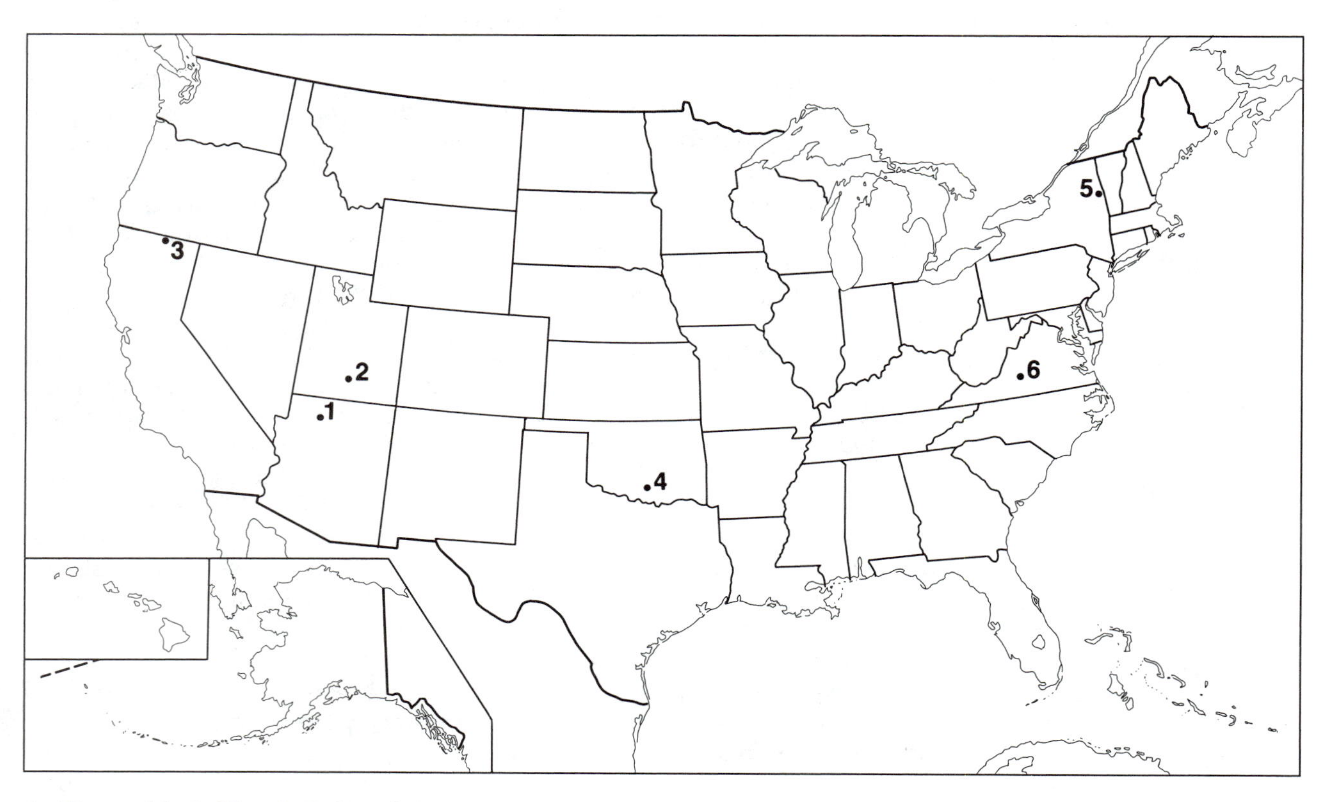

1. Figure 11–1. Kanab Point, Arizona.
2. Figure 11–2. Calf Creek, Utah.
3. Figure 11–3. Tulelake, California.
4. Figure 11–4. Springer, Oklahoma.
5. Figure 11–5. Silver Bay, New York.
6. Figure 11–6. Andersonville, Virginia.

11

Faults and Fractures

INTRODUCTION

As you have seen in earlier chapters, and will be reminded by this chapter's maps and discussions, we can learn much about faults and fractures from systematic analysis of an area's topography, as it is displayed on a topographic map.

We are only scratching the surface when we delineate what we believe to be a fracture or fault on a map. We cannot stress enough the importance of classifying the fault-associated features, particularly scarps, that so often develop along fault traces. Such classification is important because it permits us to determine roughly when the faulting took place.

Geology tells us the relative age of a fault's displacement in pre-topography time (i.e. before the present topography was formed). But it is the geomorphology depicted on the topographic map that tells us the "when" of the *more-recent* geologic past. It is important to know whether a given scarp, for example, is the direct product of fault displacement or of prolonged differential erosion.

Therefore, do not shy from such concepts as "obsequent fault-line scarp," thinking that such matters are of interest only to the academician or graduate-degree candidate. Whether you are prospecting for petroleum or trying to establish an area's recent history of crustal stability, such concepts are very important!

Kanab Point, Arizona

(Figure 11–1 in back of book, C.I. 80 feet)

This map dramatically illustrates what we meant in Chapter 1 when we said, "An excellent rule of thumb: when you see a straight topographic feature, or line of features, especially if it stands in contrast to its surroundings, *think steep*."

The area shown is part of the Grand Canyon of the Colorado River. To the north of the canyon rises the Kanab Plateau. It is lower than, and west of, the Kaibab Plateau, which is the north-rim plateau visited each year by countless tourists. The plateau to the south of the river is the Coconino Plateau.

The streams, large and small, combine to form a recognizable dendritic pattern. This pattern is usually developed in either homogeneous bedrock or areas of horizontal-to-nearly horizontal strata. The stair-step topography of this plateau area is clearly displayed at points **A, B,** and **D**.

The Colorado River, its deep inner gorge, its many tributaries, and the numerous benches on the higher ground, all command attention. To the geologist, however, the most eye-catching features on this map are the nine linear and curvilinear tributaries that combine to delineate line **EC**. It is too long, too continuous, and too nonresistant to be other than a significant nearly vertical fault or fault zone. The nine streams are subsequent streams.

Calf Creek, Utah

(Figure 11–2 in back of book, C.I. 40 feet)

It is easy to generalize. For example, does this look like an area of nearly horizontal strata, a plateau?

No, of course not. Plateaus, whether they resemble the Appalachian or the Colorado, are supposed to have flat uplands and valleys with pronounced slope-scarp-slope-scarp profiles, caused by sequences of resistant and nonresistant formations exposed in the valley sides. That sort of thing. But this area does not have that kind of topography. It certainly does not resemble the plateau area of Figure 11–1 (Kanab Point, Arizona, in back of book).

So much for generalizations. This map area is also in the Colorado Plateau, but it is part of one of that province's most breathtaking sections, the Canyon Lands. In fact, a natural arch is reported to be at **B,** and a natural bridge at **C**. Perhaps this is not slope-scarp topography because there is not enough variation among the resistances of the exposed stratigraphic units. Perhaps there are cliffs that are not apparent at first glance.

The most eye-catching features in this area are the numerous, straight-edge, narrow tributary canyons that slash across the upland like razor gashes through a sculptor's clay. On the one hand, the area in general presents an unexciting dendritic drainage pattern, whereas on the other, in addition to being startlingly straight, these canyons are grouped into parallel sets.

Are these fault traces? Are the linear valleys subsequent valleys? They certainly look like faults, though they could equally well be joints. Whether faults or joints, the valleys themselves are undoubtedly subsequent. The others, the many forming the dendritic pattern, are insequent streams. They simply "branch."

Though far from being eye-catching, the most fascinating thing about this map is the bizarre way the contours behave in certain places. From the very beginning we are taught that contours never cross each other. But points **A, K,** and **L** are but a few of the places on this map where contour lines *do* seem to cross. Here are some other fascinating points:

1. At point **H** there is a single contour line.
2. Immediately to the west, between **G** and **D,** there are *four* contour lines (line **H** branches!).
3. Between **F** and **D** there are *eight* (more branching).
4. Between **E** and **D** there are *eleven* (more branching).
5. Nearby, at **J,** the contours do other strange things.

These are simply landforms that we have little opportunity to see on topographic maps, because they are exceedingly rare. The ground slopes northeastward from **E** to **D,** a descent of over 360 feet. The 11 contours that represent that slope converge toward the southeast, until at **H** they are atop one another in a *vertical cliff*. That cliff is also over 360 feet high.

Contour **I** indicates the shallow valley of a northwest-flowing tributary. When it reaches **J**, the lip of the canyon wall overlooking the main river, the water of this tributary falls over a cliff that is over 300 feet high.

There is something about the lithology of these rocks that encourages the development and preservation of high vertical cliffs. Thick shales, interbedded with one or more thinner, resistant sandstone units, would tend more toward the missing slope-scarp profiles mentioned above. A thick sandstone unit, with a shale at its base, would offer ideal conditions for development of this topography.

According to the *AAPG Geological Highway Map of the Southern Rocky Mountain Region,* these are thick Jurassic sandstones.

Tulelake, California

(Figure 11–3 in back of book, C.I. 20 feet)

The mapping of faults on topographic maps involves three distinct procedures:

1. Recognition of the trace of the fracture.
2. Determination of whether fault displacement has occurred, and if possible, the direction of relative displacement.
3. Classification of the fault-associated scarp (if one exists) as either fault scarp or fault-line scarp.

In many cases it is not possible to succeed in all three.

This map area is an outstanding example of one where we *can* do all three. It lies in the famous Klamath Falls, Oregon area, noted for its faults and scarps. However, we will study this map with the aim of interpreting, and not of recognizing what we already know is there.

First of all, should we see these as *possible* faults? Or does what we see permit us to state flatly that these *are* faults? Is there any possibility that the many linear and curvilinear scarps in this area are something else, something utterly different from fault-related features?

There is a lake a few miles to the west. Could these possibly be the shorelines of a once-more-extensive lake, perhaps a Pleistocene lake? There is some good evidence against this. Such shorelines would not rise and fall, as does the base of scarp **EF,** in the southwest. Point **F** is about 180 feet higher than point **E**. Besides, where is the fetch that would generate the waves to cut east-facing scarp **A**? At points **D** and **G** the scarps branch in a geometric pattern, unlike the topography we usually associate with shorelines. The branching at **G** is shown in Figure 11–3a, a sketch of the area viewed from the northwest. (It is helpful to turn Figure 11–3 upside-down, to relate it to Figure 11–3a.)

The linearity and curvilinearity of these scarps, plus their remarkable parallelism, cry "*faults*!" However, when we identify these scarps as being associated with faults, we are skipping over step #1 and getting into step #2, so we must be careful. We are saying these are faults, and not mere fractures. What is more, the linearity of these fault traces tells us that their planes must be dipping very steeply.

There are two lines of evidence to support the conclusion that these are fault scarps. First, they are practically undissected. They have the appearance of being so recently formed that there has been no time for even partial destruction. Second, there are several undrained depressions and basins, such as **C,** in this map area. It would be difficult for scarp **B,** which borders basin **C,**

FIGURE 11–3a
Block diagram of part of the area depicted in Figure 11–3, viewed from the northwest.

to have developed through differential erosion, when the block that would have to be eroded more rapidly (block **C**) has no outlet, no exterior base level toward which its debris could be carried! Basin **C** must necessarily be an area of accumulation, rather than an area undergoing downcutting.

A fault scarp is, by definition, the scarp produced directly by displacement along the fault plane. The topography we are looking at in this area is the product of block faulting of the original topography, as well as of the bedrock. These scarps are geologically recently exposed fault planes.

This is an excellent topographic map on which to try your hand at mapping geology. Cover the map with a sheet of frosted Mylar or similar transparent material, and trace all the faults you can identify. Indicate up-down displacement on each. Add an **H** to denote a horst and a **G** to denote a graben.

Springer, Oklahoma

(Figure 11–4 in back of book, C.I. 10 feet)

At point **Z,** the crescent form of the nested contours is typical of those depicting the nose of a plunging fold. In this case it is a southeast-plunging anticline (axis indicated). The bend in the ridge from **N** to **P** clearly shows a corresponding slight change in the strike of the same steeply dipping, ridge-forming formation. This is an area of dipping strata (in places folded). Some of the strata are resistant and some nonresistant.

There is a fairly large fault in the map area. With care we can map it and determine its relative displacement, using this map alone. In Figure 11–3 (Tulelake, California, in back of book), we were dealing with upthrown and downthrown blocks bordered by gravity faults. Here we are going to concentrate on the geomorphic features of a sequence of unlike dipping formations, and on the manner in which those features are offset horizontally rather than vertically. Figure 11–4a, a view from the northeast, may help. (To relate Figure 11–4a to the map more easily, position the map with its northeast corner toward you.)

First, let us make a south-to-north traverse from ridge **N**. Its asymmetry is explained by the presence of a south-dipping resistant formation. After crossing a nonresistant unit (subsequent valley), we encounter narrow strike ridge **O**. It is upheld by what is apparently a thinner and somewhat less-resistant unit than **N**. Then another, thinner, nonresistant unit. And after it a third, even-narrower strike ridge, **M**.

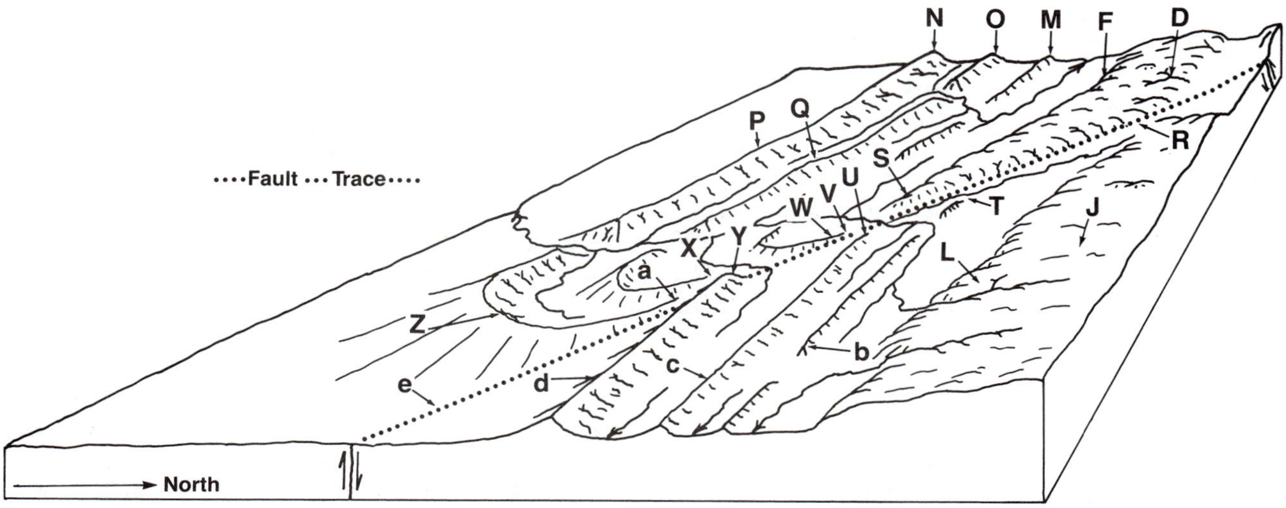

FIGURE 11–4a
Block diagram of the area depicted in Figure 11–4, viewed from the northeast.

North of **M** is a broad band of nonresistant beds, along which a large subsequent stream is meandering. Finally, north of this valley, is a thick section of more-resistant rocks (**D**), a series which is apparently most resistant in its lower part (at and near the letter **D**). This entire unit is dissected by south-flowing resequent tributaries.

In the east, starting with ridge **d,** we can make another traverse to the north and encounter the same sequence of upturned beds. The *same sequence.* We can correlate **d** with **N, c** with **O, b** with **M,** and **J** with **D**. We can *correlate* these ridges, but we cannot *trace* them from west to east across the map.

This is what we find when we go *eastward* from the **ND** traverse:

1. Ridge **N** extends eastward past **P** to the north cusp (**a**) of crescent **Z**.
2. Ridge **O** extends eastward past **Q** to point **X**.
3. Ridge **M** extends eastward to point **W**.
4. The southernmost part of upland **D,** which we indicate as **F,** extends eastward to **S,** where incidentally its strike is northwest-southeast.

Now let us go *westward* from traverse **dJ**. We can trace:

5. **d** to **Y**.
6. **c** to **U** (about 1300 feet northwest of **X**).
7. **b** to **T** (about 1800 feet northwest of **W**).
8. **L,** which correlates with **F,** to **I** (about 2500 feet northwest of **S**).

Using these points of offset, we can map with considerable precision the trace of a steeply dipping oblique fault, northwesterly across this entire sequence of south-dipping strata. In Figure 11–4a, follow the dotted line from **e** to **V** to **R** to **H**. Linear scarp **G** suggests that it continues to dot **E**. However, sharp stream-valley heads **C** and **A** make point **B** an alternative possibility, though there may only be a minor companion fault at **B**.

Finally, study the ridge crest eastward from **J**. Note the en echelon offset between divide **J** and divide **K**.

Silver Bay, New York
(Figure 11–5 in back of book, C.I. 20 feet)

This area does not boast a topography from which we can deduce a great amount of geologic information. We will find our interpretation severely limited unless we consult a geologic map. We will do that later.

For now, just what can we see here? We see a rather heterogeneous-looking area in which local relief ranges from a few hundred feet to well over a thousand. We see an area in which slopes vary from short and steep to long and gentle, with some in the high-and-steep category. We see an area of coarse-grained drainage that displays no single outstanding pattern (though many of the streams form a somewhat dendritic pattern, upon which a rather well-defined parallel pattern is superimposed).

In addition to the drainage patterns, we see a definite northeast-southwest-trending topographic grain. It is expressed in numerous places by linear steep slopes and scarps (e.g. **K, L,** and **O**). It shows in certain linear stream segments (e.g. **E** and **M**), which appear to be adjusted to some kind of structural control (fault or fracture traces).

Some of the linear scarps (e.g. **K, L,** and **O**) are so sharply defined and apparently undissected that they seem to be fault scarps comparable to those of Figure 11–3 (Tulelake, California, in back of book). Many of the others (e.g. **N**) are relatively linear, though they are more "casually" so.

Conventional symbols for swampy ground indicate that there are numerous poorly drained valleys and depressions. We may assume these to be products of Pleistocene glaciation. The presence of swampy areas gives more credence than we might otherwise allow to the possibility that some of the drainage characteristics relate to glacial deposition and/or scouring.

Might this area lie in the Glaciated Allegheny Plateau? Or the Catskills? Perhaps the Taconics or the Adirondacks? All are possible. The highly complex structures of such areas as the Taconics and Adirondacks tend to form such mixtures of massive and grained terrain.

Are the scarps fault scarps? Hardly. Whether this is Taconic or Adirondack topography, the overwhelming majority of tectonic events responsible for structure took place millions of years ago.

The topographic grain may also reflect the response of differential erosion to differences in the resistance to erosion offered by a variety of bedrock types. (Differential erosion includes differential ice-sheet scouring.)

Now, having speculated freely, let us consult a reference. The *Geologic Map of the United States* shows that the Silver Bay quadrangle is situated in the Adirondack Mountains, an area of complex crystalline geology. It is an area renowned for the control exerted over drainage lines, upland and lowland landform outlines, and lake shorelines. Control is by parallel and intersecting fault planes. The Lake Placid shoreline is famous for its alignment along fault traces.

We suggest that in preglacial periods of erosion, there was considerable differential etching in response to a variety of structural planes and rock types. Ice sheets that later invaded the area took over and continued such work. On the one hand, this smoothed and rounded topographic irregularities. On the other hand, as the ice slid along, it accentuated certain structural planes and structure-controlled topographic planes.

In some places, several specific linear features combine to define a probable major tectonic trend. For example, in the north, major trend **FD** includes linear stream segments (e.g. **I, A,** and **C**), low linear slope **B,** linear low divide **GJ,** and linear scarp **H**.

Andersonville, Virginia
(Figure 11–6 in back of book, C.I. 10 feet)

The *Geologic Map of Virginia* tells us that this area is underlain by a complex of ancient crystalline rocks, predominantly gneisses and schists. It is like the Adirondacks area in Figure 11–5 (in back of book), except this is not a glaciated area. Its topographic relief is much lower; maximum local relief barely exceeds 150 feet. In addition, this area's soils and overburden have not been stripped away. This land has remained undisturbed for great intervals of geologic time. Weathering has progressed and a differentially eroded topography has become well adjusted to the structures and lithologies of the exposed bedrock.

At a glance, the area is cut into a number of blocks and wedges. Note the near-parallelism of the most-obvious belts of linear streams and linear scarp development: **KD, VE, FG, UN, OI, WP, YZ,** and **aT**. A few exceptions, such as **XR** and **HS,** cut across the grain established by the others.

The most important question is whether the numerous linear geomorphic lines and trends are fault-associated. Based on what we can see, and what we have so far revealed about the geology, they might be. All might be. Some might be. One might be. None might be.

This is part of the Virginia Piedmont, an area in which geologically recent tectonic activity has not been rampant. Thus these are almost certainly not fault scarps. If not fault scarps, what might they be? The only other thing is steeply dipping contacts.

It is virtually impossible to make a geologic map from this small topographic map. However, our tentative and incomplete interpretation would be most helpful in directing a field investigation of the area. We should begin by delineating all linear and curvilinear stream segments and scarps as possible fault traces, and we should also indicate those that separate recognizably different topographies.

For example, line **XR** is bordered on the south by an intricately dissected upland. Its fairly steep north flank overlooks lower, more-undulating area **Q**. This topographic contrast suggests that rocks in area **Q** are different from those south of line **XR**.

Several of the northeast-trending lines separate topographies. The northwest-trending lines (e.g. **HS, AJ, BL,** and **CM**), though they may cut across various topographies, do not separate them. It would be more difficult to defend mapping such cross-grain trends as major faults, though they definitely qualify as probable fractures or fracture zones. They might also be faults with smaller displacements, however.

The geologic map shows a northwest-trending Triassic dike in the vicinity of line **HS**. Stream **HS** could be a subsequent stream developed along the outcrop of this basic intrusion. Its relative resistance could be less than that of the metamorphics that comprise the country rock.

According to the *Geologic Map of Virginia,* two major geologic contacts strike across this map near, and parallel to, lines **KD, VE, FG, UN,** and **OI**. We should consider these geomorphic features as probably fault-related or contact-related.

On the other hand, our line **XR** cuts diagonally across one of the contacts shown on the state map. In addition, that map indicates no contact in the area around lines **YZ** and **aT**. Which should we believe, the topography or the published geologic map? There is no easy answer. Any map can be incomplete or inaccurate. The contact shown on the small-scale state geologic map may be more generalized and simplified than it actually is in the field. Our line **XR** could be closer to the contact than it now appears.

Here is yet another example of that critically important fact: *we cannot assume that topography always reflects the geologic structure of an area*. In fact, at times topographic features and relations may be downright misleading. At such times the answers may be obtainable only in the field.

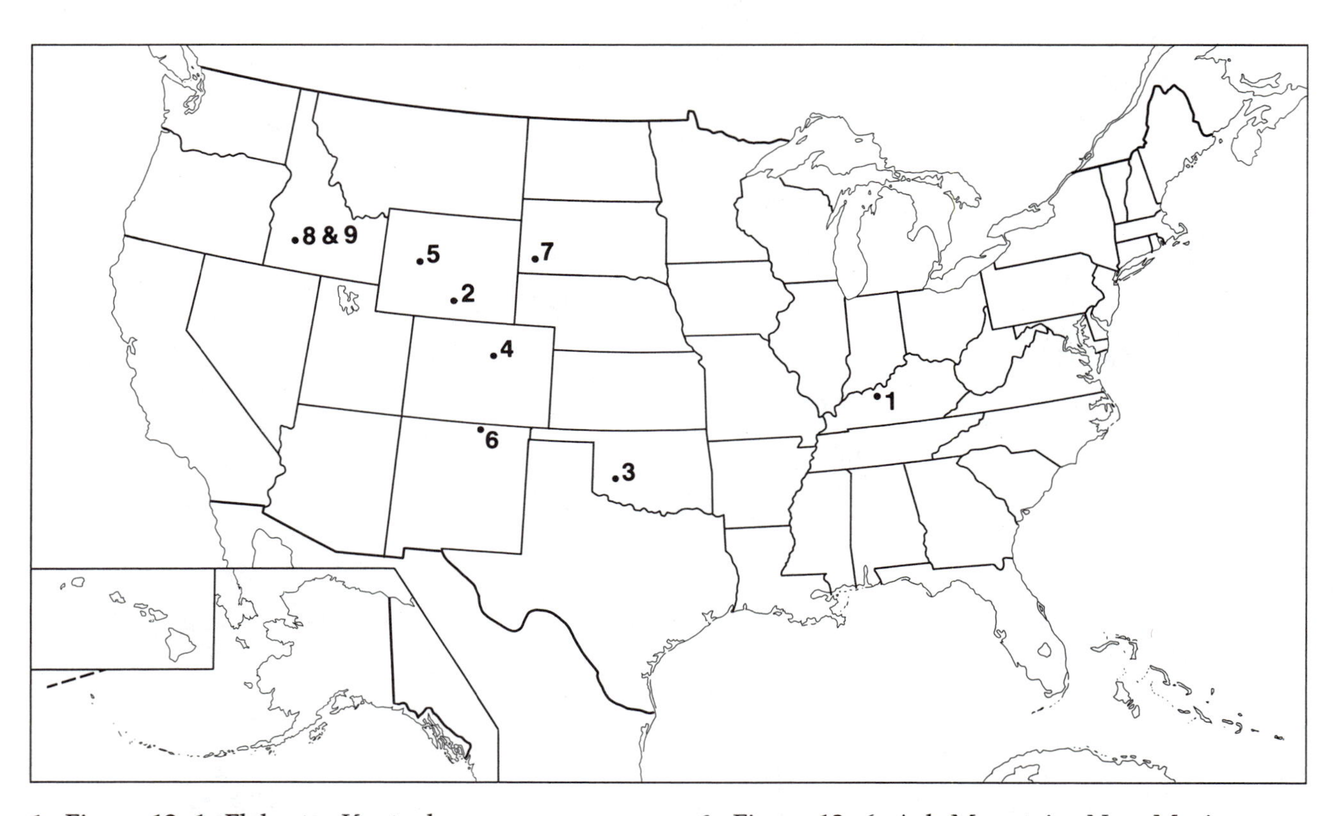

1. Figure 12–1. Flaherty, Kentucky.
2. Figure 12–2. Rawlins, Wyoming.
3. Figure 12–3. Quanah Mountain, Oklahoma.
4. Figure 12–4. Golden, Colorado.
5. Figure 12–5. Wind River, Wyoming.
6. Figure 12–6. Ash Mountain, New Mexico.
7. Figure 12–7. Buffalo Gap, South Dakota.
8. Figure 12–8. King Hill, Idaho (topographic).
9. Figure 12–9. King Hill, Idaho (geologic).

12

Anomalies

INTRODUCTION

It is extremely difficult, and sometimes misleading, to classify any assortment of topographic maps into a set of geomorphic or geologic categories. For example, Figure 7–4 (Trenton, Kentucky, in back of book) is included in the chapter devoted to karst and other depressions. Instead, we could have used it in Chapter 11, "Faults and Fractures," since one of that figure's prominent features is the topographic expression of a linear fault trace.

Having covered essential background in Chapters 1 and 2, we organized Chapters 3 through 11 by subject matter. In this chapter, "Anomalies," we did not intend to organize by a subject. We simply collected maps in which something does not quite "fit." But when we reexamined these maps, we realized that we nevertheless ended up grouping the maps by a subject.

The topography and geology represented by these maps are widely varied. But the common thread is the way prominent features either cut across other features, or are completely obscured by other features.

These are outstanding maps that offer real challenges to even the experienced interpreter. We think you will enjoy them.

Flaherty, Kentucky

(Figure 12–1 in back of book, C.I. 20 feet)

This is a difficult map. So we will begin by providing some background information from the USGS map, *Geology of the Flaherty Quadrangle, Kentucky* (Geologic Quadrangle Map GQ-229). All bedrock in this area is of Mississippian age. Dip is virtually horizontal. There are two lithologic units: one consists of limestones, including a chert member. The other consists of sandstones. Figure 12–1a is a sketch of the area, viewed from the southeast. (To orient the map to match the sketch, turn map so its upper right corner points toward you.)

As we look at this map, we must ask questions. We must be alert not to forget the routine ones, such as, "Which are the carbonate areas, and which are the sandstone?" We will later deal with that, but first let us focus on the anomaly, which can only be ridge **CH** (in Figure 12–1a, **CG**).

Why is this ridge so linear? After all, how can normal weathering and erosional processes produce such a linear feature, when the various strata they are destroying are horizontal? Yes, this is the something that does not "fit."

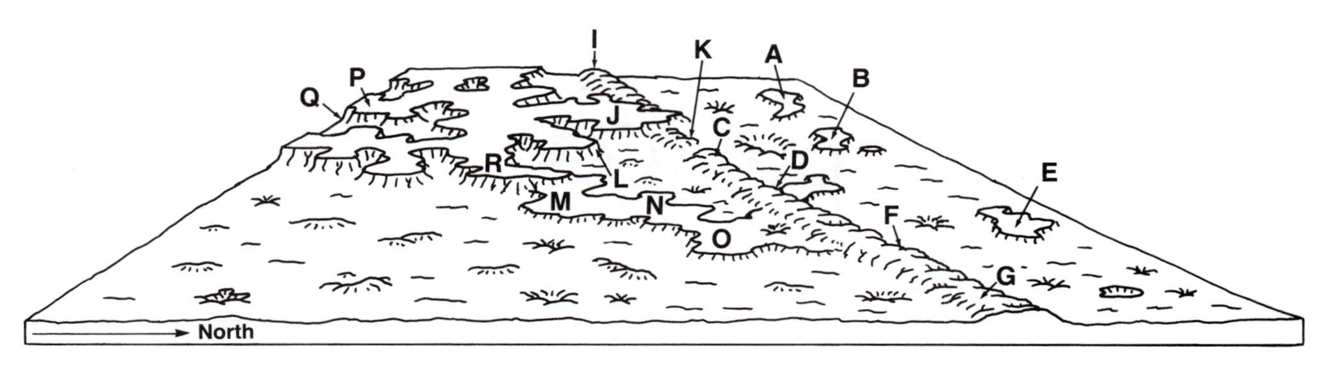

FIGURE 12–1a
Block diagram of the area depicted in Figure 12–1, viewed from the southeast.

Before we analyze the ridge's linearity, let us attempt some preliminary geologic mapping. The roughly triangular area **JRP,** in the southwest, *does* fit. This dissected upland, which must be developed in the more-resistant sandstone, exhibits precisely the type of sprawling plateau topography expected in an area of horizontal strata. The many depressions associated with it, and the large ones in the adjacent lowlands, tell us that the sandstone overlies the limestone. Further, note the fairly steep slopes (e.g. **L** and **Q**) around the edges of this upland. These support the hypothesis that the sandstone lies atop the limestone. It is a veritable caprock.

At a lower level is another, more-fragmented surface, preserved as such intermediate features as **M, N, O, A, B,** and **E**. The distribution and shape of these fragments also indicate the presence of flat-lying rock. Surface **M** actually is a ledge or shelf on the flank of upland **JRP**. Thus this intermediate surface is developed on what is apparently a somewhat-resistant unit that underlies the capping sandstone unit. Chert is reported in the limestone sequence, so we may postulate that it is the resistant layer on which these intermediate benches developed. However, there is no way we can be certain.

Now, let us return to that oddly linear and persistent ridge **CH**. It extends almost entirely across the map. We can trace it westward to point **K,** where it runs squarely into the north end of the sandstone upland. We can also recognize it at **I,** where it leaves the map. Note that this ridge varies in width, and that its crest height declines toward the east from the **FG** area.

There can be little doubt that the sandstone of area **JRP** lies atop a segment (**KI**) of whatever constitutes the ridge. We may be equally certain, since we know the lithologies, that ridge **CH** is also composed of sandstone.

Segment **CD** of the ridge is asymmetric in a way that suggests north dip. If dip is indeed to the north here, the rocks in the northern lowland must be younger than this sandstone, and those to the south must be older. Yet as we have observed, the topography north and south of the ridge is typical of nondipping bedrock. In addition, along ridge segment **FG,** the asymmetry is reversed (south dip?). In other places, the ridge profile is virtually symmetric.

Upland tracts like **JRP** are pitted with solution depressions, and such depressions abound along the flanks of ridge **CH**. The few shallow depressions near **C** and **D** are the only ones denoted along the entire upper surface of the ridge itself. Does this suggest that this sandstone is not directly underlain by limestone? Most puzzling. It is a pity that all the rocks are sedimentary. A vertical dike would come in handy here!

As we said at the beginning, this is a difficult map. We must consult our reference map once again for the answers, for few areas boast this peculiar combination of topography and bedrock geology. Yes, ridge **CH** is a sandstone, but it is not dipping. The limestones in the north and south are identical, and they too are not dipping. This ridge is not developed on a clastic dike like the much-smaller-scale ones that occur in the Dakota Badlands, though it may resemble them somewhat. However, like them, it *does* consist of a narrow band of more-resistant sandstone, flanked by less-resistant "country rock."

This is a sandstone channel fill. It was deposited in a channel that had been cut into the limestone. Later, the entire area, limestones and channel fill alike, was covered by the younger horizontal sandstone formation. Remnants of the younger sandstone are now preserved as dissected upland **JRP**.

Rawlins, Wyoming
(Figure 12–2 in back of book, C.I. 20 feet)

This map has almost surgically sculptured strike ridges, outcrop V's, asymmetries, and plunging axes. At first glance, it appears to pose no problems. The anomalous ridge in Figure 12–1 (Flaherty, Kentucky, in back of book) immediately attracted your eye. But the anomaly in this map area could be accused of trying to hide.

The simplest approach to this area is to map its dip-strike-fold structure. We will use ridge and divide asymmetry as one of our most-reliable indicators of dip. We can indicate dip by drawing arrows that point down resequent slopes. Along with the map, please refer to Figure 12–2a, a view of the area from the south.

Dips **J**, **W**, and **X** clearly define an asymmetric east-plunging anticline. Southeast dip of the next-younger resistant unit is indicated by the asymmetry at **Y**. At **Z**, we use a broken arrow to show the east dip, because here the unit forms a bench instead of a ridge. Refer back to bench **SD** in Figure 10–5 (New Enterprise, Pennsylvania, in back of book) and 10–5a. How do we know the dip here is to the east? How do we know the dip is northeast at broken arrow **K**?

We can easily extend the axis of anticline **JWX** to the eastern edge of the map. Note the geometry of each of the five homoclinal ridges: narrower and steeper-sided along the northeast-dip leg. This indicates the asymmetry of the

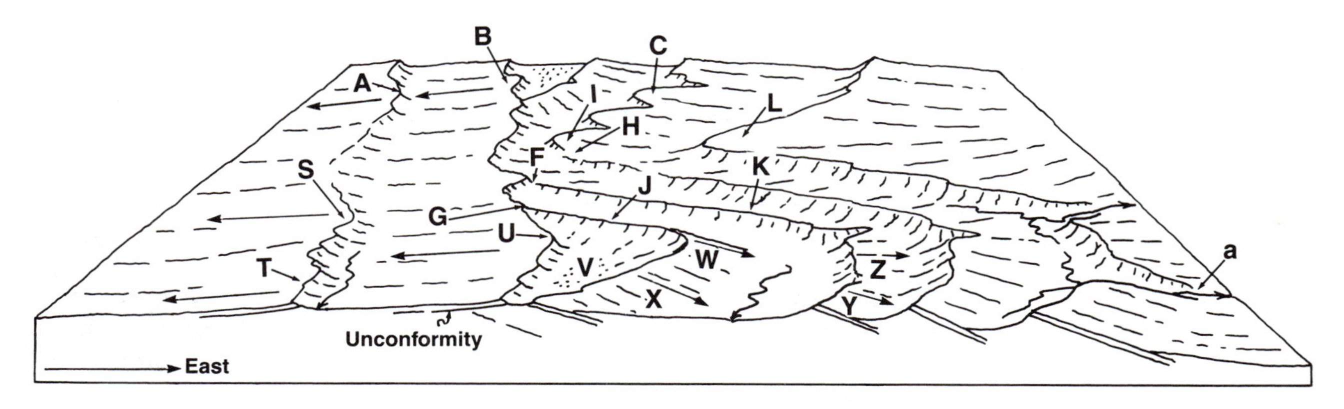

FIGURE 12–2a
Block diagram of part of the area depicted in Figure 12–2, viewed from the south.

fold (much steeper dip to the northeast than to the southeast). Can you delineate the excellent outcrop V's at **a**, **Q**, and **N**?

Look at the north-central part of the area. The much-gentler, almost cuestaform asymmetry at **I** and **C**, combined with such dips as **K**, defines an east-plunging synclinal axis at **H**. This axis continues to plunge eastward at **L**. It is still recognizable at **M**, just to the northwest of water gap **N**. Here the strike changes just enough that we can extend the axis this far eastward. Obviously it is dying out at about this point.

We can trace ridge **D** southeastward to, and past, the outcrop V's **N** and **Q**. On the other hand, secondary resistant bed **E**, which lies atop **D**, seems nearly to disappear when it gets to **O**. To the south of **O** the dip is much steeper. Erosional processes have almost, but not quite, destroyed ridge **E** in the steep-dip segment. You can still detect it at **P** and **R**.

In the west there are other strikes, other asymmetries, and other dips. But there are no axes! In this area, divides **A**, **S**, **T**, **B**, and **U** are pronouncedly of the low-dip cuesta category. Their asymmetry is apparent, though even their steeper (obsequent) flanks are quite gentle. The profile of each of these divides is much smoother and much-more rounded than the profile of the ridges that define the east dips and folds.

The homoclinal ridges in the east are apparently formed on individual resistant formations or members (such as individual sandstone beds or individual limestone beds). The cuestas in the west, however, do not suggest the presence of such individual resistant units. Instead, these landforms are of the kind more typically developed where aggregates or suites of beds are *together* more resistant than the sequences overlying and underlying them.

For example, each of these cuestas may be composed of a series of interbedded sandstones and shales. The sandstones are sufficiently numerous and individually thick enough to offer more resistance to erosion than the intercuesta lowlands. The lowlands may be underlain by sequences of shales and sandstones in which the weaker shales dominate.

Here is another possibility in the west. The entire stratigraphic section could be composed of poorly cemented units, perhaps of uniform lithology and grain size. But some parts of the section could include beds that are better cemented, or cemented by more-resistant material.

The important thing is that the west is different from the east—and not just because it lacks folding, or because its dips are uniformly gentle, or because its dips are to the west. Its *stratigraphic characteristics* are different, though we do not know precisely what they are.

By mapping the dip/fold structure and noting lithologic characteristics and contrasts, we have not only defined but also located the anomaly in this area: for some reason the geology of the west neither matches nor fits the geology of the east.

There are two possible explanations: the two blocks may be separated by (1) a *fault* or by (2) an *unconformity*. The plane separating the two blocks, whatever its nature, cannot be vertical or even moderately steeply dipping. This is because its trace (sinuous north-south dashed line), rather than being linear or nearly linear, is quite irregular. Note that ridge **JWX** disappears against this line at point **G**. Ridge **YZK** extends only to point **F**. The plane of the contact dips gently to the west. In fact, it appears to coincide with the base of the resistant unit that forms cuesta **BU**.

The strata of the western block are virtually undisturbed, with the exception of gentle tilting. Their bedding planes lie about parallel to their basal contact, so it is much easier to see this as an *angular unconformity* (Figure 12–2a) than a thrust fault. It is hard to imagine a thrust fault in which the folded and truncated "footwall" rocks are overlain by a thrust sheet that has not been deformed in any way.

Finally, note the closed depressions (e.g. **V**). Remember, this is in Wyoming, where carbonate rocks are better known for their resistance than for their solubility. This is also an area of sparse vegetation and frequent strong winds. Deflation is a far better probability.

Quanah Mountain, Oklahoma
(Figure 12–3 in back of book, C.I. 10 feet)

This is truly an area of contrasts. There are clusters of hills and rolling lowlands. There are areas of gentle slopes and low drainage density (e.g. **D** and **U**). There are intermediate areas where crenulations of the contour lines forcefully convey the intricate dissection/high drainage density (e.g. **R, L,** and **G**). There are areas of dendritic drainage (e.g. area **S**). And there are areas of linear valleys (e.g. **H** and **J**) and linear scarps (e.g. scarp **F**).

And yet, the more we study this map, the more we perceive certain *associations* as well as contrasts. We can begin to transform this topographic map into a preliminary geologic map by drawing some tentative "contacts" between the topographic "lithologies."

Look at points **N, Q, O,** and **K,** for example. Contour **N,** with its many twists, turns, and bends, is depicting a very different topography from that portrayed by contour **Q**. However, if we trace **Q** to the northwest, by the time we reach **O** we are back in an **N**-type terrain. Segment **K** lies in **Q**-type topography.

The dashed line fairly well separates the two types; we have continued it to the south, past **L,** to and beyond point **T**. A similar line **P** can be drawn a short distance to the east. You may complete both lines.

Lowland area **SU,** in the south, and lowland **D** in the north, are both **Q**-type areas. In fact, the two "contacts" we have drawn define a serpentine ribbon of **Q** type topography (belt **QMT**) that connects areas **SU** and **D**.

Why are there such geomorphic contrasts? Part of the answer seems obvious: there must be several different bedrock lithologies in an area of such striking topographic "lithology" contrasts. Surely the rock underlying mountain **I** cannot be the same rock type as that occupying lowland **D**. Whether **I** is high because it is more resistant, or because it is structurally high, is another matter. For now, the significant thing is that it is dissected in one way, and lowland **D** is dissected in another.

Could the high areas have been folded up? The topographic contrasts here are not those commonly occuring around folds and domes. In fact, these various topographies interfinger in a most-disorganized manner.

Could the high areas have been faulted up? Though the high areas are cut by many fractures and are probably faulted internally, most of their borders with lowland areas do not resemble fault scarps or fault-line scarps. We cannot say categorically that they were not faulted up at some time, but they do not appear to have been faulted up against the lowland rocks. This is an important distinction to keep in mind.

Could the resistant rocks be igneous material that intruded the less-resistant lowland rocks? Possibly.

Linear stream **B** is puzzling. The drainage patterns in the lowlands are mostly dendritic. But such a linear stream must be a subsequent stream, and the subsequent streams here are almost exclusively in the higher areas. There is also something puzzling about the lowland area south of stream **B**. This is the only lowland where we find numerous small closed contours (**C**). These are small hilltops in a lowland expanse that is far more irregular than area **D** to the east, and lowland **SU** in the south.

How many different types of topography are there in this map area? We see four:

1. Lowlands of the **D** and **SU** variety.
2. Somewhat different lowland of the area **C** type.
3. Intermediate "crenulate" areas (e.g. **R, L,** and **G**).
4. Uplands such as **I, E,** and **A**.

If you are acquainted with the geology of Oklahoma, you may have recognized this as part of the Wichita Mountains. If not, here are some facts you should know about the area's geology. Let us see how they fit the topography.

The high terrain is underlain by complexly jointed and faulted nonstratified bedrock. It is shown on the *Geologic Map of the State of Oklahoma* as Precambrian "granite." Beneath parts of the northern lowland, that same map indicates that there are exposures of Precambrian gabbro. What the map does not show, but we surmise, is that the "granite" areas include a *variety* of crystalline rock types, rather than uniformly pure granite.

Lowlands **D** and **SU,** and the serpentine belt **QMT,** are shown to be underlain by Permian clastics. The only way Permian sedimentary rocks could lie around and among Precambrian hills, in the irregular interfingering we see, would be for them to have been deposited on an extremely irregular topography. Apparently, belt **QMT** is a winding valley or pass that was filled by accumulating Permian detritus. It is a buried valley which has recently been partially exhumed.

Until proven otherwise, scarps such as **F** may be thought of as probable Permian fault-line scarps. Like valley **QMT,** they were buried by Permian deposits and have recently been exhumed. Note that scarp **F,** stream **B,** and subsequent trend **H-J** are nearly parallel.

The geologic map removes at least some of the confusion about area **C**. It shows this area (including stream **B**) also to be underlain by Precambrian granite. The topographic irregularity may be explained by the presence of highly fractured crystalline bedrock. The low topography indicates that this area must have been well eroded down (beveled). It is not difficult to picture the planation of this area during the same pre-Permian cycle of erosion that witnessed the cutting and broadening of valley **QMT**.

Many questions remain. Are the variations in terrain in the "granite" areas due to corresponding variations in bedrock composition? Or might they be explained by variations in the degree or intensity of fracturing of a single rock type in different areas? We cannot hope to answer solely from a study of the topographic map.

Consider what we have gained here. We now have a much better idea as to what questions to ask, and how to obtain data on which to base our answers.

Golden, Colorado

(Figure 12–4 in back of book, C.I. 10 feet)

Crossing this map from east to west is a stretch of the Denver and Rio Grande Western Railway. We can identify its route on this map because its construction changed the original topography:

☐ At **E** the built-up roadbed seals off a small valley.
☐ At **F** they gouged out a long, sweeping gash which is over a quarter-mile long.
☐ At **B, C,** and **K** the contours are bent in a repeated singular pattern.
☐ At **L** and **O** there are two smaller linear cuts.
☐ At **M** another small valley is sealed off.

Using these and other such features, you should have little difficulty tracing the rail line from border to border.

Humankind's hand is also apparent in the southwestern part of the map area. A narrow, high ridge extends from **H** to **P** and is broken only at **Q,** where stream **I** has cut a water gap. This ridge seems almost too geometric to be a natural landform. On the other hand, it could hardly be the remnant of some kind of dam. Its south end (**P**) is just below 6000 feet elevation, whereas its north end rises to about 6150 feet, with some intermediate segments over 6200 feet.

At both ends, the ridge is paralleled by narrow, straight, trench-like cuts. At these points, and at several places along the west side of the ridge proper, we find the word "claypit" on the published quadrangle. The linear depression about a mile north of **H** is similarly designated. Note in passing that the broad, gentle, easterly slope just south of this isolated depression is uninterrupted in any way that would suggest protruding bedrock on that surface. Ridge **PH** strongly resembles a resistant vertical dike.

Though clay has been extracted from the pits, the ridge is surely not composed entirely of clay. A steeply dipping sandstone, with which the clay beds are associated, would explain the local topography. This topography, incidentally, was created by differential erosion.

Why does this sandstone not have the slightest topographic expression north of **F**? Before we attempt to answer, we must first establish what this gently sloping smooth surface is. Could it be the top of a thick alluvial fan or plain into which the large, east-flowing, consequent streams have become incised? (In response, say, to a change of climate?)

Note something odd about this east-sloping surface. The contours seem to "match" across the valley of stream **I**. For example, contour **R** lines up almost perfectly with contour **J,** which continues northward as **A**. However, contour **JA** is exactly 200 feet *above* contour **R**! The valley of stream **I** does not separate two parts of a single surface. These are two *different* east-sloping surfaces, separated by a deep valley. Since the "surface" on which **A** and **R** rest is not a single surface, it certainly could not be the top of a single fan. These relations are shown in Figure 12–4a, a sketch of the area viewed from the southeast.

If it were a single fan surface at the elevation of surface **JA,** the top of the upturned sandstone at **P** would be about 175 feet below that fan surface. And if it were a single fan surface at the elevation of **P,** ridge-end **H** would project above it as a 175-foot-high wall!

Farther to the east, contour **N** has the same elevation as **R**. We suggest you trace out this contour, in red, across the entire map. Steep slope **D** (arrow) is a step down from the upper surface (**JA**) to the lower surface (**PRN**).

Now, back to sandstone ridge **PH**. To remove ridge **PH** as a ridge to the north of claypit **G,** we must either bevel it or cover it. Better yet, picture the upturned beveled edge of the sandstone as being covered unconformably by a thin blanket of debris. This debris has spread over a nearly flat, gently sloping erosional surface: a *pediment*.

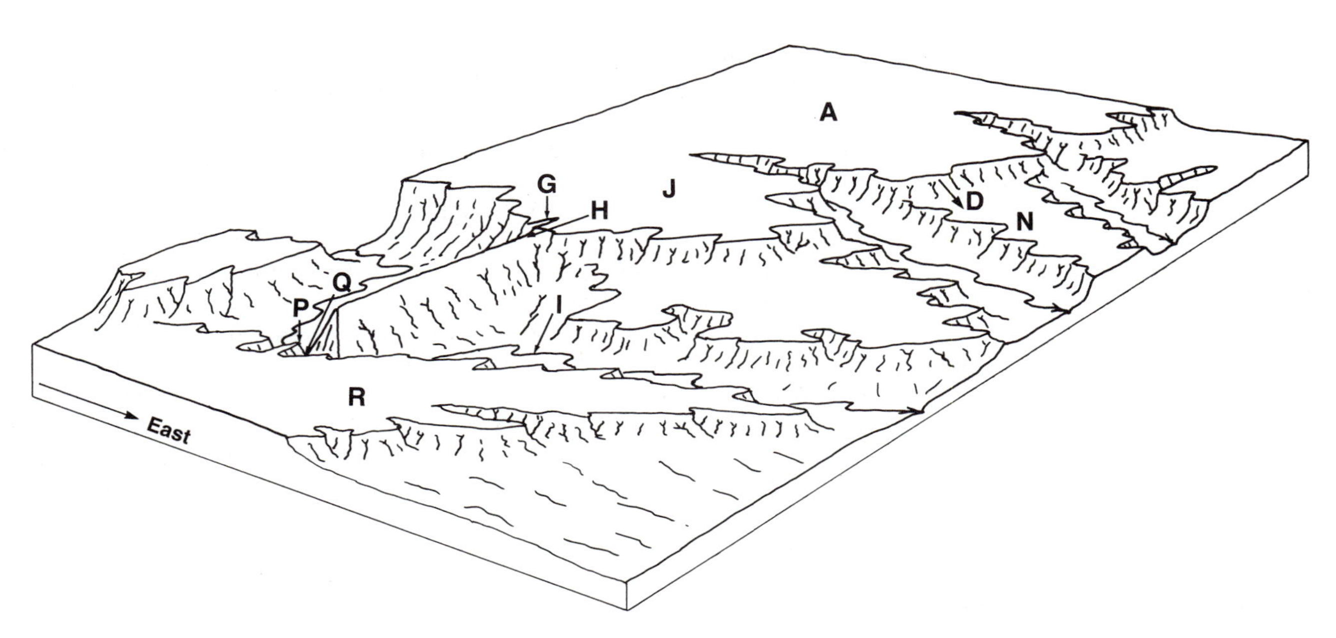

FIGURE 12–4a
Block diagram of part of the area depicted in Figure 12–4, viewed from the southeast.

What we have here is a broad, veneered pediment (**GJA**). At one time it must have cut across the entire map area. It beveled both resistant and nonresistant upturned strata, including sandstone ridge **PH**. The thin blanket of alluvium evenly spread atop this pediment has completely obscured the underlying bedrock, and all topographic irregularities developed on it.

At some later time, for reasons unknown, the streams were rejuvenated. This set the stage for partial destruction of the original pediment, and the cutting of a lower, younger pediment (**PRN**) 175–200 feet below the first. Present surface **PRN** consists of remnants of this younger pediment. It is a surface on which another thin layer of detritus had apparently been spread during its formation. It is this layer of material that covers the upturned sandstone at and south of **P**.

Later rejuvenation and downcutting on a large scale have created the present deep valleys and considerable destruction of the two pediments. Incidentally, water gap **Q** was cut by stream **I** after it had been superposed down through the pediment veneer into the underlying, steeply dipping strata. Ridge **PH** is not an exhumed ancient ridge. It is a latter-day ridge, produced by the differential erosion that accompanied the postsuperposition downcutting by stream **I**.

Wind River, Wyoming
(Figure 12–5 in back of book, C.I. 20 feet)

If you are able to visualize and comprehend the topography and geology of Figure 12–4 (Golden, Colorado, in back of book), you are ready for this map. The topography here does not resemble that depicted in Figure 12–4, but there are numerous comparable elements in the geomorphic and geologic histories of both areas.

In Figure 12–4, the most-apparent anomaly is the small, steep-dip ridge **PH,** developed in the limited area from which the unconformably overlying material was removed. In this area of Wyoming, the extensive expanse of differentially eroded underlying strata is the topographic norm. As depicted in Figure 12–5a, a sketch of the area viewed from the southeast, gently sloping table **AB** is the anomalous feature.

The structure is simplicity itself: a homocline, a structure in which dip is uniform in one direction with very slight variations in magnitude. The north-northwesterly strike is apparent at many places. It is indicated by numerous linear contour lines (e.g. **P, O, G,** and **R**) and by linear subsequent streams (e.g. **T**). Divide asymmetries and uniform easterly slopes (e.g. **S, Q, M,** and **F**) proclaim east-northeast dip.

Both the outline configuration and the extremely gentle slope of surface **A** contrast with the geomorphic grain and landform assemblages of the remaining map area. Three arrows indicate the slope direction of surface **AB**. Smaller but similar gentle slope remnants are slopes **L, C,** and **D**.

Note how the north end of resequent slope **F** literally runs into the side of gentle-slope remnant **D**. A similar relation exists between resequent slope **I** and gentle-slope surface **AB**. The inclination of slope **I** also contrasts with that of slope-remnant **L**. These relations are shown in Figure 12–5a.

To visualize the position of today's topography in time, you must first understand *remnant* as we use it here. A remnant is a small remaining part of a once far-more-extensive whole. If we have correctly identified the features here, if these are indeed remnants of a former extensive surface, then we must imagine that time when most (all?) of this map area resembled surface **AB** and its companions. If you turn back to Figure 12–4 and mentally fill in the deep valleys that now exist, you will reconstruct just such a broad and simple topography.

This is an area of dipping strata, and some of the strata are much more resistant than others. The production of such a broad surface must have entailed long intervals of weathering and erosion, large-scale pedimentation. As in area 12–4, some fragmental debris (not necessarily thick) was strewn upon this gently sloping surface, creating an unconformably overlying blanket. Consequent streams crossed the sloping surface of this alluvial blanket from west-southwest to east-northeast.

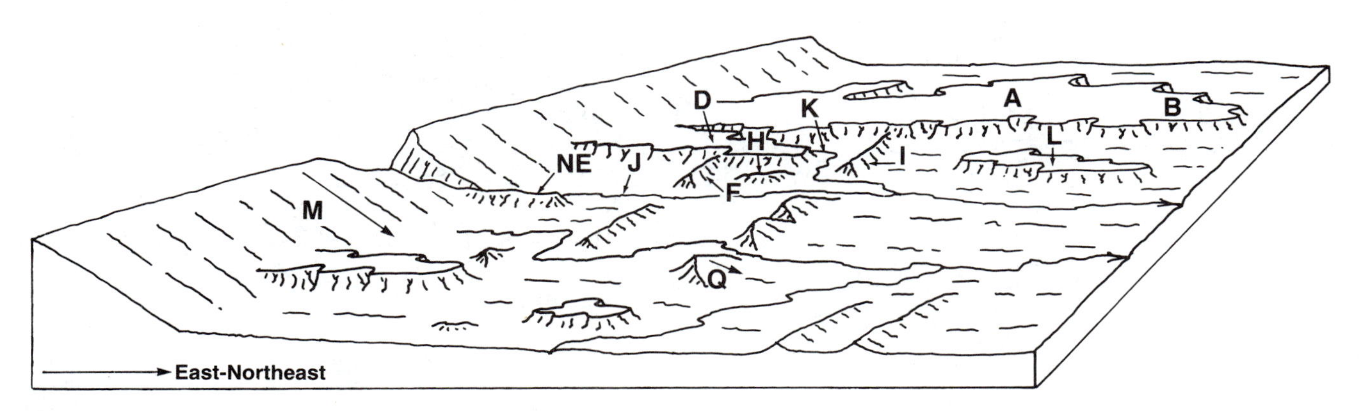

FIGURE 12–5a
Block diagram of part of the area depicted in Figure 12–5, viewed from the southeast.

Widespread rejuvenation of these streams superposed them through the blanket, and through the unconformity into the underlying dipping strata. After the streams penetrated the unconformity, the differential etching process began. It transformed the previously beveled bedrock into a topography replete with hogbacks, subsequent valleys, and water gaps. In this topography, areal growth was accomplished at the expense of the pediment surfaces, which concurrently were being destroyed.

The process is well advanced here, but not quite complete. Surface **AB** and its associates are transient features. As their destruction progresses, the area will increasingly take on the characteristics of hogback **S,** asymmetric hill **Q,** and subsequent linear stream **T**.

Before we leave this area, note the odd east-northeast-trending linearity of narrow divides **NE** and **H,** and the linear streams **J** and **K** which parallel them. These may be topographic reflections of minor cross-faults in the area.

Ash Mountain, New Mexico

(Figure 12–6 in back of book, C.I. 40 feet)

There is a similarity between this area and that of Figure 12–5 (Wind River, Wyoming, in back of book). In each, the prevailing topography is dominated by the response of a sequence of dipping formations to the processes of differential erosion. In both areas, the anomalous feature is a high, sloping bench (**l**).

Please refer to Figure 12–6a, a sketch of the area viewed from the north. Some topographies can be sketched from any direction, but some cannot. The

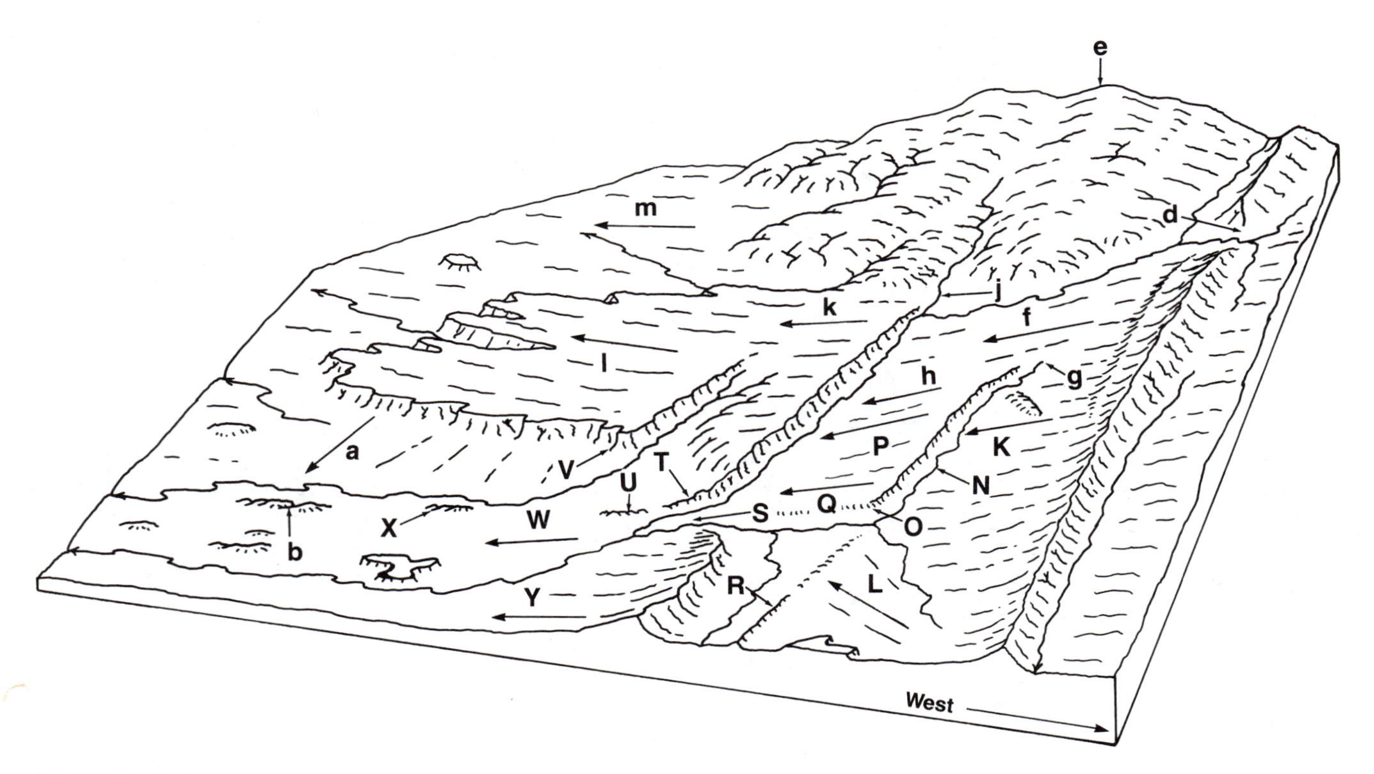

FIGURE 12–6a
Block diagram of part of the area depicted in Figure 12–6, viewed from the north.

only direction from which we could meaningfully show the topography here was from the north. To place the map in proper relation with Figure 12–6a, rotate map so north is toward you.

Although bench **l** is the highest, largest, and most outstanding of the sloping surfaces in this area, it is far from the only one. The more we study this map, the more gentle slopes we see. We have labeled several of them (slopes **C, D, E, F, M, L, K, W,** and **m**). We have divided one into four segments: **f, h, P,** and **Q**.

This is another area in which strike ridges disappear beneath sloping surfaces. We saw it in the Golden, Colorado area (Figure 12–4 in back of book), and in the Wind River, Wyoming area (Figure 12–5 in back of book).

In the present area, the relation between strike ridge **R** and surface **fhPQ** is extremely interesting and revealing. Segmented ridge **R** encounters surface **Q** at point **O**. Note the linearity of west-facing scarp **Oi**. This tells us that, along that line, the steeply dipping, resistant, ridge-forming rock unit **R** is protecting the adjacent segment (**PQ**) of slope **fhPQ**. Look at contour bend **N** and its companions. A linear subsequent stream has eroded headward (southward from **O** toward **g**) along the nonresistant strata that lie immediately west of unit **R**.

Slope **fhPQ** is graded to Slope **W** and virtually continues eastward as slope **W**. Drainage in the vicinity of **W** is toward lowland **c**. Its topography is one of dissection and erosion, and not of thick alluvial accumulation. Slope **W**, therefore, must be an erosional slope (pediment), like nearby slopes **a** and **Y**. They may be covered by debris, as many pediments are. If surface **W** is erosional, it then follows that surface **fhPQ** cannot be the top of a thick alluvial pile. Once we establish this and understand why it is so, we can look at the other gentle slopes and classify them too as erosional.

As we have seen, slope surface **fhPQ** is being destroyed by stream **N** and its tributaries. But destruction may at times be a constructive process. After all, while the stream **N** system is destroying slope surface **fhPQ,** it is creating a newer, lower pediment, **K**. Pediment **K** is younger than pediment **fhPQ,** part of which it has replaced at a lower level. Pediment **L** is, apparently, also younger than pediment **fhPQ**. (Why?)

Pediment **m,** in the southeast, is a present-day pediment, still growing. Like all pediments, it is growing headward or upslope as the nearby mountain front is being eroded back. Pediments **a** and **Y** are also present-day pediments, as is pediment **F** in the north. Note that as pediment **a** grows headward, the north flank of remnant **l** is being destroyed.

What about pediment **E,** which is really a continuation to the south of pediment **F**? It is growing to the west, upslope, as scarp **B** retreats. At the same time, it is being destroyed by steams **H**. Destructional processes working at **B,** creating new additions to surface **E,** do not know what is going on downslope, nor do destructional processes operating at **H** know what is happening at the pediment's head!

All of the pediments we have discussed are beside or among the homoclinal ridges or hogbacks, except slopes **m** and **a**. The **PQ** segment of pediment **fhPQ** is bordered by units **R** and **T**. But the upper part of that slope, **fh,** truncates the resistant unit **R** (which forms ridge **R**) and the strata immediately to the east and west.

On closer examination, we see that ridge **V**'s relation to high slope remnant **l** (the oldest, highest pediment) is similar to that between ridge **R** and the lower (younger) pediment **fhPQ**. The relation between unit **R** and the **PQ** segment of **fhPQ** is also present where ridge-forming unit **T** borders the head of slope **k** (along scarp **Tj**). This a steeper westward (headward) extension of slope **l**. It may be that the combined resistance of units **V** and **T** is responsible for the steeper bench slope in the **k** area.

Production of erosional surfaces such as **L, fhPQ,** and **l** requires considerable intervals of time during each of which local base level remains constant. We see this area as having experienced cycles of stability, interrupted by periods of drainage rejuvenation. It was necessary to destroy most of the level **l** erosional surface to initiate development of surfaces such as **fhPQ** and **a**. Likewise, surface **fhPQ** must be, and is being, destroyed so that surface **K** may be established and developed.

Mountain Mass. To the southwest, mountain mass **e** is unlike either the strike ridges and valleys or the several pediment slopes. It appears to be simply a large mass of "material" which lies upon the strata and structures that made the topography to the north and northeast. We cannot see enough of **e** to tell what it is, and the geologic maps of the region are of little help. For example, the 1932 *Geologic Map of the United States* (reprinted in 1960) shows Tertiary intrusive rock in this area, whereas the 1967 *AAPG Geological Highway Map of the Southern Rocky Mountain Region* places Eocene clastics both in this area and to the east of ridge **Tj**.

Structure of Steeply Dipping Beds. Let us now turn to a completely different matter, the structure of the steeply dipping beds. In water gap **I,** a clearly defined outcrop V (dashed lines) indicates moderately steep east dip. Immediately north of point **S** (the south end of the ridge that extends south from gap **I**) the contour bends define a curve (dashed line) in the descending ridge crest. This is the north half of another outcrop V. The south half of this V is depicted by the contour bends at **U**. Ridge **Tj** and ridge **IS** are en echelon, rather than end-to-end. This suggests a transverse fault through gap **S** (and perhaps another through gap **I**). Note how curved divide **T** continues around past point **U**. Small linear hills **X** and **b** might be strike ridges. Since this is an area of regional east dip (e.g. east-dip asymmetry at **Z**), the possibility of local east strike merits our attention.

Steep east dip is seen also at water gap **G,** which cuts through ridge-trend **R**. To the south, ridge **R** is too symmetric to indicate dip to either east or west.

There are three water gaps (**A, J,** and **d**) in the other large ridge. It is obvious that the dip along this hogback is extremely steep. (This is evident from the ridge's narrowness in comparison to its height and from its steep-sided, nearly symmetric profile.) In all three of these water gaps, there are only steep west-pointing outcrop V's. West dip.

Using the dips established from the outcrop V's we have examined, we may draw the axis of a long, tight, well-defined anticline along the main north-south valley that separates the two main hogbacks. Perhaps we should draw it a bit to the west of the center of the valley (e.g. through letter **M**), since the "fold" is asymmetric (steeper west dip).

What about those massive, resistant rocks farther to the west and northwest? There is nothing like them in the east. If we tell you that there are no strike faults mapped in this area, can you then explain these confusing relations? This is not an anticline, despite the east dip along one ridge and west dip along the other. The various geologic maps agree. From the northwest corner of the map, to and including ridge **T,** the bedrock units consist of:

- □ Western highland: Precambrian crystalline rocks.
- □ Western hogback and flanking lowlands: Pennsylvanian clastics.
- □ Ridge **U** and nearby strata: Cretaceous formations.

In other words: all of the dipping formations we perceive to be west-dipping as well as east-dipping (on the basis of outcrop Vs) actually dip to the *east*! Those forming the western hogback must be overturned. No wonder we were confused by those "west" dips!

Just to the north of this area, the Precambrian is faulted up against the Pennsylvanian. It is possible (although apparently unreported) that the fault extends southward into this area. However, examine the topographic boundary between the massive western mountains and the adjacent lowland/ridge topography to the east. Nothing tells us to designate this as a fault contact rather than a normal nonconformity, or vice versa.

As a matter of fact, where *should* we draw the Precambrian-Pennsylvanian contact? Perhaps just west of letter **A**, along the base of the east-facing scarp that borders the Precambrian highland belt.

Buffalo Gap, South Dakota

(Figure 12–7 in back of book, C.I. 10 feet)

This area lies on the southeast flank of the Black Hills Uplift. Regional dip is to the southeast. This is clearly indicated by the pronounced asymmetry of the main north-northeast-trending homoclinal ridge, which cuts across the center of the map. Four arrows on its resequent flank denote the dip (**K, X, o,** and **k**).

The resistant formation underlying this ridge is folded into a south-plunging anticline at the south edge of the map. At **i** the dip is to the south; at **h** it is to the southwest. The asymmetry at **g** defines the west flank of an adjacent south-plunging syncline in the very corner of the map. Note that the main ridge is cut at three places (**Z, p,** and **n**) by water gaps. Most of the map area is depicted in Figure 12–7a, a view from the southeast.

Overlying the main ridge-forming unit is a second resistant bed. It is dissected into shorter, more-numerous segments (e.g. **M, Y,** and **m**).

Note the linear western border **RC** of the broad western lowland and the flatiron-like spur end near **C**. These attest to northeast strike, and probable uniform southeast dip throughout at least the western half of the map area. Similar dip east of the main strike ridge is indicated by linear hill **L**, the line of knobs **N, O,** and **P,** and linear contour trend **cQ**. Southeastward dip is also shown by some other features that you may be able to detect.

So far we have noted nothing anomalous about this area. Nonetheless, there are several extremely anomalous and mystifying features. They just do not seem to reach out and grab our attention!

Anomalies. We do not expect to find northwest-southeast-trending ridges in an area of southeasterly regional dip, but in this map area there are several such ridges. These surely merit our consideration. Not only do they trend in a direction contrary to strike, but they are closely associated with gaps in main ridge **Kk**. It is not easy to accept this association as a matter of chance.

Before we can discuss the anomalism of these ridges, we must identify them. Whatever they are, they have been subjected to extensive erosion. Their outlines are irregular, and in places they are even fragmented.

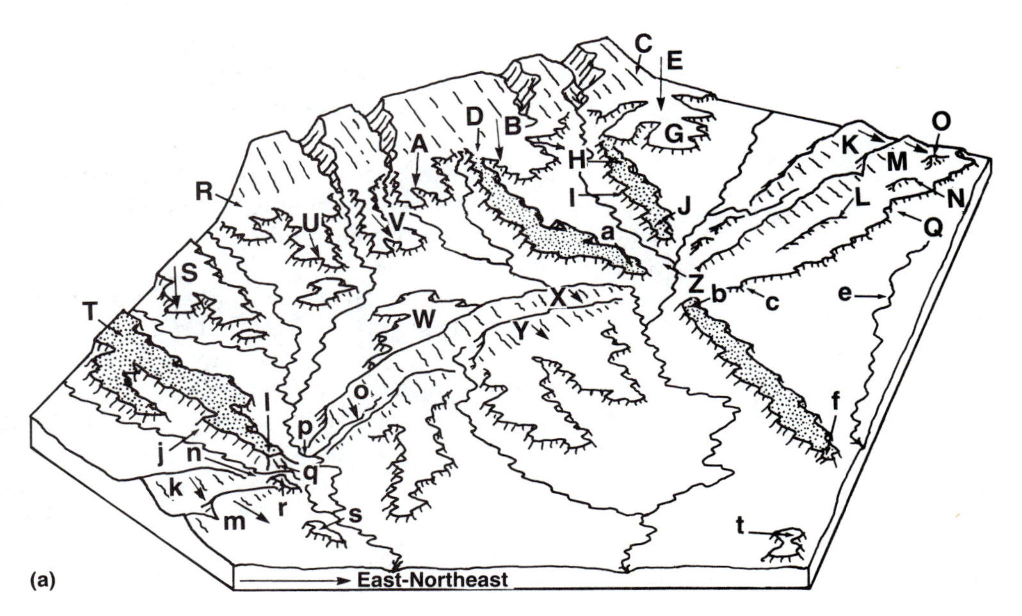

FIGURE 12–7a
Block diagram of part of the area depicted in Figure 12–7, viewed from the southeast.

We can trace the first one southeasterly from **D** to **a** (near the center of the map). Here it divides the largest of the gaps in ridge **Kk** into a double gap (**Z**). One-third mile farther southeast is the northwest end of another segment (**bf**), which is over a mile long. Note particularly that the entire segmented ridge **Da-bf** descends gradually from a high at **D** to a low at **f**. Also note that, unlike the several strike ridges, this "ridge" is flat-topped. It is an elongate, narrow, somewhat sinuous or serpentine "table." To the southeast are two small, isolated flat-topped hills (**t** and **u**). These could be additional isolated remnants of this surface.

The second sloping flat-topped surface (**HJ**) lies to the north of ridge **Da**. Its southeast end seems to be aiming toward the northern division of gap **Z**.

The third elongate, sloping "table" is surface **Tl**, in the southwestern part of the map area. Like surface **Da**, surface **Tl** ends in a water-gap area. In fact, it is responsible for there being two gaps (**p** and **n**) in what is actually a single large break (between **o** and **k**) in the main **Kk** hogback. Looking back to large gap **Z**, we see that the east end of bench **Da** breaks that gap apart in much the same way. The only significant differences between divided gap **Z** and paired gaps **p** and **n** are:

1. At **l**, the bench stands higher above the two gap-cutting streams than bench-end **a** stands above its nearby streams.
2. Whereas the bench at **a** merely drops off at **Z**, bench **Tl** is terminated at **q** by a uniformly sloping facet. This facet strongly resembles a flatiron or similar dip slope. (Refer to Figure 2–5 to see the resemblance.) In fact, slope **q** "fits" almost perfectly as an isolated fragment of resequent slope trend **ok**.

Small, southeast-trending hills **r** and **s** appear to be remnants of what at one time was a much-longer **Tl** ridge.

A broad belt of sloping surfaces extends northeasterly, parallel to and adjacent to the west border **RC** of the large western lowland (**S, U, V, A, B, E, G**). These are obviously the remnants of a single extensive surface. The surface must have resembled surface **cQe** in the east, but it must have been much more widespread. Of particular importance is the fact that **Da** is actually a finger-like extension of remnant surface **B**. Bench **HJ** is "almost" separated from remnant surface **G** by a shallow sag or saddle (**F**). Similarly, bench **Tl** is another preserved finger of this regional sloping expanse.

Back in Time. By mentally turning back the geomorphic clock, we can reconstruct the former sweeping, low topography (Figure 12–7b). The present surface remnants are the mere remains. When we go back in time, we find to the west of ridge-trend **Kk** an erosionally graded surface. It slopes toward the two broad gaps (**Z** and **pn**) in the ridge, which at that time was much lower. For example, ridge segment **Xo,** which now rises up to 200 feet above the floor of gap **p,** stood a mere 50 feet or so above adjacent lowland segment **W**. **W** was then part of the regional lowland surface.

At that time, as today, southeast-flowing streams must have drained through the two large water gaps. If these main drainage lines were graded ("mature" in the Davisian sense), they would have flowed as meandering streams along well-defined, relatively broad floodplains. The sediments strewn on those floodplains would have been permeable. As time passed, conditions may have favored cementation of those materials by minerals which were deposited by rising groundwater (caliche). On such matters we can only speculate.

If our general assumptions are correct, the valley-floor detritus became relatively resistant to weathering and erosion. By contrast, the entire stratigraphic section underlying the area to the southeast of line **RC** is nonresistant (except for the three resistant units forming ridges **Kk, M,** and **L**). Convincing

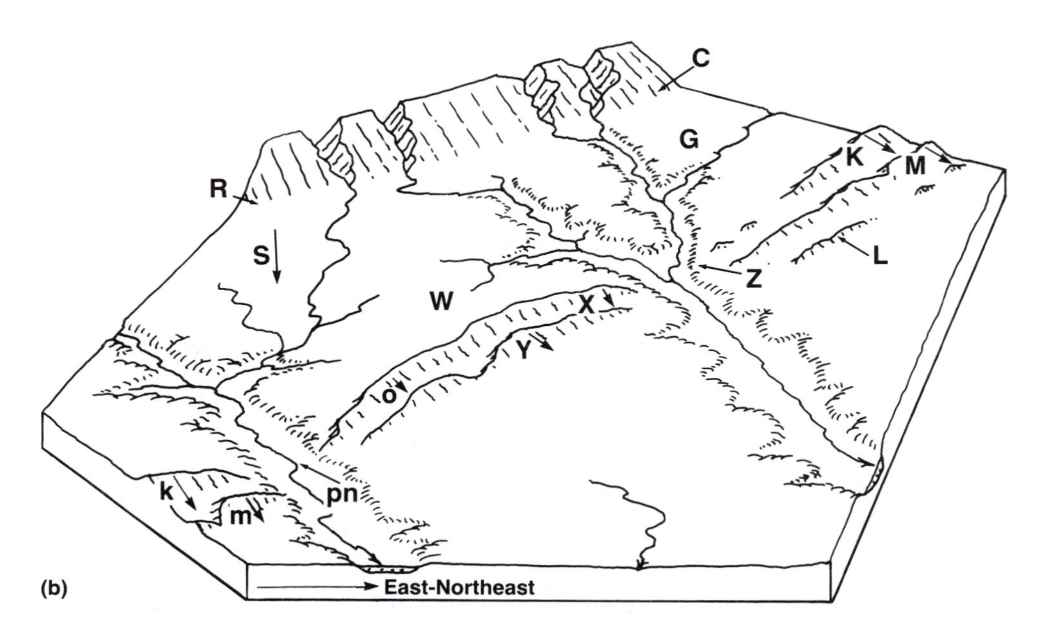

FIGURE 12–7b
Block diagram of part of the area depicted in Figure 12–7, viewed from the southeast, as it may once have appeared.

evidence of this nonresistance is the virtual perfection of the former regional erosional surface that we have just reconstructed.

For reasons unknown, possibly a climatic change, the entire area's drainage was rejuvenated after the creation of the gently sloping erosional surface. The present topography is the product of that later downcutting, downwasting, and backwasting. This erosion interval saw most of the former major interfluve tracts destroyed. At the same time, most of the serpentine floodplain belts were being preserved. It is these preserved floodplains that constitute our *anomalies*.

Two Explanations? In this discussion we have proposed one possible explanation. We feel that it fits the facts, as revealed by the topographic map.

Another possible scenario calls for lavas to have flowed down the former valleys. They would have blanketed the valley floors with a layer of igneous rock more resistant than the rock underlying the surrounding low-lying interfluvial expanses. Rejuvenation, as in our other hypothesis, would have subjected the entire area to accelerated differential erosion. Ultimately this would have led to the topographic inversion we now have in the area: the former valley floors standing as sinuous, narrow, southeast-sloping, flat-topped ridges.

The *Geologic Map of the United States* does not show any Tertiary extrusive rock in this area. Also, the flat-topped remnant surfaces are fairly well defined by relatively steep border slopes (e.g. **I, d,** and **j**), but they do not display the pronounced scarp/slope border we would expect with such a true caprock as lava. So, our second hypothesis does not appear to work.

One final word about dip-slope **q**. It is a dip-slope because the overlying nonresistant beds were open to attack from the southeast, as well as from both sides. Their removal exposed the top of the resistant unit. On the other hand, the same resistant unit that forms dip-slope **q** now protects bench **Tl,** and the nonresistant bedrock beneath its gravels, from further attack from the southeast. Thus, upstream from its former water gap, the former stream's valley-floor debris is preserved as a bench, whereas downstream it is almost completely destroyed.

King Hill, Idaho

(Figure 12–8 in back of book, C.I. 40 feet, and Figure 12–9 in back of book, *Reconnaissance Geologic Map of West-Central Snake River Plain, Idaho*)

This section is different from the others we have studied so far. It is accompanied by a geologic map of nearly the same scale and of virtually the same area.

Originally we intended to use the geologic map for a show-and-tell analysis of the King Hill topographic map. However, as we studied the topographic map, we realized that we had discussed and analyzed most of the geomorphic/geologic features and relations depicted here.

Therefore, we decided to move over and let you take the wheel. We will give you some necessary geologic information, and then list the things you are to do, and in what order.

Essential Facts. From youngest down to oldest, the mapped rock units depicted in Figure 12–9 are listed in Table 12–1.

The bar-and-ball symbol denotes the downdropped side of a normal fault. Faults are dotted where concealed. The small double-barb lines crossing the faults are called "tie bars." They indicate that the rock units on both sides of

TABLE 12–1
Mapped Rock Units in Figure 12–9

Symbol	Rock Unit	Epoch
Qls	Landslide material	Recent
Qbf	Fan material	Middle Pleistocene
Qbb	Lava flows	Middle Pleistocene
Qtg	Basin fill	Lower Pleistocene/Upper Pliocene
Tb	Basalt	Middle Pliocene
Tbs	Sediments	Middle Pliocene
Tiv	Tuffs/lavas	Lower Pliocene

the fault are the same in that area. (For example, in the southeast, a tie bar shows that areas **Y** and **Z** are occupied by "basin fill" **QTg**.)

The geologic map is a guide—a guide *only*—since it is in places inaccurate and/or incomplete. With its help, you can make your own geologic map of this area.

Preparation. The first step is to attach a transparent overlay, such as a frosted Mylar, to Figure 12–8. We suggest that you use several different pencil colors, such as blue for streams, black for structure (e.g. faults and dips), and brown for stratigraphic contacts.

Drainage. Begin by tracing the numerous large and small streams (e.g. those over one inch long) that you can delineate on the topographic map. This is important because drainage is the only feature common to both maps. The only way you can transfer a mapped fault to the topographic map is to use as guides the streams it crosses or otherwise relates to. Similarly, the only way you can transfer a scarp trend is to use the streams as reference.

Structure. Next turn to the structure. Keep the following points firmly in mind:

1. Proceed slowly.
2. The two maps do not cover precisely the same area, so be careful.
3. No matter how completely you delineate faults, ascertain dips, and establish contacts, there are going to be some (many?) places where you may be uncertain or confused.
4. In some places, the topography will *not* reflect the geology.
5. You will map many things that *only* the topographic map can provide.
6. If you add any geologic information to what has already been compiled, you are *making a contribution*!

Figure 12–9 shows only three dips (**A, C,** and **D**). Slopes **A** and **C** (Figure 12–8) coincide with their indicated dip directions. We have marked point **D** on Figure 12–8, in case you wish to transfer the dip symbol. Since these slopes are on lava surfaces, can we assume that they are virtual, if not exact, dip slopes? What do you think of slopes **H, O, Q, R,** and **T**?

Note that most of the mapped faults that you can identify on Figure 12–8 are represented by linear and curvilinear scarps and steep slopes. In many cases the scarps do not have the same curvature, or continue along the same lines, as their corresponding geologically mapped traces. *Believe the topography*!

For example, mapped fault segment **G** may lie along topographic line **G,** but what about its extension to **F**? At point **F** on the topographic map there is nothing but a continuous, gentle southeasterly slope. On the other hand, steep linear scarp **E** (Figure 12–8) does not have a geologically mapped structural counterpart, though the base of scarp **E** appears to be the trace of an east-west-striking fault downdropped on the north. This fault, incidentally, would extend through **G**. (That is, if the fault-associated scarps in this area of Tertiary and Quaternary rocks are fault scarps, and not fault-line scarps. What do you think they are? Why?)

We have indicated many other interesting scarps such as **M, L, B, K, S, U,** and **V**. According to the geologic map, these are not fault-associated. What do you think? One of the most striking is scarp **N**. Incidentally, it faces north, whereas its apparent southeast extension, scarp **P,** faces south.
gate closed contour (**J**) can be. Is this not a low, north-facing scarp?

Contacts. You may find it far more difficult to sketch in the contacts than to map the faults. If you try to add contacts, you will probably discover that in places they do not closely follow the corresponding contacts shown on the geologic map. Elsewhere you will encounter mapped contacts that you cannot "find" in the topography. When this happens, *do not* transfer them! This is *not* an exercise in transposing data from one map to another. The geologic map will help direct you to faults and contacts that you can find on the topographic map. But, as we said, you must use the geologic map solely as a *guide*, and not as an infallible source. The geologic map was made by human beings engaged in the same activity that you are now undertaking: interpretation.

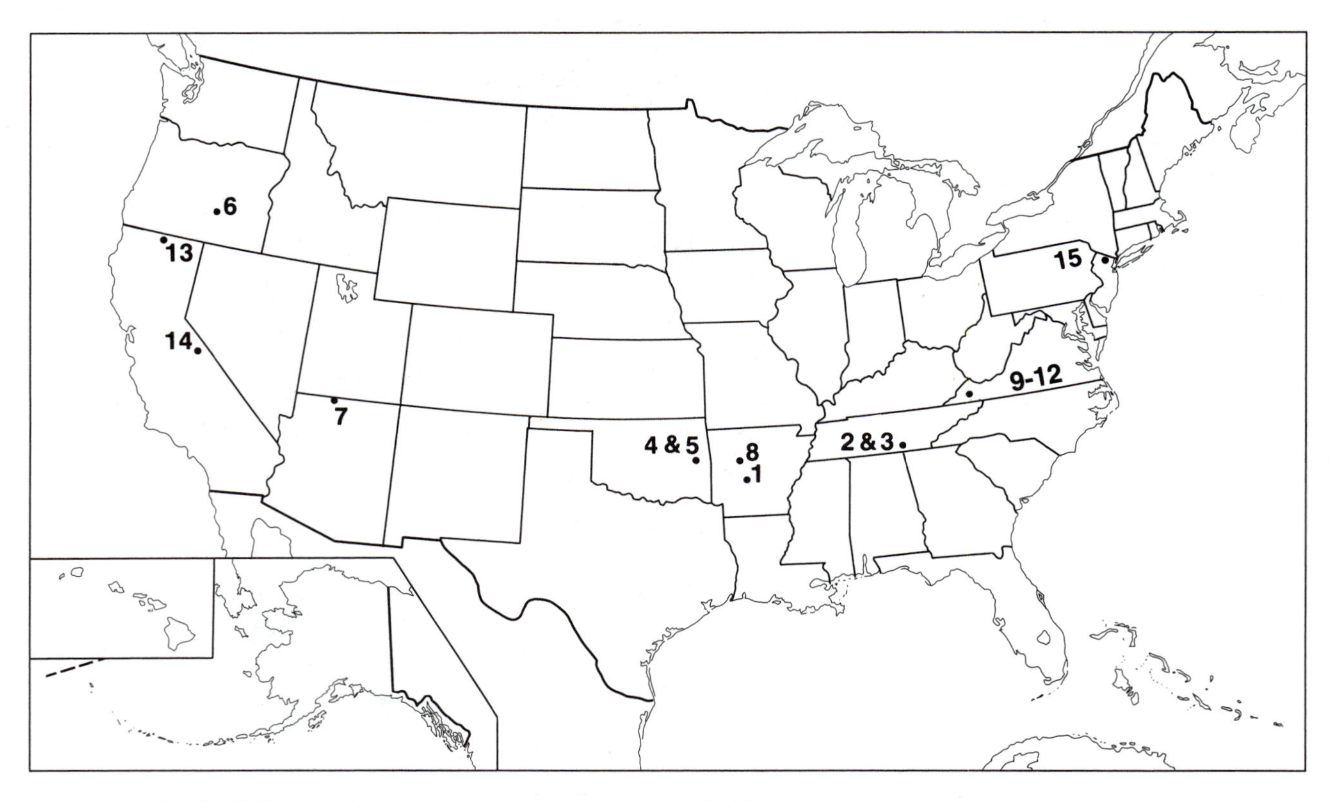

1. Figure 13–1. Cabot, Arkansas.
2. Figure 13–2. Chattanooga, Tennessee.
3. Figure 13–3. Chattanooga, Tennessee.
4. Figure 13–4. Fort Smith, Arkansas/Oklahoma (topographic).
5. Figure 13–5. Fort Smith, Arkansas/Oklahoma (geologic).
6. Figure 13–6. Alec Butte, Oregon.
7. Figure 13–7. House Rock Spring NW, Arizona.
8. Figure 13–8. Damascus, Arkansas.
9. Figure 13–9. Hilton, Virginia.
10. Figure 13–10. Mendota, Virginia.
11. Figure 13–11. Brumley, Virginia.
12. Figure 13–12. Hilton, Virginia.
13. Figure 13–13. Bray, California.
14. Figure 13–14. Mt. Tom, California.
15. Figure 13–15. Newark, New Jersey.

13

Problems

INTRODUCTION

Your authors have been working with, studying, teaching, and interpreting topographic maps for over 65 years, if you sum our experience. Despite such prolonged familiarity, which might have bred contempt or apathy, we still are filled with awe when we rediscover, time and again, just how much information is conveyed by what one student called "those funny little squiggly lines."

Each map in this final group presents its own questions, its own unique topography, and its own unique combination of structures, lithologies, and geologic/geomorphic histories. We have tried to embrace a great range of variables, while keeping the number of maps to a minimum.

We have also tried to resolve as many of the problems and answer as many of the questions as possible. At times we have been forced to do what all map interpreters (and all geologists) must do: conclude that, from the evidence available, there may be two or more possible answers, but we cannot decide. In our opinion, it is far better to designate a feature with the choices, and a question mark. It is better to say, "it is either an angular unconformity or a fault," than it is to unwittingly use a fault contact to illustrate a "typical unconformity"!

Cabot, Arkansas
(Figure 13–1 in back of book, C.I. 10 feet)

If you turn back to Figure 10–9 (Cato, Arkansas, in back of book), you will see a homoclinal ridge that is bent back on itself in the form of a hairpin. The geometry of that ridge, particularly its asymmetry, indicates that it is a homoclinal ridge. The overall structure is that of a west-plunging syncline.

In this map area we have three major homoclinal ridges (**N, F,** and **I**). They seem to be trying, without much success, to complete similar bends. Two of them are shown in Figure 13–1a, a sketch of part of the area viewed from the south. Ridge **N** almost makes it, getting around to northwest dip (arrow) at **L**. At **O,** a small segment of a low ridge indicates northeast strike (possible northwest dip, indicated by broken arrow). This same minor ridge-forming unit forms a low strike ridge at **D**.

Ridge **F** is truncated abruptly just south of **Q**. Here the dip is more to the west-northwest. The geology is obscured by highway construction. Ridge **I** is cut off just south of **T,** where dip is due west. If we draw a nearly straight

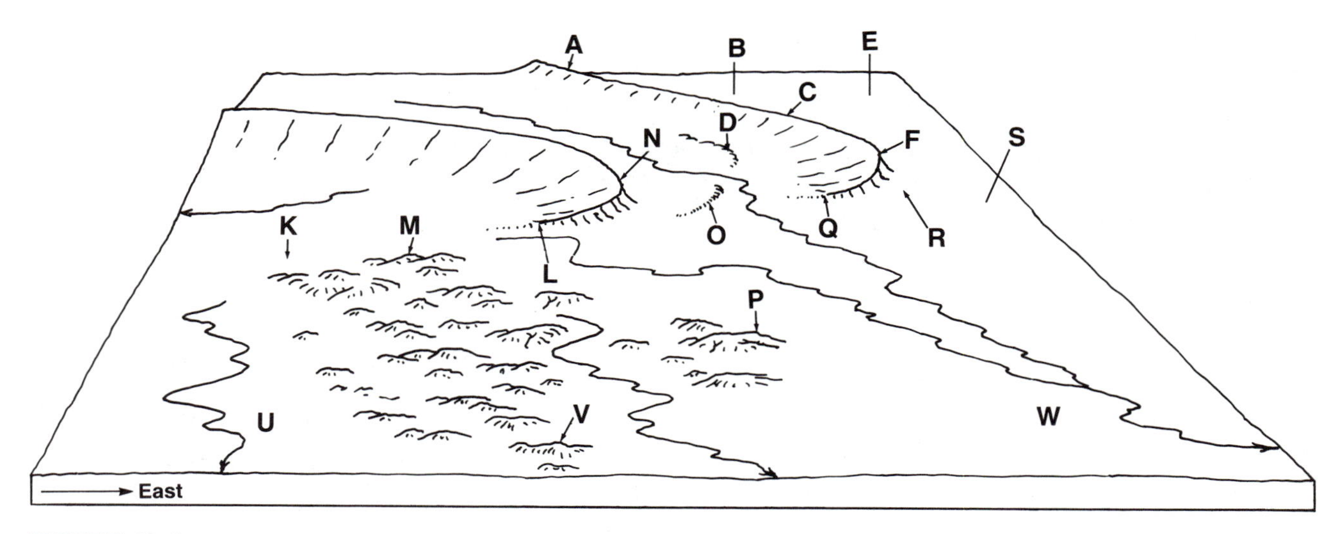

FIGURE 13–1a
Block diagram of part of the area depicted in Figure 13–1, viewed from the south.

line from **K** to **S,** we will divide the map into two halves. The northern half has parallel ridges and subsequent valleys. The southern half, though mostly featureless lowlands (e.g. areas **U** and **W**), contains some hilly ground (area **MPV**). In this hilly area a dendritic drainage pattern reflects a lack of structural control.

A major criterion for recognizing faults is the abrupt truncation of topographic features, particularly features that are adjusted to bedrock lithology and structure. These three strike ridges are examples. Line **KS,** therefore, may be designated a possible fault trace.

We shall return to line **KS,** but first let us consider the strike ridges in more detail. Throughout its length, ridge **N** maintains an asymmetry consistent with west and southwest dip. Ridges **F** and **I** are not so consistent. In the vicinity of **C,** ridge **F** is virtually symmetric. Along segment **A,** the asymmetry is reversed. If segment **A** were isolated from the remainder of this ridge, it might appear to be the strike ridge of a resistant, northeast-dipping formation.

Along segment **A,** the dip is definitely steep. In fact, it may even be overturned. Ridge asymmetry here may be the response of erosional processes to a series of thin beds, rather than to a single thicker unit. So we should not deduce dip direction on the basis of such a single local geomorphic feature.

Similarly, the topography at **G** denotes steep dip, but it tells us nothing about dip direction. The dips at **F, Q, I,** and **T,** however, do tell us that the dips at **A, C,** and **G** are to the southwest, regardless of whether they are steep, vertical, or overturned.

Though the dip directions along the homoclinal ridges merit our consideration, it is line **KS** that presents the problem. Exactly what is the geologic nature of this line? Is it truly a fault trace, as the topography suggests? If we had nothing but the topographic map to refer to, it would be reasonable for us to designate this as a possible fault trace.

Does the topography offer alternate possibilities? It does, so long as we refuse to consider other information we might know about the geology of this part of Arkansas. For example, topography such as that in area **MPV** need not be developed on flat-lying sedimentary rocks. It could be that the bedrock consists of deeply weathered, relatively nonresistant, massive crystalline rock.

Could line **KS** be the border of a large intrusive? It could, though its linearity is suspect.

The *AAPG Geological Highway Map of the Mid-Continent Region* reveals that the entire north half of the map area consists of Pennsylvanian strata. They are primarily if not entirely clastics (sandstones and shales). The south half is underlain by Tertiary and Quaternary Coastal Plain/Mississippi Embayment sedimentary rocks.

So, is line **KS** a former shoreline? What material underlies flat-floored triangular embayment **RES**? It appears to be an extension of featureless lowland **W,** rather than a continuation of dissected subsequent lowland belt **BE**. Area **J** also resembles areas **W** and **RES**.

Deformation of the Pennsylvanian formations is pre-Tertiary. Thus we know that the Tertiary has an unconformable relation with the underlying Pennsylvanian strata throughout the area south of line **KS**. If hilly area **MPV** is Tertiary bedrock, the broad flat low areas (e.g. **W** and **U**) are probably blanketed by debris of Quaternary age.

What then, is line **KS**? The topographic map, though it provides the question, cannot supply the answer. The geologic map, though it identifies the bedrock units and shows line **KS** to be a "contact," does not answer the question either.

It is a contact, but what kind? It seems to separate topography developed in the folded Pennsylvanian bedrock from that developed in probable Tertiary (area **MPV**). However, the Quaternary lies upon, and effectively hides, many other details.

The Tertiary almost certainly has not been faulted down into contact with the Pennsylvanian, but that does not rule out the possibility (only a possibility) that line **KS** is the trace of a steeply dipping older fault. Pre-Tertiary differential erosion might have produced the truncated ridges before any Tertiary sediments were deposited.

On the basis of our incomplete knowledge of this area's geology, all we can do is speculate. We must accept that we simply do not know the answers.

We have not even asked all the questions. For example, **G** and **H** are two major parallel strike ridges. Why is there but one ridge at **I**?

Can you explain the "broken" contours in the areas marked by small **x**'s? Why are these odd topographic features restricted to ridge **I**?

Chattanooga, Tennessee

(Figure 13–2 in back of book, C.I. 100 feet)

This is a remarkable map. For one thing, the topography has over 1000 feet of relief. For another, the geology west of line **OD** is unlike that to the east. Also, there is something most strange about segment **UR** of the main river.

Let us start with the area west of line **OD,** an area we will designate **OAD**. There are two noteworthy things about this area. First, it is dissected by a system of streams forming a dendritic pattern. Second, this is a plateau area. It is devoid of significant variations in upland elevation, and it is bordered by steep, scarp-like slopes (e.g. **B** and **M**). The extremely irregular outline of the western front, and the dendritic drainage, are typical of areas of nearly horizontal resistant caprock.

The area east of line **OD** has the same topographic relief. But it displays a pronounced linear grain, best represented by stream **QF** and scarps **PE, SH,** and **UK**. This trend is also revealed by smaller topographic features, such as ridge **V** and subsequent stream **W**.

By now you are familiar with the structural implications of topographic and drainage linearities. You can deduce that this eastern area is one of numerous, steeply dipping structural planes. Whether they are dipping bedding planes or faults we must yet determine.

Scarp **PE** could be a fault-associated scarp. Northwest dip (note arrow) is suggested by the asymmetry of slope and dissection characteristics along segment **IJ** of this front. Across valley **QF,** the abrupt and virtually undissected scarp **SH** contrasts sharply with the dissection and gentler slopes to the southeast. These asymmetries reveal dip to the southeast, as indicated by arrows **G** and **T**.

Valley **QF,** therefore, appears to be one of four things:

1. A graben.
2. A horst (possible, if scarps **PE** and **SH** are obsequent fault-line scarps).
3. An erosion-widened subsequent valley developed along a fault zone.
4. A breached anticline, in which case stream **QF** would also be a subsequent stream.

By similar reasoning, scarp **UK** may be either of two things. It could be a fault-associated scarp. Or it could be a scarp that is linear because the dip of the resistant caprock is locally steep. The contours indicate no apparent asymmetry to support the latter possibility. However, features such as linear ridge **V** and stream **W** convincingly testify to the steep dip of something along this front. Thus linear upland **SHKU,** like valley **QF,** may have more than one possible structural explanation. It may be a horst, a graben, or a flat-floored syncline with upturned edges.

Whatever the structure beneath valley **QF** and uplands **OAD** and **SHKU,** we do know one thing. The rocks that underlie the uplands are resistant, and those beneath the lowlands are nonresistant.

When we say the lowlands are developed over nonresistant rocks, we are not saying that all valleys here are valleys because resistant rock is absent. For example, why do valleys **L** and **N** exist? They exist simply because streams have cut through the resistant plateau caprock. These are not subsequent valleys! Given enough time, stream **C,** which is currently cutting into the caprock surface, may eventually carve a valley as deep as **L** or **N**.

Granted, downcutting in these two valleys has undoubtedly penetrated the resistant caprock and entered the nonresistant underlying formations, but the nonresistant rock beneath the caprock had nothing to do with the creation of these valleys as valleys.

Let us assume that stream **QF** follows a generally linear trend, regardless of its valley's structure, and despite its serpentine bends. We will assume it follows a linear trend because it developed along a belt of nonresistant bedrock. Therefore, can we also assume that, like most subsequent streams, it developed by headward growth? We can visualize it, then, as a small tributary, beginning at confluence **Q** and growing headward toward the northeast. It is important that you understand this scenario for stream **QF,** since we shall refer back to it in our discussion of Figure 13–3 (also Chattanooga, in back of book).

The Meandering Stream in a Canyon. The final feature we will discuss is the fascinating incised, meandering course **UR** of the main river as it cuts westward through the eastern plateau area, upland **SHKU,** and its extension to the southwest. Figure 13–2a is a sketch of this canyon viewed from the south.

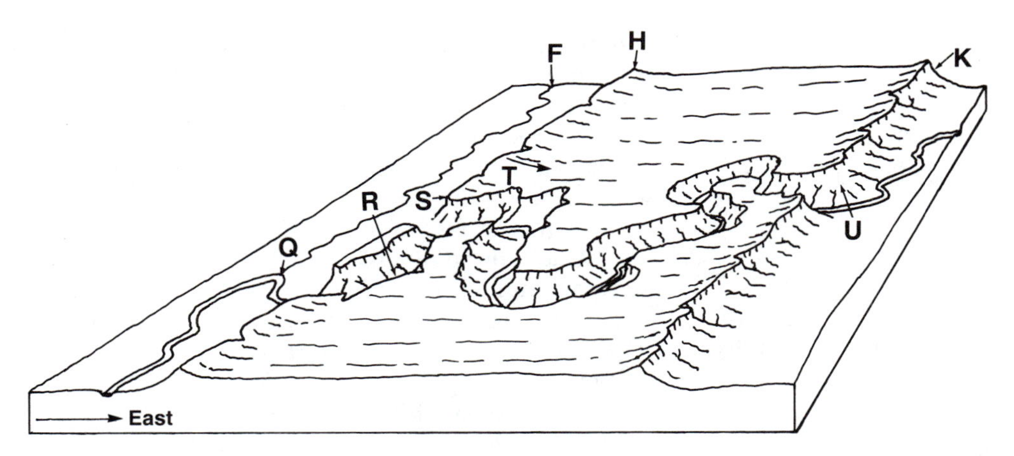

FIGURE 13–2a
Block diagram of part of the area depicted in Figure 13–2, viewed from the south.

This is clearly an anomalous feature. It is anomalous because, though it flows in a canyonlike valley, its course is typical for a stream in a broad floodplain. Many streams exhibit such conflicting characteristics. Their incised meanders are usually explained as the product of rejuvenation, from regional tilting or a climatic change.

This stream is unlike them, however. It flows from a lowland underlain by nonresistant bedrock, into and across an upland maintained by resistant caprock, and then back into another lowland developed on nonresistant bedrock. In Chapter 9 we studied several other streams that cross through high, resistant barriers, but they do not *meander* through them!

Is it possible that this river, antecedent to block faulting which uplifted plateau area **SHKU** across its path, was flowing on a broad floodplain prior to said uplift? From the contours we can see, such a history is possible. That would be an example of straightforward, independent map interpretation. We cannot claim that such an event happened, but we must acknowledge it as a possibility.

However, this area lies in southeastern Tennessee. Its last tectonic deformation was late in the Paleozoic, so the likelihood of antecedence fades into the swirling clouds of fantasy.

This river is the Tennessee, one of the most mystifying and provocative rivers in the United States. This is but one of its many anomalous stretches. Various hypotheses have been advanced to explain serpentine canyon **UR**. That fact alone attests to its being a bona fide anomaly.

Chattanooga, Tennessee
(Figure 13–3 in back of book, C.I. 100 feet)

The southwest corner of this map area lies immediately northeast of the area covered by Figure 13–2 (also Chattanooga, in back of book). Southwest-flowing stream **YP** is the headwater section of the stream we designated **QF** in Figure 13–2. Similarly, scarp **XO** is an extension of scarp **PE**. Scarp **ZS** is an extension of scarp **SH**. Scarp **aM** is a continuation of scarp **UK**.

In Figure 13–2, main dividing line **OD** separates a western dendritically dissected plateau terrain from an eastern area that is partly plateau with a definite topographic grain. In Figure 13–3, division line **aM** separates a southeastern lowland (area **aMf**) from the entire northwestern upland area (**aMA**). Lowland **aMf** is remarkable for its closely spaced, parallel, low ridges (e.g. **N**) and linear subsequent streams (e.g **W**). The northwestern corner of area **aMA** displays a modified dendritic drainage pattern. However, a general northeast-southwest-trending grain intrudes (e.g. streams **Q** and **B**).

From the few contour lines that are mapped in lowland area **aMf,** we cannot determine much about the details of its structure. All we can tell is that there are numerous steeply dipping structural planes (bedding and/or fault). These are responsible for the parallel ridges and streams.

The ability of a few contours to convey geologic as well as geomorphic information is extremely well illustrated in the southeastern part of this map area. By contrast, this same lowland area is ideal for demonstrating the limitations of trying to extract specific geologic data from a topographic map.

For example, this map cannot tell us (but a more-detailed geologic map can) that a certain southeast-dipping Cambrian dolomite crops out at points **b, c, d,** and **e**. And this map cannot tell us (but a geologic map can) that these outcrops are separated by a series of steeply dipping reverse faults. Area **aMf** is part of the Valley and Ridge geomorphic province. Within the confines of this small, triangular area, there are at least 13 such southeast-dipping thrust faults. They all strike parallel to the topographic grain, as they should, since they are largely responsible for it.

As interesting as the thrust area may be, it is the *plateau* area that is important to us. Its topography will permit us to improve our interpretation of Figure 13–2. From what we could see on Figure 13–2, linear valley **QF** appeared to be a breached anticline. However, we could not rule out the possibility of a fault or two playing a major role in creating this landform. In addi-

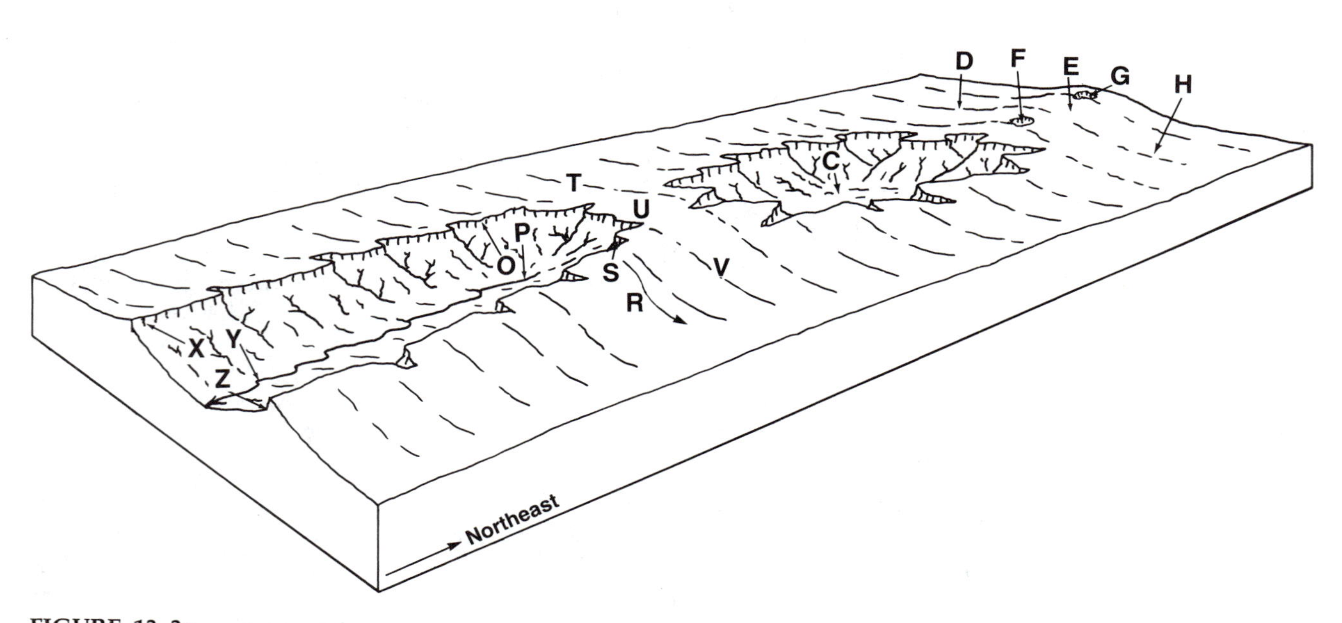

FIGURE 13–3a
Block diagram of part of the area depicted in Figure 13–3, viewed from the southeast.

tion, we postulated the headward growth of subsequent stream **QF** as the main process in carving that linear valley.

In this map area, the valley of stream **YP** presents the same criteria that were available to us in Figure 13–2: two infacing scarps (**XO** and **ZS**), ridge asymmetry (e.g. dip arrow **R**), and uniform valley width. All our conjectures about valley **QF** carry over to valley segment **YP**.

There is one significant difference, however. Unlike segment **QF,** this valley terminates at a well-defined head, **P** (Figure 13–3a). (Figure 13–3a is a view from the southeast of the **Z**-to-**E** area on the map.) Immediately upstream from headwall area **P,** there is a continuous upland surface, from the northwest side (**T**), to a high point (**U**), and back down to the southeast side at point **V**. We see the same up-and-over continuity across **DEH** and **IJK**. The latter arch continues to the map border near point **L**.

This smooth, linear, northeast-trending topographic high is best explained as an anticlinal ridge, which is in the process of being breached. This permits us to state with more assurance that valley **QF** (Figure 13–2), with its extension **YP,** is an anticlinal valley. We can also say that stream **QF,** with its head **YP,** is a subsequent stream.

What is the meaning of large depression **C** (Figures 13–3 and 13–3a)? And the smaller ones (**F** and **G**)? Depression **C,** which falls squarely along the axis of the anticline, is over five miles long and two miles wide! It is definitely not a backyard variety of sink. Yet, except for solution, how can we explain this feature?

Depression **C** is attention-demanding. Yet of even greater importance to understanding valley **YP** (and its downstream continuation **QF**) is divide **U**. This surprisingly narrow divide separates depression **C** from valley head **P**. This divide is being attacked from both sides.

It is the kind of geomorphic feature about which the eminent W. M. Davis reportedly commented during field trips, "It's a good thing we came here today. Tomorrow this may be gone."

The question is, of course, what will happen to depression **C** when divide **U** has been removed? It will simply become a new headward extension of valley **YP**. And while divide **U** is being destroyed, depression **C** will enlarge to the northeast as well (as it will in other directions, of course). Depressions **F** and **G** will also grow larger. These depression enlargements will necessarily diminish the size and life of the divides that separate them (e.g. the neck between point **D** and crest **E**). Eventually, depression **F** as well as depression **G** will also be incorporated into the ever-lengthening valley we now designate **YP**. After the removal of divide **U**, the valley will become **YC**. Later it will become **YF**. Then **YG**. And so on.

We originally envisioned that stream **QF** (Figure 13–2) had extended itself as a subsequent stream northeastward, from its beginning as a small tributary at confluence **Q**. The headward valley growth by solution that we have just described is a little different, is it not?

Does this mean that our concept of stream **QF**'s headward growth was faulty? It does not. There is nothing in Figure 13–2 to suggest the presence of soluble bedrock beneath the plateau's caprock. Our original hypothesis, based on what we could glean from that map, was perfectly reasonable. All we have done, by incorporating the landforms of this map into our thinking, has been to add another dimension. Now we can say that valley **QF** (Figure 13–2), like its headwater segment **YP,** apparently grew headward by the *combined* processes of solution, and headward stream and slope erosion.

Fort Smith, Arkansas/Oklahoma

(Figure 13–4 in back of book, C.I. 100 feet, and

Figure 13–4a in back of book, C.I. 50 meters, and

Figure 13–5 in back of book, *Geologic Map of Oklahoma*)

Situated just west of the Oklahoma-Arkansas state line, this map area is underlain by gently dipping Pennsylvanian strata. The large rivers are the Arkansas (**J**) and the Canadian (**P**). Letters designate features on Figure 13–4. Numbers designate features on Figure 13–5. (A few letter references appear on both maps, e.g. rivers **J** and **P**.)

Our geologic map, Figure 13–5, shows the distribution of five formations (**Pa, Pb, Pc, Pd,** and **Pe**). It does not denote their relative ages or their lithologies.

What does the topographic map tell us about these formations and their structures? Asymmetries such as **H,** and those at **B** and **M,** reveal gentle local dips to the southwest and west. (Transfer letters **B** and **M** to Figure 13–5.) These dips tell us that the general north-south trends of the formations on Figure 13–5 reflect a regional dip to the west. It is from old (**Pa**) in the east to young (**Pe**) in the west.

High terrain in the northeast (e.g. **K** and **L**) tells us that unit **Pa** is at least in part a relatively resistant formation. This resistance is also reflected in the cuesta at **H** and the high ground at **Q** and **S**. (Transfer these two letters to Figure 13–5.) Some nonresistant members are revealed by topographically low **Pa** areas such as **C, I,** and **E** (indicated on Figure 13–5).

Note (Figure 13–5) that area **H** is bordered on the northwest by fault **5** and on the southeast by fault **7**. The dip at **H** is to the southwest (toward **C**). So the resistant members of **Pa** must be in the lower part of that formation. The less-resistant upper, overlying part must occupy lowland **C**. Areas **I** and **E** bracket area **H,** and are topographically lower. In addition, the top of **Pa** is mapped at points **3** and **6**. But upsection from **C**, it lies far to the southwest, at **13**. It follows that areas **I** and **E** must be underlain by the upper part of **Pa**. It then follows that area **H** is faulted up (a horst) between areas **I** and **E**.

Along scarps **F** and **G,** resistant rocks stand topographically above nonresistant rock areas. From this we conclude that these scarps, which separate cuesta **H** from lowlands **E** and **I,** were developed by differential erosion. They were not produced directly by fault displacement. We see these therefore as fault-line scarps. Since they reflect structure (block **H** is topographically *and* structurally high), we classify them as resequent fault-line scarps.

There is a remarkable coincidence between fault **11** and contour segment **A**. However, keep in mind that (1) a *contour* is the intersection of the ground surface and a *horizontal* plane, but (2) a *fault trace* is the intersection of the ground surface and a *dipping* fault plane. Fault trace **11** probably lies along the *base* of the low scarp, whereas contour **A** runs along the *face* of the scarp. If we had only the topographic map, and believed there to be a fault here, we would map it along dashed line **O** (Figure 13–4), instead of along contour **A**.

How should we classify the southeast-facing scarp that looks down upon fault trace **11** (dashed line **O**) from contour **A**? The rocks on the north side (**Pc**) are younger than those to the south (**Pb**). So, offset must be up on the south and down on the north. A simple fault scarp would face to the north, so this must be an obsequent fault-line scarp.

Note the detailed correlation between the topography along scarp **B** and outcrop pattern **1**. The caprock along this scarp must be the basal resistant part of formation **Pd**. Formation **Pc** must be a nonresistant unit occupying the lower part of the obsequent slope. These same formations can be recognized at **M** and **N,** and at numerous other places to the south of the Canadian River (**P**). (Transfer lettered point **N** to Figure 13–5.)

Between north-facing scarp **F** (along fault **5**) and south-facing scarp **A** is unnaturally curvilinear south-facing scarp **D**. This feature is just as indicative of fault control as any other scarp in this area that has already been mapped as fault-associated. However, according to the geologic map, it is associated not with a fault but with an unfaulted contact (**4**). That map indicates line **4–2–12** to be virtually a strike line (north dip of appreciable steepness). From **4** to **2** this dip is shown to replace the fault (fault **11**). From **2** to **12** it is shown to complement it.

Regardless of what the geologic map indicates, there is an excellent possibility that a continuous fault or fault zone extends from **5** to at least **9**. This is like the fault a few miles to the south, which is continuous from **7** to **8**. Gentle southwest dip causes the similarity between contact **10** and contour segment **A**.

Elsewhere on this topographic map you will recognize other mapped faults (e.g. fault **15** at scarp **X**). However, in other places you will see no topographic evidence of some mapped faults (e.g. fault **14** near point **W**). And in yet other places, the topography strongly suggests faults, but they do not appear on the geologic map (e.g. at points **R, T, U,** and **V**).

Many features and relations remain undesignated and undiscussed. To get the most from these maps, you should make your own independent interpretation. So, as in Chapter 12, proceed to develop your own map. Begin by placing a sheet of Mylar or similar material over the topographic map (Figure 13–4). Trace the three rivers (in blue). Then, on the overlay:

1. Indicate dip and strike wherever possible.
2. Identify the mapped faults that you can "see" in the topography. (Trace in red.)
3. Draw in those mapped faults or segments that have no topographic expression. (Use a different color, or lines of red dots.)
4. Map linear and curvilinear scarps, slopes, or valleys that you believe to be fault controlled, but which have not been mapped as faults. If you can determine displacement, show up and down sides. (Use a different color.)
5. Map as many stratigraphic contacts as possible. This will be difficult in some areas and impossible in others. (Use another color.) Do not be disturbed if you can only partially succeed. (You cannot just put your overlay over the geologic map and trace off the contacts, because the two maps are at slightly different scales.)

Figure 13–4a (also at 1/250,000) illustrates the importance of contour interval. It gives further meaning to our earlier statement that when we "interpret" a topographic map, we are interpreting the topography *as it is depicted* on that map. Though Figure 13–4a is an excellent map, its contour interval of 50 meters (164 feet) renders it considerably harder to interpret than Figure 13–4 (C.I. 100 feet).

Alec Butte, Oregon
(Figure 13–6 in back of book, C.I. 10 feet)

Despite the apparent simplicity of the geology and topography, this map poses some perplexing questions:

- ☐ Why are there undrained depressions (**1, 2, 3,** and **4**) in this small area?
- ☐ What is the nature of the linear and curvilinear scarps and slopes (e.g. **D, A,** and **M**)?
- ☐ What is the significance, if any, of these scarps and slopes being mapped as undissected?
- ☐ Why does scarp **R** face to the east, whereas the northwestern extension of this same topographic trend is west-facing scarp **O**?
- ☐ Why does transverse stream **I** cut across the well-defined northwest-southeast topographic grain? It not only flows across the grain; it actually cuts water gaps through four topographic obstacles (gaps **L, G, H,** and **B**). Why does it do this instead of avoiding such features?
- ☐ Is there any connection among these features?

Let us proceed cautiously and see what we can come up with. Instead of beginning with our first question (the depressions), let us consider the scarps. They strongly resemble the fault scarps in the Tulelake area of northern California (Figure 11–3 in back of book). If these are fault scarps, does that let us tie together the other features?

Yes. Geologically recent block faults could have produced such scarps. The lack of dissection can easily be explained by the brevity of post-faulting time available for dissection. It could also explain why the linear scarps and slopes "come and go," descending and losing identity at their ends.

For example, we can trace scarp **J** southeastward only to point **K**. Immediately beyond this, there is no suggestion of such a topographic or structural feature. A fault with diminishing throw southward (from over 50 feet at **J,** to zero at or near **K**) would produce this topography. Block faulting would also explain the undrained depressions.

But what about east-facing scarp **R** and west-facing scarp **O**? Imagine a line from point **F** southeast to point **P,** to indicate an up-on-the-east fault. Could you visualize the fault losing its throw at or near **P**? Then, imagine continuing the fracture southeastward to **S,** but with up-on-the-west throw, the throw increasing southeastward from zero at **P** to a maximum near **R**. Would you have, as a consequence, an east-facing fault scarp such as the one we have here (scarp **R**)? Have we not just described a *scissor fault*?

How might faulting relate to stream **I**? If the scarps were appreciably dissected, they might be seen as fault-line scarps. In this case stream **I** could be superposed across a partially exhumed fault topography. However, the scarps are not dissected, which weighs against their being fault-line scarps. They are also associated with numerous nearby depressions. It is difficult to propose an exhumation that permits downcutting to continue below the local external base levels. Deflation could be involved, but these depressions are huge features. So it is not likely.

Let us turn back the clock, prior to block faulting. A northeast-flowing stream, the "ancestral **I,**" crossed this area from **N** to point **C** and beyond. Then the block faulting occurred, progressing so slowly that the stream was able to maintain its course by cutting virtual canyons through the rising blocks. The depressions are simply down-dropped and/or down-tilted blocks. They lie just far enough from the stream and its tributaries so they are not drained by them.

On the other hand, if the stream had not maintained its course, there would also be large, undrained depressions at **E** and **Q**.

House Rock Spring NW, Arizona
(Figure 13–7 in back of book, C.I. 40 feet)

This area lies to the north of that depicted in Figure 10–4 (Jacob Lake, Arizona, in back of book). The *AAPG Geological Highway Map of the Southern Rocky Mountain Region* shows that the Permian limestone displayed in the Figure 10–4 area also forms the surface bedrock here, in the upland bordered by scarp-slopes **OB** and **TMF**.

The Colorado Plateau is renowned for its monoclines (such as those of Figure 10–4) and block faults. It is not unusual to discover that a fault's throw is replaced, along strike, by monoclinal folding. We seem to have an example of such structural transition here.

Scarp **TM** has all the characteristics of a fault-associated scarp. It is curvilinear. It is steep. And it is closely paralleled by drainage divide **SJ**. In some ways divide **SJ** resembles remarkably divide **AE** in Figure 10–1 (Cambridge NE, Kansas, in back of book). Similarities include the pronounced asymmetry of divide **SJ,** and the uniformly gentle westerly slopes, which appear to be resequent or even dip slopes. Since **S** is higher than **J,** it is unwise (and incorrect) to assume that line **SJ** is a strike line. Thus we have indicated probable dip (arrows **R** and **I**) to the northwest, rather than due west.

Scarp **OB** also appears to be fault-associated. The offset or jog near **A** introduces the possibility of a fault belt or zone, rather than a single steeply dipping fault. It looks like an en echelon structure. Note that this scarp, unlike scarp **TMF,** does not have a nearby accompanying divide. In fact, the gentle upland in this western area also slopes toward the northwest. We interpret this slope as a reflection of gentle northwest dip (arrows **P** and **C**). Thus we see block **OBFMT** as a single tilted fault block.

1. If both border scarps are either fault scarps or resequent fault-line scarps, the block is a *horst*.
2. If they are both obsequent fault-line scarps, it is a *graben*.
3. If one scarp is an obsequent fault-line scarp, and the other is either a fault scarp or a resequent fault-line scarp, the block is merely a *tilted fault block*.

We suggest you draw a simple cross section to show these three possible structure/topography relations.

Cut deeply into the upland surface are many northwesterly flowing streams. They generally parallel dip direction. Are these consequent or resequent streams? Until we know the age of the faulting and its relation to the stripping of the limestone, we cannot even guess.

It is interesting, though, that some streams, like **QH,** not only flow obliquely to dip direction, but also flow in anomalously linear valleys. These streams must be adjusted to some north-south structural control. A possibility is subsidiary faults that parallel the main border faults.

Classification of the border scarps, upland streams, and upland surface is a complex task. You may do this solely from what you can see on this map, plus the sparse information we have provided. Or, you may consult references to the area, and improve your chances of correct interpretation.

What is the nature of east-facing slope/scarp segment **MF**? For a short distance north from **M,** this feature is steep enough to be a scarp. But over most of its length it is more moderately inclined and should be thought of as a *slope*. Like scarp segment **TM** this slope is accompanied by a nearby parallel divide (**LE**). Like divide **SJ,** divide **LE** is asymmetric in profile.

We can "see" the trace of the border fault associated with scarp **TM,** but no such fault trace is suggested by slope **MF**. Instead, this slope segment looks like a dip slope (or a near-dip resequent slope). We would map a fault along line **TM,** but not extend it northward beyond **M**. Instead, we would map a questioned anticlinal axis along line **LE,** with gentle plunge to the north-northeast.

We propose a classification of resequent for these east-dip slopes (e.g. **D**). Perhaps this will give you ideas about classifying scarps **OB** and **TM**.

Consider a couple of minor topographic features, slope **U** in the east and slope **N** in the west. These slopes, particularly slope **N,** meet their respective fault-associated scarps (e.g. **OB**) at a sharply defined angle. Are these structure-controlled surfaces, or do you have some other explanation? What is the significance, if any, of depression **K** in the east?

Finally, we come to this map area's most obvious and tantalizing feature: the sinuous, meandering course of canyon **MG,** cut through upland **OBFMT** by the main west-flowing stream. Is this the gorge of an antecedent stream? Is it the result of superposition through a now-destroyed angular unconformity? Or could this be the product of some other series of events?

This problem can be resolved only if we know the answers to the other major questions—those about scarps **OB** and **TM** and slope **MF;** about the broad, gentle slopes (e.g. **R**); about the northwest-flowing streams; about linear streams such as **QH;** about the asymmetry of divides **SJ** and **LE;** and about possible east dip **D**!

There are too many things about this area and its geomorphic history that we do not know, and that this map cannot tell us.

Damascus, Arkansas

(Figure 13–8 in back of book, C.I. 20 feet)

From their appearance in this limited area, we cannot tell whether stream **B** is tributary to stream **O,** or vice versa. On the other hand, we can see the discordance between stream **B** (and most of stream **O**) and the geologic strike or grain. Strike is slightly to the north of east in the center of the map (ridge **W**), and in the north (ridge **N**). Strike is somewhat south of east in the south (ridge **d**). Note that divide **W** is parallel to divide **N,** but not to divide **d**. (See Figure 13–8a, a sketch of the area viewed from the west.)

Before we speculate on what produced this discordance between the structure and the two major streams, let us look more closely at the structure. Though interrupted by the gap cut by stream **B,** the asymmetry of divide **AN** is preserved. This asymmetry is clearly the topographic response to the gentle southeast dip of a resistant formation. Dip is indicated by three long southeasterly slopes (arrows **C, J,** and **M**).

An independent indication of southeast dip is a well-defined, if sinuous, outcrop V. It is the pattern of sharp contour bends along the curved descending line **DF,** and its counterpart across the stream, descending curved line **IEG**. (Both are defined by dashed lines.)

It is both interesting and significant that dip slopes **C, J,** and **M** are so long, whereas dip slope **K** is noticeably shorter. It is cut off by the scarp that

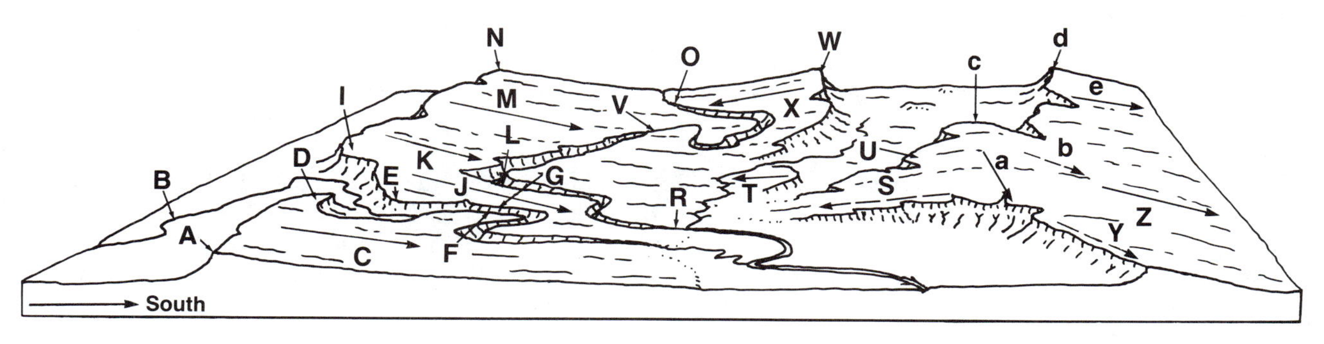

FIGURE 13–8a
Block diagram of part of the area depicted in Figure 13–8, viewed from the west.

drops down to bend **L** of stream **O,** a drop of over 150 feet. Along most of its length, such as at stretches **V** and **R,** stream **O** is not so deeply incised. We shall refer back to these relations.

The asymmetry of divide (cuesta) **W** shows north dip (arrow **X**). This, together with south dip **CJM,** defines a gentle, broad syncline. It extends across the entire map area, its axis passing through points **R** and **V**.

The dip along divide **d,** however, is back toward the south (arrow **e**). This places another axis, an anticline, between ridges **W** and **d**. The shape of the closed contours at **f,** and the asymmetry of that hill, indicate southwest dip (arrow) at this point.

Refer to water gap **BK** in Figure 9–1 (Cumberland, Maryland, in back of book). You will again see how the outcrop V's point both east and west. They combine to disclose the anticlinal structure of the main ridge in that map area. The structure is revealed by lines of sharp contour bends (**EB-BC** and **DK-KF**).

Similar lines defined by contour bends are featured on our present map. Two especially important ones are identified by broken arrows **Y** and **S** (not dip arrows). These arrows designate *apparent* or *component* dip directions. Apparent dip **S** is not far from true local dip (direction shown by nearby arrow **T**). True dip direction near **Y** is shown by dip arrow **Z**.

The erosional pattern around topographic prominence **c** is confusing. We might imagine a gentle dip, as shown by broken arrow **b,** at right angles to "strike" **Uc**. However, line **Uc** is *not* the strike. It cannot be, since **U** is at least 40 feet lower than **c**. Look at dips **Z** and **T** and component dips **Y** and **S,** and consider that line **Uc** must be a component dip toward the northwest. Therefore, point **c** must be on the anticlinal axis. Thus dip from point **c** is very close to, if not exactly, due west (arrow **a,** the *plunge*).

Streams. Let us return to stream stretches **V** and **R**. Along these, stream **O** is not deeply incised into the surrounding terrain. There are no deep valleys along these segments. This is because the resistant formation responsible for dip slopes **J, M, T,** and **X** is structurally low along the trend **VR**. Remember that **VR** is the approximate trace of the synclinal axis. The resistant unit is near stream level at **R,** and not much above stream level at **V**. Along trend **VR** the entire interfluve area is topographically low.

Our final problem is the discordance between major streams **B** and **O,** and the geologic structure and its topography. Could these be antecedent streams? Or are they more probably superposed? If the latter, from what depositional or erosional surface? When confronted by such questions and such topography, we cannot use the topographic map to provide all the answers.

Except for the valleys of the main drainage lines, the topography of this area is remarkably adjusted to geologic structure and lithology. This tells us that the main streams have followed their present courses for a long time, during which the overall topographic-geologic adjustment became firmly established.

If you consult geologic references to this area, you will learn the age of the bedrock, the age of the folding, and whether there are unconformably younger strata nearby—rocks through which these streams may have been superposed. Or could they merely have been superposed from a concordant sequence of overlying nonresistant strata?

More Puzzles. Before leaving this map, consider the following:

1. What stratigraphic relation exists between cuesta divide **Q** and divide **N**?
2. Is **P** a dip slope? A resequent slope? Both?
3. What is the dip direction at **H**?
4. Is the rock at **H** older than, younger than, or the same age as that at **N** (assume no strike faults)?
5. Is the resistant unit at **f** older than, younger than, or the same age as that at **d** (assume no strike faults)?
6. Is the resistant unit at **f** older than, younger than, or the same age as that at **H**?
7. How many resistant formations can you identify in this map area? Label them chronologically, starting with 1 for the oldest.

Hilton, Virginia

(Figure 13–9 in back of book, C.I. 20 feet)

This map and the following three are all from the Kingsport–Gate City–Clinch Mountain area of southwesternmost Virginia. This is an area where the topography is intimately adjusted to the geology. It is also an area where some streams are in flagrant disharmony with the geology, and with their adjusted neighbors as well.

We will study one of these disharmonic streams in Figures 13–10 and 13–11. In Figure 13–12 we will use the adjustment of the topography to the geology to solve a geologic problem. However, since we cannot identify the anomalous until we have established the normal, we will concentrate on the topographic-geologic union so beautifully displayed on this map.

The *Geologic Map of Virginia* shows six sedimentary formations cropping out in this map area, ranging in age from Lower Paleozoic to Mississippian. Some are thick and some are thin. Some are homogeneous and some heterogeneous. Some are resistant and some are nonresistant. One is soluble and the others are not. They all dip in the same direction at about the same angle. They are (though not in chronological order):

Unit 1. Relatively thin, fairly resistant sandstone.
Unit 2. Thicker, more-massive, more-resistant sandstone-conglomerate unit.
Unit 3. Thin, extremely nonresistant mudstone.
Unit 4. Extremely thick sequence of interbedded thin shales, siltstones, and limestones.

Unit 5. Thick carbonate sequence.
Unit 6. Nonresistant, thin-bedded shale unit, considerably thicker than Unit 3, which it does not directly overlie or underlie.

The strike is obvious, being well defined by ridges **AB** and **PO,** valley **CL,** linear tributary belt **UN,** and generally by the east-west meandering course of stream **V**. Our first problem is to determine dip direction. The best clues are the asymmetry of ridges **AB** and **PO,** and the south-pointing V's (dashed lines) in water gaps **S** and **I**.

The V's are not essential, however. Ridge **AB**'s asymmetry consists of more than greater steepness of the north flank in contrast to gentler slope of the south. The north flank is also concave in profile, whereas the south flank is uniform. The north flank is much-more intricately dissected; its drainage density is much higher, its texture much finer. All in all, the north flank's topography is one produced by rapid downcutting and backwasting. It is the product of an oversteepening typically found on slopes that are developed on nonresistant rocks that are capped by a resistant rock layer.

In contrast, the uniform south-flank slope suggests stripping of overlying, nonresistant beds from the top of a dipping, resistant formation. This is either a dip slope, or very close to one; dip may be slightly steeper than slope. (Compare this to Figure 8–5b, which shows a greater discordance between dip and resequent slope than seems to exist in this area.)

Note the wider spacing of the parallel and subparallel streams that flow down the south flank of ridge **AB**. This need for larger (and hence fewer) drainage basins is partly due to the greater resistance of the rocks on this flank.

Everything we see along ridge **AB** points to south dip. Its great height means that it must be developed on resistant Unit 2, which underlies the entire south flank and the crest of the ridge. The south flank (the resequent slope) is drained by resequent, south-flowing streams.

The north flank (the obsequent flank) is drained by obsequent streams. It is developed on a thick sequence of nonresistant rocks. As a whole, they are certainly not carbonates. Unit 4 is the only candidate for this position. This unit reportedly contains some limestone interbeds. But apparently they are too few and too thin to develop sinks that the 20-foot contour interval can show. The high drainage density, small basin size, and oversteepening all identify a rock sequence with high shale content. Thus, by elimination, this must be Unit 4.

Since we have assigned Unit 2 to ridge **AB,** ridge **PO** is easy to identify. It must be capped by the sandstone of Unit 1. The sandstone occupies its south flank and forms the outcrop V's at gaps **S** and **I**.

The large karst area south of ridge **PO** must be the outcrop area of Unit 5. Unit 6 occupies strike valley **CL,** including all but the very top of the obsequent (north) flank of segmented ridge **PO**. Its contact with the caprock (Unit 1) is at or near point **H**.

But where is thin Unit 3, the nonresistant mudstone? This unit is reportedly not in contact with Unit 6, which underlies lowland belt **CL**. So Unit 3 must outcrop either north of ridge crest **AB** or south of ridge crest **PO**. If it underlies Unit 2 (crest **AB**), it is beyond our grasp. Nothing north of ridge **AB** suggests a separate nonresistant unit in the general Unit-4 area (all of Unit 4 is nonresistant).

Might it overlie Unit 1? The several linear subsequent streams (e.g. **J**) in narrow, straight, lowland band **UN** stand out in sharp contrast to both the

TABLE 13–1
Summary of Formations.

Unit	Description	Age	Direction
Unit 4	Extremely thick sequence; interbedded thin shales, siltstones, and limestones	Lower Paleozoic	North
Unit 2	Thicker, more-massive, more-resistant sandstone-conglomerate	↓	↓
Unit 6	Nonresistant, thin-bedded shale		
Unit 1	Relatively thin, fairly resistant sandstone		
Unit 3	Thin, extremely nonresistant mudstone		
Unit 5	Thick carbonate sequence	Missis-sippian	South

adjacent ridge **PO** (Unit 1) and the disorganized hills, knobs, and depressions of the Unit 5 limestone area to the south. These linear streams have obviously developed along a strip of upturned, nonresistant, nonsoluble bedrock which we will designate Unit 3. The formations are summarized in Table 13–1.

Geomorphology and Streams. Except for the narrow, 200-foot-deep wind gap at **Q,** the crests for ridge **PO** are remarkably uniform. Might stream **D** once have flowed southward through this gap to join stream **V** at or near **R**? If so, why does stream **D** now turn eastward to flow through water gap **S**?

Why does meandering stream **V** not wander across a flat floodplain instead of following its serpentine valley? It may be pertinent that valley **CL** contains a series of benches (e.g. **E, F, G,** and **M**). Each is a remnant of what must have been a continuous, nearly flat lowland floor, extending along belt **CL**. If we can picture stream **V** meandering across a floodplain, 200 feet above its present channel, we can also envision a similarly higher erosional surface in nearby shale belt **CL**. (Compare to Figure 12–7—Buffalo Gap, South Dakota, in back of book, and Figures 12–7a and 12–7b. These depict another such surface in a much more arid setting.)

Stream **V** is an incised river. Could its incision (rejuvenation) have set off a chain reaction of rejuvenation in its many tributaries? The extreme nonresistance of Unit 6's shale explains the broad, gentle erosional slopes like **K** along lowland **CL**. As time passes, low surface **K** and its counterpart around **S** will grow more extensive as the terrace remnants are more completely consumed by erosion. Finally the remnants will be destroyed, and nothing will remain but the new broad, low surface. It will be a replica of the one that is now being chewed up.

Before we move on to our next map, consider incised meander bend **T**. It swings northward to encounter sandstone Unit 1 in gap **S,** at the point of the outcrop V. If this meander were again rejuvenated, as it cut down it would encounter that sandstone in its channel. If it cut into the resistant sandstone, could it escape the sandstone's grasp? Try to visualize this event. We shall refer back to this during our analysis of Figure 13–11.

Exercise. Draw a north-south topographic profile along the line through points **G** and **U**. Show the relation between the geology and topography by adding the six formations to the profile.

Mendota, Virginia

(Figure 13–10 in back of book, C.I. 20 feet)

The geomorphic units of Figure 13–9 (Hilton, Virginia, in back of book) extend northeasterly across this area. But our labeling is different from that area to this area. The labeling on each map is as follows (Figure 13–9 before slash/ Figure 13–10 after slash): ridge **AB/AB** (same), valley **CL/RH,** ridge **PO/SJ,** and valley trend **UN/TI**.

The six formations identified on Figure 13–9 are designated by the same numbers on Figure 13–10 (refer to Table 13–1).

In Figure 13–9 we were confronted with a geomorphology/geology correlation. Our interest here is the anomalous course of a single stream, major river **V**. What is anomalous about stream **V**?

In Figure 13–9, this stream flows along a belt of nonresistant (soluble) limestone. Its meandering course parallels the strike of that limestone, nearby ridge **PO,** shale belt **CL,** and more-distant ridge **AB**. There is nothing noteworthy about a subsequent stream that flows along the strike of upturned, nonresistant strata. So, very little was said about stream **V** in that map area. Except that it is *incised*.

Note the contrast of stream **V**'s course in Figure 13–10 (also depicted in Figure 13–10a, a view of the stream valley from the southwest). At **G,** stream **V** leaves the limestone-mudstone area. It flows through gap **F** in ridge **SJ**. Continuing across the shale lowland to **D,** it encounters the base of the dip slope (or resequent slope) of sandstone **2**. It then turns and flows obliquely (**C**) along lowland **RH** to **P**. Here it undercuts the obsequent slope of ridge **SJ** (capped by sandstone **1**). At **P** it turns back along strike to the northeast, to gap **F**. It continues back through that gap into the limestone-mudstone lowland (at **Q**), along which it had been flowing in the first place.

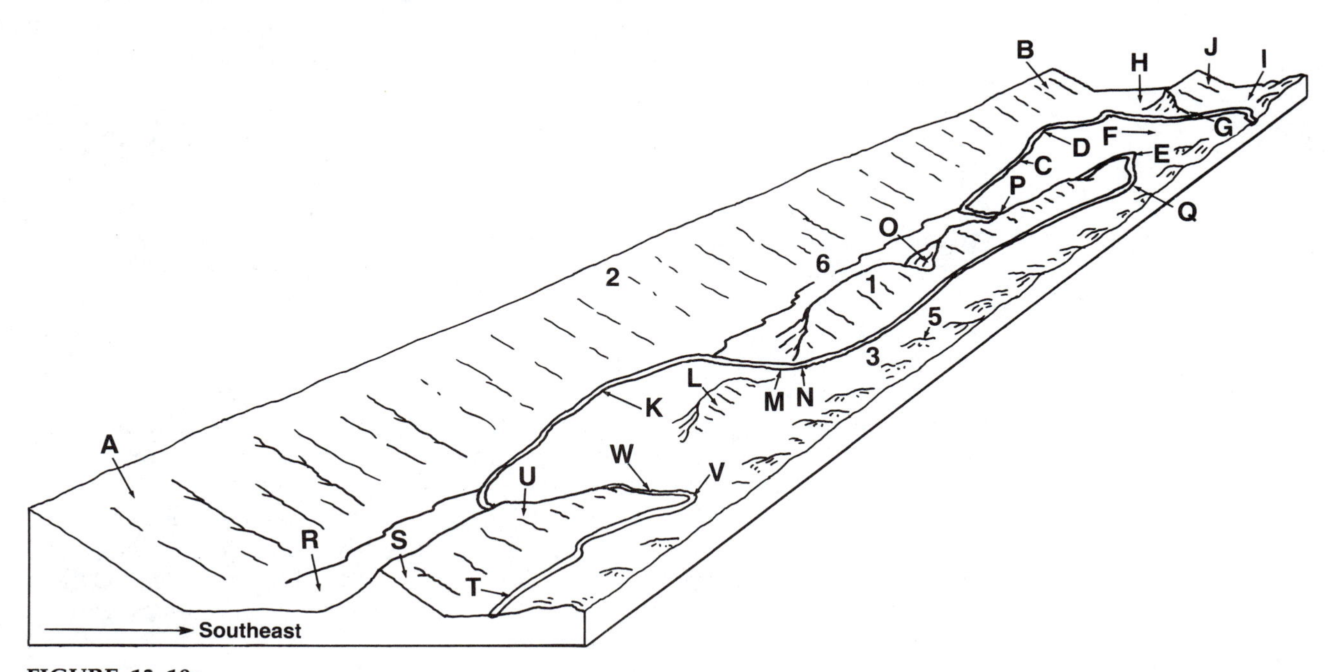

FIGURE 13–10a
Block diagram of part of the area depicted in Figure 13–10, viewed from the southwest.

We refer to the ballooning of the stream across area **FDPF** as the "Mendota Hernia." There is nothing anomalous about the ballooning per se. The stream was able to sweep laterally across the nonresistant shale, once it got through the sandstone barrier. What is anomalous is the stream's leaving the easy going of the limestone and mudstone, *to cross a resistant obstacle,* to enter the shale. To further complicate matters, the stream turns around and returns to the carbonate lowland *through the very same gap*! This is like the little child who has run away from home, only to return when nightfall approaches, while a lonely owl salutes the stillness.

From **Q,** the stream continues along strike—actually, along the narrow outcrop band of the Unit 3 mudstone—to point **N**. Here it turns yet again, back into and through another gap (**M**) in ridge **SJ,** into shale belt **RH**. In this shale area, stream **V** has swept out another broad floodplain, across to **K** and back to gap **W** (this time, not the gap through which it entered!). Just outside of gap **W,** at point **V,** stream **V** turns southwestward. It continues its original subsequent course, which it left twice and returned to twice, within this small map area.

Is stream segment **MKW** another **FDPF**? Yes and no. Each is a course that progresses through limestone-sandstone-shale-sandstone-limestone. However, in one case the stream flows through two separate gaps. In the other case, the stream flows into and out of the shale area through the same gap.

Can we formulate a single hypothesis to explain both anomalies? Or must we conclude that they were produced by significantly different sequences of events? Perhaps we should select one of the two features, explain it, and then try to explain the other in the same way.

In Figure 13–9, at point **T,** stream **V** swings over against the sandstone of ridge **PO**. The greater resistance of the sandstone may deter the stream from further lateral migration to the north. But the possibility exists that the stream will cut laterally into the sandstone, and eventually through it. It is not very thick. If it can cut through the sandstone and into the nonresistant shale, it will be like the running back who has penetrated a wall of blitzing linebackers: full speed ahead! Herniation!

The in-and-out course through gap **F** (Figure 13–10) could have developed by just such lateral cutting. In this case, it was through sandstone **1** into shale **6**. This is a possible explanation.

What about area **MKW**? In this instance we must contend with ridge segment **L,** which separates gaps **M** and **W**. Hilltop **L** and ridge crest **U** are both more than 300 feet above the stream (where it flows through gap **W**). Though we may still not know how the stream got across the sandstone into area **FDPF,** one thing we do know is that it has cut down at least 300 feet since it accomplished that entry.

Let us return to area **FDPF**. What would happen at point **F** if stream **V** were rejuvenated? At **G** and **E,** the stream would be incised into the sandstone, which it now crosses just below water level. As downcutting progressed at these two points, it would not be progressing at point **F,** would it? Over time, a small hill would appear at **F**. It would "grow" as more downcutting occurred at **G** and **E**. Could this be how isolated hill **L** came into existence? After rejuvenation, stream **V** would flow northwestward through "gap **G**" and back southeastward through "gap **E**," the counterparts of present-day gaps **M** and **W**.

Does this mean that loop **MKW** and gaps **M** and **W** were produced by such a sequence of events, but at an earlier time? Maybe. We cannot be certain.

There is another possibility. Please refer once again to meander impingement **T** in Figure 13–9. We have already considered that meander loop **T** might continue northward laterally, through the sandstone and into the underlying shale. What would happen at **T** if, instead of migrating laterally north, the stream, responding to rejuvenation, were merely to cut down? Would it not then become lodged in the sandstone? And, if downcutting continued, would it not eventually cut through the sandstone, into the underlying shale?

We may now proceed to Figure 13–11. Perhaps it will contribute to our understanding (or to our confusion).

(However, one parting question: how would you explain wind gap **O**?)

Brumley, Virginia

(Figure 13–11 in back of book, C.I. 40 feet)

This map displays the same geomorphic and geologic units that we studied in Figures 13–9 (Hilton, Virginia) and 13–10 (Mendota, Virginia, both in back of book)—except one. Unit **4** lies to the north of this map area.

Three small triangular hills **N, P,** and **S** and gap **O** reinforce some of our speculations regarding the anomalies in Figure 13–10. The north-south asymmetry of hill **N,** combined with its uniform south slope, identify it as a small flatiron-like hill or cockscomb. It is the type often developed by segmentation of a homoclinal ridge or hogback. Its south slope is a resequent slope, as are the south slopes of strike-ridge **J** and hogback **CF** (note dip-slope arrow **D**).

The outstanding thing about hill **N** is not its production by differential erosion. The notable point is that hill **L,** to the north of valley **M,** is upheld by the same sandstone that forms hill **N**. Valley segment **R** separates sandstone-capped hill **Q** from hills **P** and **S,** which are also developed on the same sandstone. It is the same way that valley segment **M** separates hill **L** from hill **N**.

Unlike linear, subsequent lowland belt **HE,** which is developed along the nonresistant shale of Unit **6,** valley segments **M** and **R** are cut down into the south slope (resequent or dip) of a sandstone ridge. These are definitely not subsequent valley segments! The geomorphic/geologic relations in this critical area are depicted in Figure 13–11a, a sketch of the area viewed from the southwest.

Stream **U** flows in shale along these segments (**M** and **R**), but it is not "adjusted" along the shale. It did not extend its head along this band of nonresistant rock. Instead, it simply cut down through the dipping sandstone into the underlying shale.

How did stream **U** become established along segments **M** and **R**? How was it able to get as far north as **O** and **I**? The only way stream **U** could have cut valley segments **M** and **R**, as confined as they are, was for it to have cut vertically from directly above. That is obvious.

Thus, the question changes: how did it come to be above **M** and **R** in the first place? It could only have migrated laterally northward to **M** and **R** at a time when the limestone lowland surface was much higher. It would have been at or above the tops of present hills **N** and **P,** when the limestone (**5**) and mudstone (**3**) still lay upon the sandstone that now forms these hills.

Consider valley segment **TG**. The limestone terrain to the south does not rise very high above the stream. Along this segment the stream is not flowing in the sandstone, but in the mudstone and lowermost limestone that overlie the sandstone. Yet even slight downcutting would lower the stream into the sandstone. We can imagine area **NPS** resembling area **TG** prior to the downcutting that produced valley segments **M** and **R** and the three triangular hills.

If downcutting continues, what will happen to hills **N, P, S, L,** and **Q**? They will not all suffer the same fate. Hills **L** and **Q** are capped by isolated

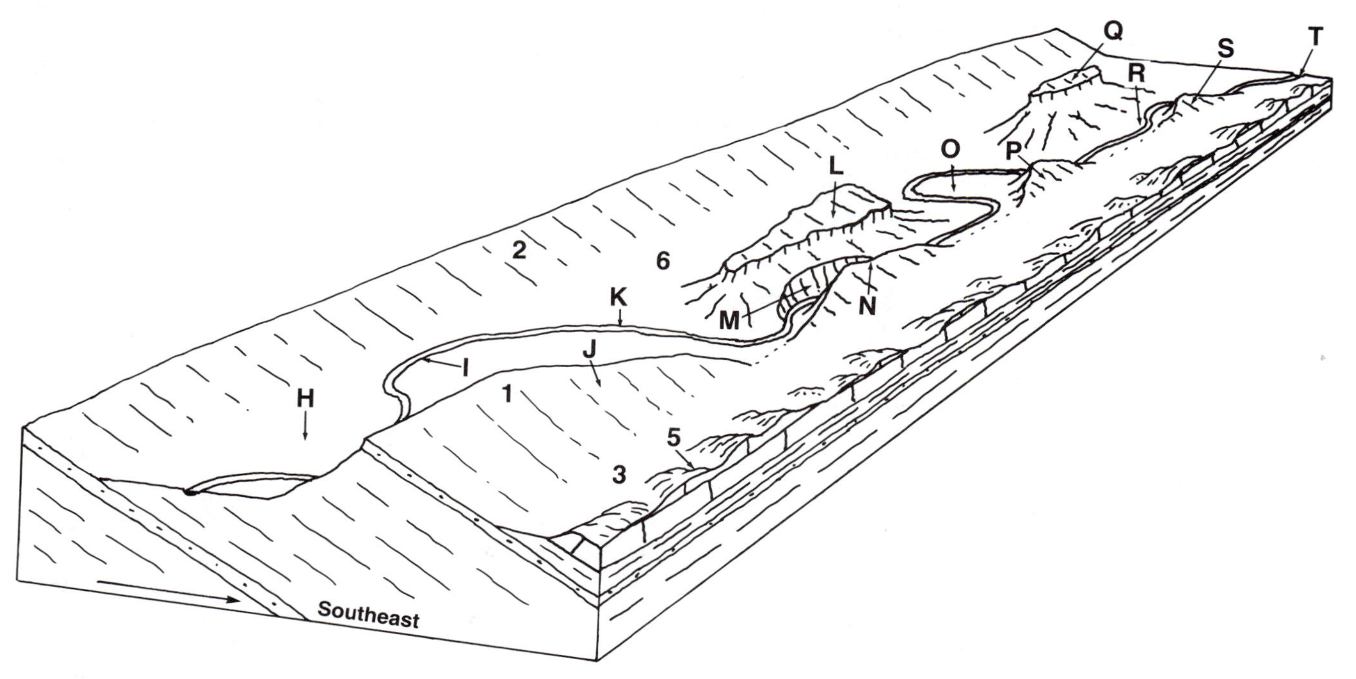

FIGURE 13–11a
Block diagram of part of the area depicted in Figure 13–11, viewed from the southwest.

outcrops of sandstone. But hills **N, P,** and **S** are the edges of dipping sandstone that continues underground to the south, beneath the limestone lowland. Downwasting, backwasting, and lateral corrasion will destroy hills **L** and **Q**.

On the other hand, lowering valley **RM** and the limestone lowland to the south, accompanied by much-slower lowering of hilltops **N, P,** and **S,** will result in the growth of these three ridge segments. In time, they will be joined into a single homoclinal ridge or hogback, comparable to ridge **CF** in the east and ridge **J** in the west.

The important thing is this: with remnant hills **L** and **Q** destroyed, and with a continuous hogback created along the line **NPS,** the entire segment **ROMK** of stream **U** will be flowing along the shale belt. It will be north of that future sandstone hogback, and separated by that ridge from the limestone in which it originally flowed.

Can you see how these observations of the present, and visualizations of the past and the future, afford an understanding of the geomorphology depicted in Figure 13–10?

Hilton, Virginia

(Figure 13–12 in back of book, C.I. 20 feet)

This map area lies in the same general region as Figures 13–9, 13–10, and 13–11 (Hilton, Mendota, and Brumley, Virginia, all in back of book). But this area's stratigraphic units are all of *Cambro-Ordovician* age, much older than the surface units in the other map areas.

Let us use this map as a basis for constructing a geologic map. We will achieve some success, but experience some fustrations. Let us see how far we can go.

In certain parts of the area we can see dendritic drainage patterns. But the area as a whole displays a well-developed grain, trending about east-northeasterly (apparently, the regional strike). Using similarities within areas, and differences among areas, we can divide this map into several lithologic bands or belts.

The most outstanding is narrow band **JK,** across the middle of the map. It rises to the same general level as the broader belt **PL** immediately to the south. But it nevertheless stands well above lowland **IH,** which borders it on the north. So it is apparently of medium resistance. What sets band **JK** apart is its unique dissection. Were it distributed over a broader area, it would undoubtedly be cut by a dendritic system of closely spaced streams (high drainage density; fine texture). Shales and shaly strata typically display such drainage characteristics.

There are two belts that reveal soluble bedrock. These are lowlands **IH** in the north and **OT** in the south. Limestones are certainly soluble in Virginia, so we may classify these as probable limestone areas.

Area **PL** is quite broad, since its topography extends southward to include hilly trend **SN**. Within area **PLNS,** however, we have what must be a narrow band of steeply dipping nonresistance, etched out by subsequent stream **QM**. This could be a steeply dipping strike fault, or a particularly nonresistant, thin member of the formation underlying this rolling upland.

Interestingly, upland **BCG** in the north is also cut by a thin band of nonresistance. It has been selectively eroded by subsequent stream segments **D** and **E**. Note that upland **BCG** exhibits a modest amount of solution (e. g. sinks **A** and **F**), and that similar depressions occur in the south (at **R** and at **S**). The relief, drainage density, slopes, and overall topographic "signature" of upland **BCG** are remarkably like those of upland **PLNS**. Both uplands are bordered on the south by limestone lowlands.

What Do We Know? These are our observations, and a few tentative geomorphology/geology suggestions. What do we *not* know? What do we *think* we know?

We know that this area is underlain by a thick sequence of steeply dipping strata that strike east-northeast (unless the area is cut into parallel slices by a series of steeply dipping faults that strike that direction). We know that there are two bands of soluble rocks, most likely limestones. We know there is one narrow belt of (probably) shale or a shaly rock. And we know that there are two bands of fairly resistant, nonshaly, largely nonlimestone bedrock. The latter display topographic characteristics too "weak" to be caused by thick sandstones or conglomerates. We cannot rule out some limestone members in these strata, in view of the few depressions in each.

We do *not* know the dip direction! Hence we do not know the relative age of each stratigraphic unit. According to the *Geologic Map of Virginia*, there are only three geomorphically mappable stratigraphic units that crop out in this area. They are, from oldest to youngest: (1) a shale unit, (2) a thick dolomite sequence with some thin limestone and other interbeds, and (3) a thick limestone unit. That map tells us that all the strata dip in the same direction. (For the present, we will not divulge what that direction is.)

In a normal sequence, dip should be from oldest (shale) to youngest (limestone): downdip/upsection. This distribution is true in the southern part of the map, where we have what must be the shale (**JK**), dolomite[1] (**PLNS**),

[1]Dolomite is much less soluble than limestone. This accounts for the paucity of sinks in area **PLNS** (Figure 13–12). Its tendency toward fairly high permeability (many fractures) is typically reflected by low drainage density and coarse drainage texture. It is also more resistant than limestone.

and limestone (**OU**), arranged north-to-south. We cannot "see" south dip in the topography (e.g. asymmetries). But we can see it in the distribution of geomorphic units.

However, we run into a problem in the north. Northward from the shale (oldest) is the limestone (youngest) and, beyond that, the dolomite (of intermediate age). The line separating shale **JK** from limestone **IH** must be the trace of a fault, along which the south block has moved up, and the north block down. We only need the one fault. The relative positions of the northern dolomite and limestone terrains are explained by the south dip in that area. We have established that the dip in the south is to the south.

We cannot determine the dip direction of the fault plane. But we can tell from its linear trace that its dip is steep. This may be a steeply north-dipping normal fault, or a steeply south-dipping reverse fault. Whichever it is, the north-facing shale scarp that overlooks the fault trace must be a resequent fault-line scarp. The reasons are: (1) the faulting is too old for the scarp to be a fault scarp (displacement almost surely occurred late in the Paleozoic); (2) the shale topography overlooks the terrain underlain by the nonresistant limestone; and the scarp faces the down-dropped block, as did the long-since-destroyed original fault scarp.

Exercise. Draw a north-south topographic profile and geologic cross-section along a line through points **D** and **T**.

Bray, California

(Figure 13–13 in back of book, C.I. 100 feet—solid contours, and 50 feet—dashed contours)

The *Geologic Map of the United States* shows the exposed bedrock to consist of Late Tertiary and Quaternary volcanics and interbedded continental deposits.

We are not concerned with the two large circular mountains **h** and **c,** obviously volcanic cones. More outstanding, from the viewpoint of geomorphology/structure/history, are the clearly defined, east-facing, north-trending scarps, such as **gO** in the west. Two of these scarps are depicted in Figure 13–13a, a sketch of the eastern part of the area viewed from the south.

Three of the scarps (**ji, ob,** and **KM**) extend into the cones. In fact, scarp **KM** is restricted to the flank of cone **c**. On the other hand, segment **of** of scarp **ob,** and the southernmost segment of scarp **ji,** lie in the lowlands to the south of cones **c** and **h**.

Scarp **lF** is perhaps the most fascinating. Along segment **lT** it rises to 100 feet, though most of that segment is lower. Parts of segment **TS,** on the other hand, rise to over 250 feet. A short, low, east-facing scarp is indicated by the single contour along line **DC**.

Except for knowing this to be an area of young volcanic materials, this topographic map is all we have with which to decipher the geology. What does it tell us?

It tells us this is not an area of extensive erosion. Contour bends that indicate lowland drainage lines (e.g. bends **Q** and **a**) also indicate that these streams do not occupy deep valleys. The map also tells us that this is an area of widespread extrusives. The prevalent crenulated character of the contours (e.g at points **A** and **P**) shows a terrain of disorganized roughness typical of such areas.

Linear "valley" **bf** (dashed line) is a valley, but must all valleys be created by downcutting streams? Valley **bf** is asymmetric in profile, having one ex-

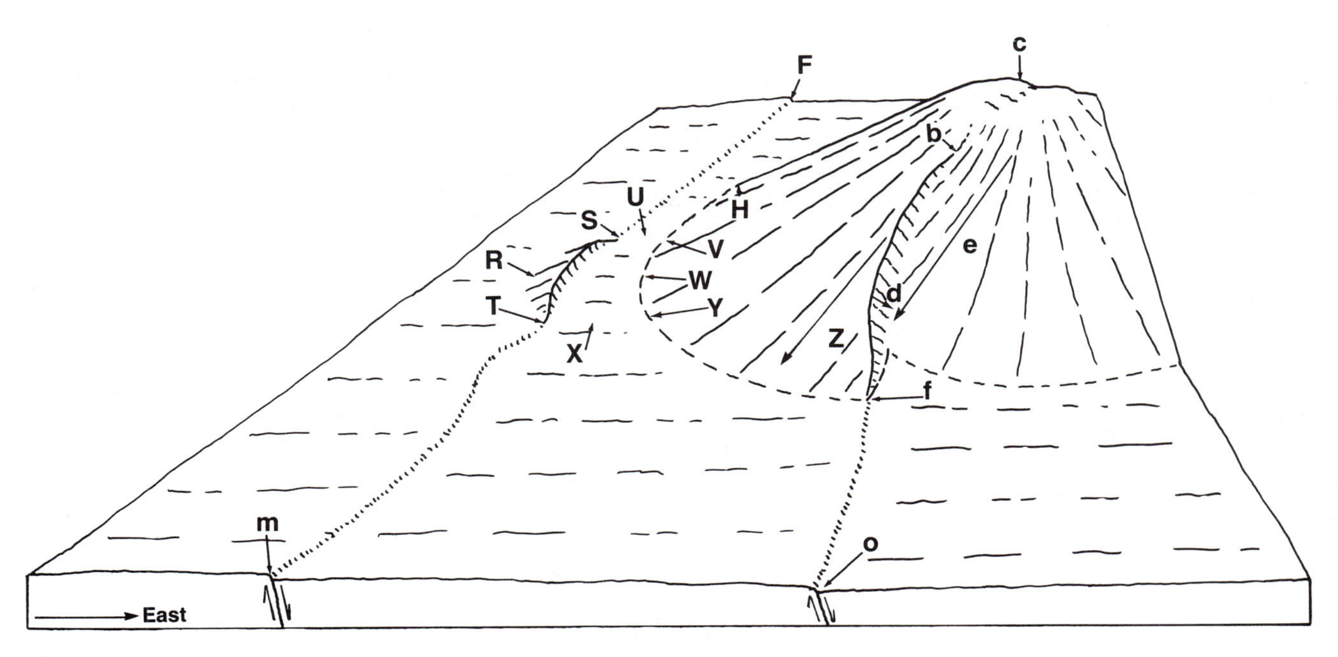

FIGURE 13–13a
Block diagram of part of the area depicted in Figure 13–13, viewed from the south.

tremely gentle slope **e** in contrast to its steep opposite slope **d**. How can such asymmetry be explained as erosion, when this is the only large valley on the entire southern flank of the cone?

On the north flank of cone **c** is an "almost valley," **KM**. Immediately to the west and east are parallel slopes **I** and **N**. We believe both are original slopes. Thus we do not consider this "almost valley" to be a valley. Its steep slope **L** must be a fault scarp, if our interpretation of slopes **I** and **N** is correct. Slopes **I** and **N**, then, are apparently offset segments of the same, once-continuous flank of the cone. Similarly, we identify slopes **e** and **Z** as once being a continuous slope. They too have apparently been offset by a fault. Steep slope **d**, like **L**, is a fault scarp.

There are convincing arguments against any of the prominent scarps being caused by erosion: (1) the youth of the bedrock; (2) its overall homogeneity and the pristine appearance of the scarps; and (3) the apparent lack of deep dissection throughout the entire area, including the flanks of the cones. It is for these reasons that we designate all of them as fault scarps. We will now pose several questions. The answers we can obtain only from the topography.

1. Why is the topographic offset (throw?) along fault scarp **ob** so small in the lowland (**of**), and so great (about 200 feet) on the flank of cone **c** (scarp **fb**)?
2. Why does fault **KM**, which has a scarp (throw) up to 200 feet on the flank of cone **c**, have no discernible extension into the lowland north of **M**?
3. Why is segment **TS** (immediately adjacent to cone **c**) so much higher than any other segment of fault scarp **lF**?

Note parallel, gentle, northwesterly slopes **k** and **n**. These appear to be parts of a once-continuous slope that has been offset along segment **m** by fault **lT**. Similarly, the large oval area **SBEJG** in the north (an area of extrusive rocks) is offset along fault-segment **SF**. The important thing for us to visualize is what this north-south-trending zone (**lF**) must have looked like before the faulting.

We believe at that time only linear ridge **TS** existed! It did not stand as high above the narrow valley **XU** as it does now (east-facing scarp **TS** was then about 100 feet lower). It nevertheless rose about 300 feet above the lowland to the west of **R,** as it does now. Ridge **TS** was not formed solely by the fault displacement that produced scarps **lT** and **SF**. It was merely lifted higher along its eastern side by that faulting.

The curvature of the western flank of ridge **TS** (around **TRS**) nearly parallels the curvature of adjacent flank **YWVH** of cone **c**. Along fault segment **TS,** it is the west side that is uplifted. The west sides of faults **fb** and **KM** are also uplifted. Is ridge **TS** a small part of the once-exposed lower flank of cone **c**? (The remainder of its base has long since been buried by more-recent extrusive and continental deposits.) It is extremely odd that the only high feature along trend **lF** happens to be ridge **TS,** and that it is so shaped, and that it lies where it does!

Consider this: (1) if this area has been intermittently faulted, and (2) if periods of faulting were separated by intervals of thick accumulation (accumulations that could have buried the low-lying fault scarps, but not the scarps on the cone flanks), and (3) if the last event was the faulting that produced the present low fault scarps—might such a composite trend as **lF** have been created?

We do not know the geomorphic-structural history of this area. We have only noted features and relations depicted by the contours. We have set forth a highly tentative hypothesis which is in harmony with the map.

A truly fascinating area!

Mt. Tom, California
(Figure 13–14 in back of book, C.I. 80 feet)

According to the *Geologic Map of the United States,* the western upland is underlain by crystalline rocks of the Sierra Batholith. The gently sloping surface to the northeast of reference line **Da** is a lava plain, designated on the Mt. Tom quadrangle as "volcanic tableland."

As you can see by the V-bends of the contours, the entire east-facing mountain front **TKABU** is stream-cut. Immediately to the west of divide segment **TKA,** a well-defined cirque (**R**) and trough (**RS**) attest to local Pleistocene alpine glaciation. Lateral moraines **V** and **W** tell us that a small valley glacier from the south once extended, as a piedmont glacier, into the eastern lowland. Note that a fan (**X**) has built out from the tip of this abandoned glacial tongue.

We cannot determine from the map whether the most challenging features in this area are related to Pleistocene glaciation. These anomalous features consist of several streams and segments, principally **DM, XM, FL, MN, IN,** and **NQ**. Note that stream **DM** is the only one that flows along axial trend **Da** of the lowland. Note further that, instead of following the axial trend along its entire length, at point **M** this stream angles off easterly as segment **MN**. It flows along a narrow gorge which increases in depth to its confluence **N** with stream **INQ**.

Downstream from confluence **N,** stream segment **NQ** flows parallel to basin-axis segment **Ma**. **NQ** is also restricted to a deep, narrow canyon. These relations are shown in Figure 13–14a, a sketch of the area viewed from the southeast.

Stream behavior in the **MNQa** area stands out as particularly anomalous. Compare the course of stream **INQ** with those of stream **FL** and its companions to the northwest. All of the latter streams are, like **IN**, obviously consequent streams. They developed on the pristine lava slopes. The difference between them and **IN** is that, unlike **IN**, they continue to flow down that slope all the way to the lowland floor (axial line **Da**). In fact, the normal courses of these streams help us recognize the courses of segments **MN** and **NQ** as anomalous.

The crenulations displayed by contours **Z** attest to the extrusive slope extending southwestward to line **Ma**. The slope does not lie solely to the northeast of canyon **NQ**. Like stream segments **IN** and **MN**, stream segment **NQ** has cut a canyon into the lava surface. The remarkable thing is that, if that particular canyon segment were not there, slope **P** would simply continue, to line **Ma**, as slope **O**.

Several things tell us that the young volcanic tableland has undergone minor fault displacements and stream adjustments: (1) the linearity of scarp **J**; (2) the linearity of scarp **H**; (3) the linearity of nearby parallel stream segment **G**; and (4) the linearity of stream segment **EF**. To the northwest of **D**, along an extension of line **Da**, linear scarp **C** is also highly suggestive of faulting.

None of these features and relations explain why stream **DM**, instead of continuing to and beyond **a**, turns into the sloping volcanics area and cuts its way to confluence **N**. But they do indicate that this is an active area, tectonically and climatically.

It is not easy to produce an explanation. It would have to address the following questions:

1. Why does stream **IN** not continue southward to point **Y**?
2. Why does stream **IN** turn abruptly at **N** and flow to **Q**, along the *side* of (instead of *down*) broad slope **PO**?
3. Why does stream **DM** not continue southeasterly to **a**, instead of turning and flowing into the rising lava "plateau" surface?

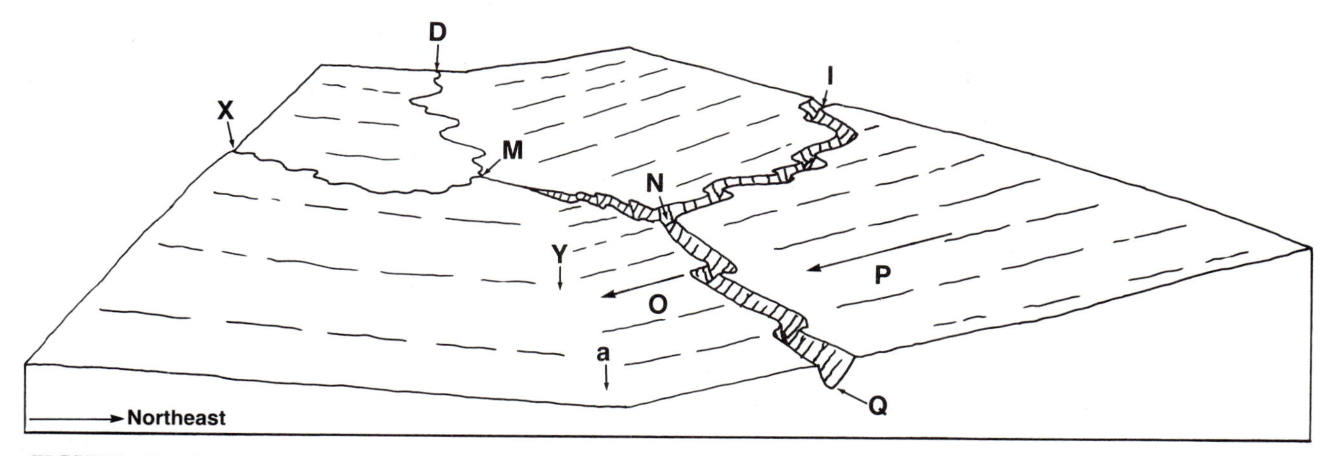

FIGURE 13–14a
Block diagram of part of the area depicted in Figure 13–14, viewed from the southeast.

If surface **MNQa** were unlike the lava topography to the north, we might suspect an erosional slope, perhaps even an exhumed lava surface. In your mind, fill in the bottom of the lowland, up to the level of the surface on which the letters **N** and **Q** are drawn. You can visualize line **NQ** as the topographic axis of the basin then, just as **Da** is now. Under such conditions, segment **NQ** would have simply downcut, to become anchored in the resistant volcanics. This would occur as the valley fill to the southwest of that line was being removed, and while volcanic slope **MNQa** was being exhumed. If that is what actually happened, then segment **MN** is merely a short, superposed stream segment.

There are disturbing things about such a hypothesis. Surface **MNQa** resembles an *original* volcanic slope far more than a surface from which overlying material has been stripped. Also, throughout the entire lowland area **UBDa,** there are no landforms to suggest that downcutting and removal are in progress, or ever were. It is possible that such slopes as **UBDV** are broad pediments. But why are there no remnant benches from a higher basin floor?

Is there another explanation? If not superposition, could this be the product of antecedence? Could a once-more-horizontal volcanic surface have been tilted upward to the northeast? Could segment **NQ** have developed along a lowland floor, with **MN** as a normal tributary, and could these segments then have become incised as the land rose about them?

Yes, it is possible. But we just do not know. Perhaps that is why this is such a great map!

Do you have any suggestions?

Newark, New Jersey

(Figure 13–15 in back of book, C.I. 100 feet, and Figure 13–15a in back of book, *Geologic Map of New Jersey*)

This area offers a marvelous variety of geomorphic and geologic features and relations. The bedrock ranges from Precambrian to Lower Paleozoic to Triassic. Lithologies (see Figure 13–15a) include gneisses and other crystalline rocks (**p€**,) shales and limestone (**€O**), coarse clastics (**S**), redbeds (**Tr**), diabase (**Trd**), and basalt (**Trb**). The terminal moraine of a Wisconsinan ice sheet also snakes across the map from northwest to southeast. The moraine is shown by dashed lines. Normal faults are indicated by hachures. Reverse and thrust faults are shown by conventional triangular barbs.

A small portion of the Great Valley of the Valley and Ridge Province (triangular Cambro-Ordovician area) occupies the northwest corner of the map. The Reading Prong (**p€**) of the New England Upland Province extends southwesterly across the west-central part of the map. The southeastern half of the map shows a portion of the Triassic Lowland.

Although we can recognize in the topography much of the geology shown on the geologic map (Figure 13–15a), there is nevertheless a definite limit to what we can interpret directly from the topographic map (Figure 13–15). For instance, we can differentiate between resistant and nonresistant rocks. But there is no way we can point to a particular ridge and declare, "This is a gneiss," or "This is a basalt." In a few places we can ascertain dip direction (e.g. at **B** and **C**). We can draw many contacts. But we cannot readily distinguish among concordant contacts, unconformities, normal faults, and thrusts.

Since the scale of the two maps is virtually the same, the easiest way to match the topography with the geology, or vice versa, is to trace the geologic map onto a transparent sheet and lay it over the topographic map. However, that is neither instructive nor rewarding. We suggest an alternative:

1. Set the geologic map aside. Lay a transparent sheet, such as thin frosted Mylar, over the topographic map.
2. Wherever you can, indicate strike and dip directions.
3. Draw as many contacts as possible. You will have to leave many of them incomplete. For the most part, these will be lines separating resistant and nonresistant formations. Though you will not be able to differentiate among them, some of the contacts will be normal, but some will be faults, and others will be unconformities.
4. When you have compiled as much geologic information as possible from the topographic map, remove your overlay and adjust it over the geologic map.
5. Compare your incomplete geology with the published geology. Remove the overlay and set it aside.
6. Place another sheet of Mylar over the *topographic* map.
7. This time, using the geologic map as a guide, try to recognize all mapped contacts and faults on the *topographic* map. Carefully trace them onto the overlay, using conventional symbols to designate fault traces, both normal and thrust. Note that in some places you cannot recognize the moraine, while in other places it effectively obscures the bedrock geology. For example, around **A** it hides the trace of the Ramapo Fault, the main fault separating the Reading Prong from the Triassic Lowland. Be sure to trace as much of the moraine as you can.
8. As a final step, you may wish to trace the entire geologic map (Figure 13–15a) and lay this tracing over the revised geologic map you have just made as Step **7**, before removing the latter from the topographic map. This way you can appraise your recognition of the known geology on the topographic map.

Among other things, there are three special varieties of slopes and/or scarps in this map area—obsequent, resequent, and fault-line. (You must determine whether they are *obsequent* fault-line or *resequent* fault-line.) In addition, there are anticlinal mountains and ridges, synclinal valleys and ridges, and homoclinal valleys and ridges.

We trust you can identify them!

Appendix: Getting Acquainted with Topographic Maps

INTRODUCTION

A topographic map is remarkably like a text page. To understand the meaning of the text, we must first know the meaning of each word and the function of each punctuation mark. This in turn demands that we be able to recognize each word and punctuation symbol. All of this relies upon a working acquaintance with the alphabet from which we can assemble the individual words. You cannot run before you learn to walk, or walk before you learn to crawl.

We want to teach you how to *interpret* topographic maps, but we cannot do that unless you are conversant with the "alphabet" and have learned how to recognize the "words." This appendix will give you that background, if you lack it. We turn first to the "alphabet letters" that combine to form topographic maps. These include the elements of direction, distance, scale, and contour lines.

DIRECTION

Suppose you plan to drive from St. Louis to New Orleans. You locate both cities on a highway map and note that you will be traveling from near the "top" of the map to near the "bottom" of the map. You will be driving "down" to New Orleans.

But in what *direction* will you be driving? Highway signs indicate north, south, east, and west, yet they do not indicate up, down, over, left, or right. If you are at all familiar with highway maps, you know that when you hold the map in "reading position" (when the place names and marginal information are right-side-up), you are looking from the "bottom" of the map toward the "top," or from the south toward the north (Figure A–1). North is toward the "top" of the map. East lies toward your *right,* west toward your *left,* and south toward *you*.

Once you have clearly established in your mind this basic map orientation, you can proceed to designate intermediate directions, such as northwest (upper left), and southeast (lower right). Most directions fall between these

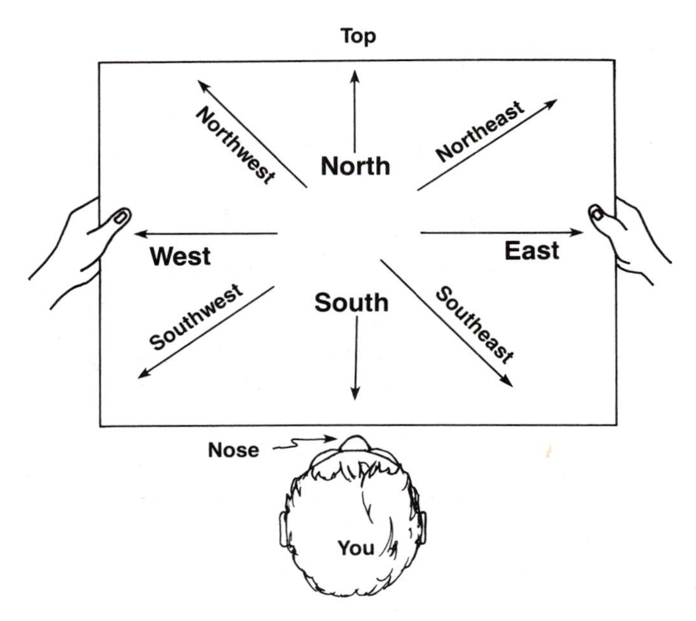

FIGURE A–1

special reference directions. For example, the direction from point **A** to point **B** in Figure A–2 is "just a little bit to the north of east" or "quite a bit south of northeast."

Bearing

During a single hour, the minute hand of a conventional analog clock sweeps through one complete circle of 360 degrees (360°). In moving from 12 to 3 o'clock (from north to east, Figure A–3), the hand passes through 90°. Suppose we place a protractor on our map and find that direction **AB,** that "little bit to the north of east," is in fact 6° to the north of east. (It is also 84° to the east of north, clockwise.)

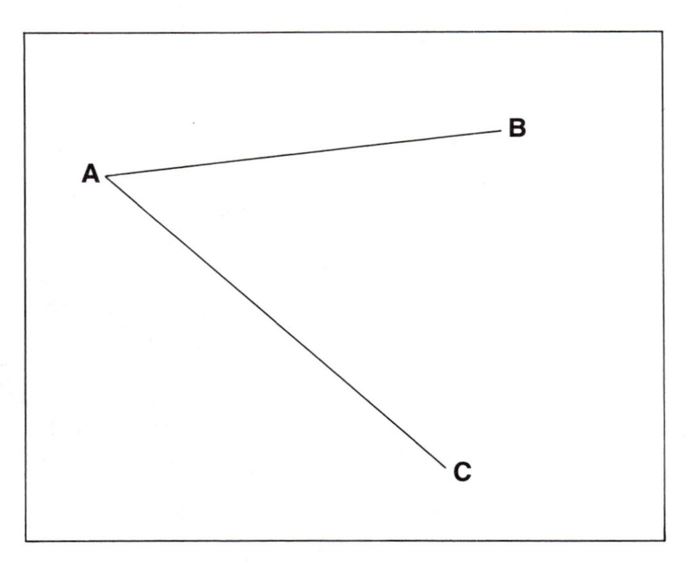

FIGURE A–2

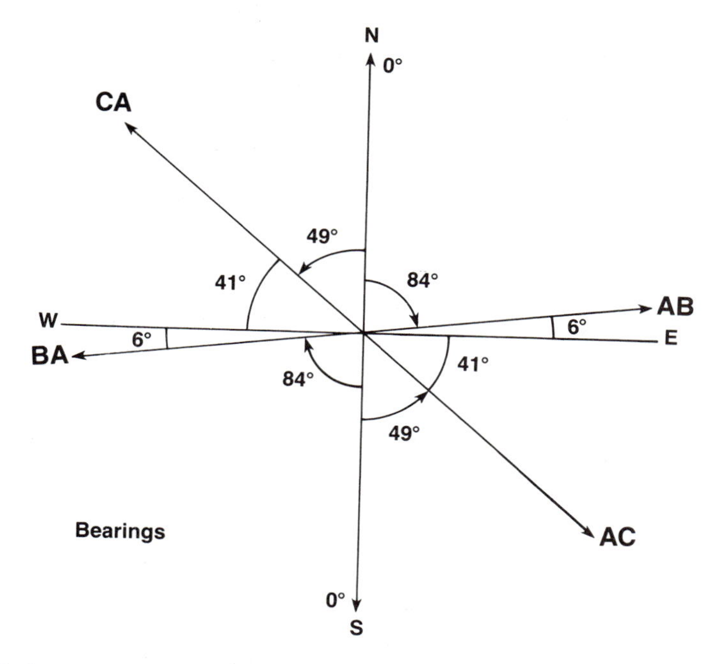

FIGURE A–3

How may we describe such a direction? To avoid confusion, we select north as our primary direction of reference and state that direction **AB** is "north 84 degrees east," which we write N84°E.

Suppose now that we wish to describe the opposite direction, from **B** to **A** (direction **BA,** Figure A–3). This direction is 6° to the south of due west, or 84° to the west of south (clockwise). In this case, we select south as our reference direction, and say that this direction is "south 84 degrees west," or S84°W.

The two directions we have considered (**AB** and **BA**) have both been measured clockwise from their respective principal directions of reference (clockwise 84° from north; clockwise 84° from south). Now look at direction **AC** (Figure A–3), which is generally from northwest to southeast. More precisely, this direction, measured with a protractor, is 41° (clockwise) to the south of east, and 49° (counterclockwise) to the east of south.

Now, should we designate this as a *clockwise* direction—east 41 degrees south, or E41°S—as we designated **AB** and **BA**? Or should it be a *counterclockwise* direction—south 49 degrees east, or S49°E? Standard practice dictates that we use only *north* and *south* as our principal directions of reference. So, this direction is S49°E. Accordingly, its opposite, direction **CA,** is N49°W.

Such directional designations are called *bearings*. North and east are separated by 90° of arc (the northeast quadrant), as are south and east (southeast quadrant), south and west (southwest quadrant), and north and west (northwest quadrant). Since every bearing must fall within one of the four quadrants, no bearing of more than 90° is possible.

For example, if we measure 96° clockwise from north, our line will fall 6° to the south of east, in the southeast quadrant. We designate this direction S84°E. It is essential that you understand bearings, since *strike* and *dip* directions are usually given in terms of bearing.

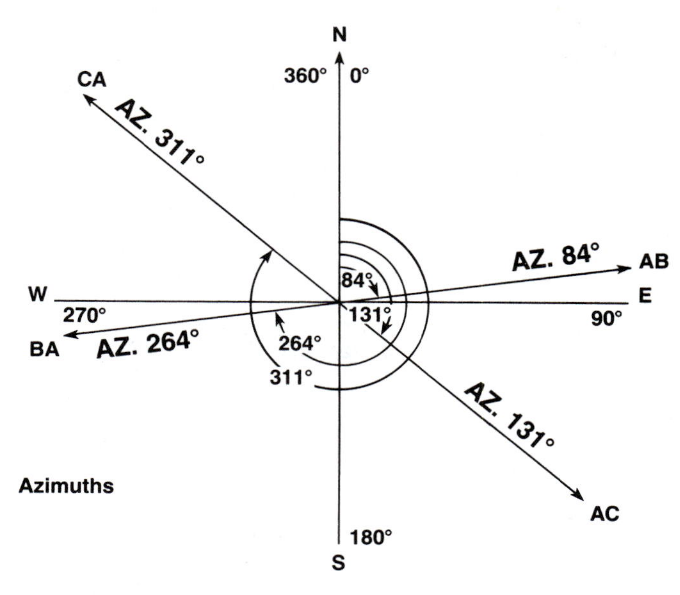

FIGURE A–4

If you are not very familiar with specific directions or degrees, it may at first appear to you that measuring directions in 1° increments is very precise and "scientific." In many instances, however, far greater precision is required. When such is the case, we may divide each degree into 60 minutes (60′) and each minute into 60 seconds (60″). Thus an exactly measured direction may be recorded as something like N42°04′33″W. Though it is important that you know how to designate map angles in minutes and seconds, rest assured that, in most map interpretation studies, accuracy to within the nearest degree, or approximated half-degree, is more than adequate.

Azimuth

The other method of measuring direction is by *azimuth* (Figure A–4). Like bearings, azimuths are stated in degrees, minutes, and seconds. Unlike bearings, however, azimuths are not measured clockwise and counterclockwise from north and south reference directions. Instead, all azimuth directions are measured *clockwise* from *north only*. Due north is 0°, east is 90°, south is 180°, and west is 270°. The numbers continue to increase from 270° toward the north, to 360°. (Due north can be 360° as well as 0°, just as 12 o'clock midnight is both the beginning and the end of a day.)

To review, compare the ways in which our four directions are indicated as bearings and azimuths in Figures A–3 and A–4. The bearing and azimuth values are summarized in Table A–1.

TABLE A–1
Bearings and Azimuths for Directions in Figure A–3 and A–4.

Direction	Bearing	Azimuth
AB	N84°E	84°
BA	S84°W	264° (84° + 180°)
AC	S49°E	131° (180° − 49°, or 90° + 41°)
CA	N49°W	311° (270° + 41°, or 360° − 49°)

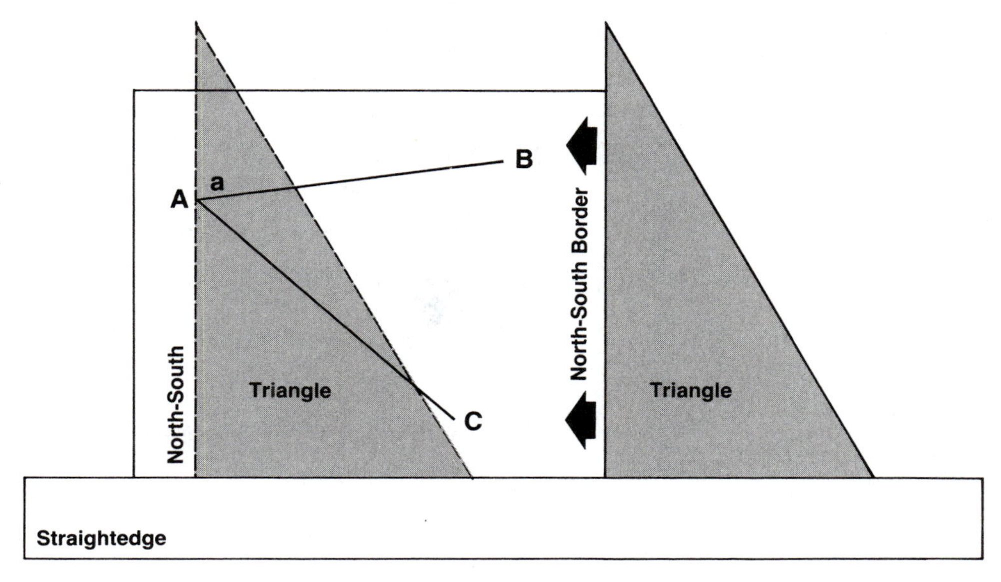

FIGURE A–5

Measuring Bearing and Azimuth—One Method

Just how do we measure the angles on Figure A–2? This map has no north-south or east-west reference lines passing through points **A, B,** or **C**. We could draw north-south lines through these points and measure the intersecting angles directly, but how do we draw north-south lines on such a map?

The easiest way is to lay the edge of a triangle along either the right or left border of the map (the borders are north-south lines). Then lay a ruler or other straightedge along another of the triangle's sides (Figure A–5). Next, slide the triangle over until the side that you lined up with the north-south map border now lies upon the desired point (e.g. **A**). Now draw a vertical line along the triangle side through point **A**. This line is parallel to the map border, a north-south line. You now have a north-south line through **A**. (We could also draw north-south lines through points **B** and **C,** but that would be wasted effort, as will be explained.)

All we need, in addition to lines **AB** and **AC,** is the north-south line we just drew through point **A**. Using a protractor, we can now measure the angle between the north-south line and line **AB**. From this measurement, we can determine both the bearing and azimuth of direction **AB,** and compute them for opposite direction **BA**. Then, by measuring the angle between the north-south line and line **AC,** we can determine the bearing and azimuth of direction **AC,** and compute them for opposite direction **CA**.

Exercise. Using the procedure outlined above, determine the bearings and azimuths of directions **BC** and **CB**. Use Figure A–2.

Measuring Bearing and Azimuth—A Better Way

There is another method for measuring bearings and azimuths that is less complicated and does not require the time-consuming drawing of north-south lines

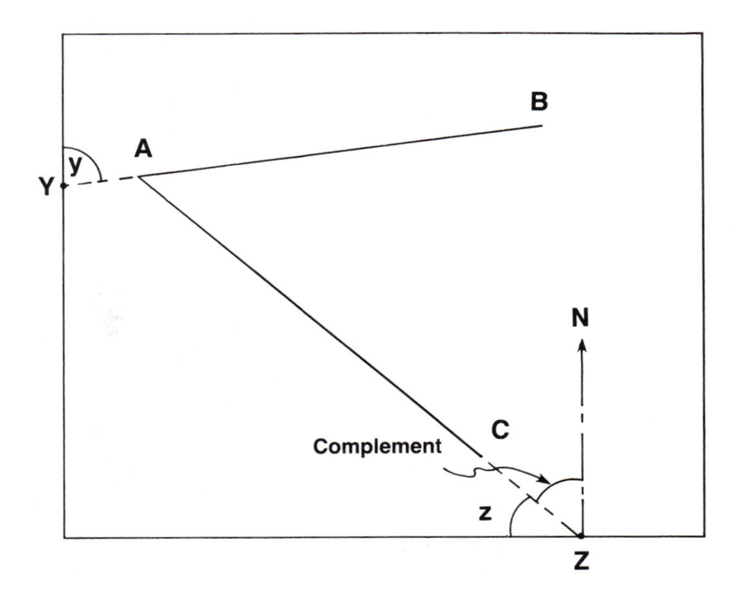

FIGURE A–6

on our map (lines which later might create confusion). We suggest you adopt this second method.

In the previous method, we constructed a north-south line so we could measure the angles between it and the map lines for which we needed directional data. Had we drawn 10 such north-south lines across the map, the angles of intersection between them and our other lines would all have been the same. In other words, it makes no difference which north-south line we use. For that matter, east-west lines make 90° intersections with north-south lines. So, we could just as well have measured the intersections between our map lines and an east-west line, and by simple arithmetic computed the bearings and azimuths.

Before we constructed any lines, we already had two perfectly good north-south lines—the east and west borders of the map. We also had two east-west lines—the north and south borders. All we must do to obtain the bearing of direction **AB** is to *extend line* ***AB*** *until it intersects one of the map borders*.

In Figure A–6 we have drawn such an extension of line **AB,** to the left. It intersects the west border at point **Y**. The acute angle **y** at **Y** is equal to the acute angle **a** at point **A,** Figure A–5. The extension of line **AC** to the southeast intersects the south border at point **Z**. The acute angle **z** at **Z** is 41°. The north-south line **ZN** through point **Z** makes a 90° intersection with the east-west line, so the 41° acute angle **z** is the *complement* (totals 90°) of the angle line **AC** would make with a north-south line. We can also use this intersection to compute the azimuth of direction **AC**.

Exercises. The following are several practical exercises to be sure that you understand the material we have covered so far. If you do not understand it, you are not going to comprehend the material that follows!

Determine the bearing and azimuth of each of the directions in Figure A–7, and enter your results in Table A–2.

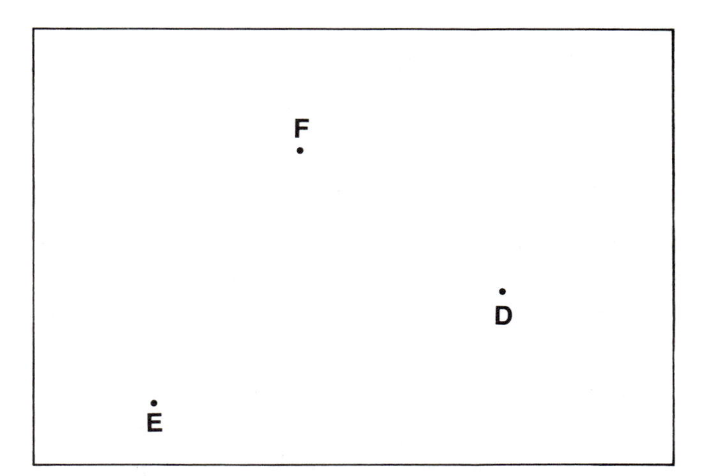

FIGURE A–7

TABLE A–2
Bearings and Azimuths for Directions in Figure A–7.

Direction Line	Bearing	Azimuth
DE	______	______________________________
ED	______	______________________________
DF	______	______________________________
FD	______	______________________________
EF	______	______________________________
FE	______	______________________________

Magnetic North

So far we have dealt with *true north,* or *geographic north,* which is the "north" directed toward the North Pole of the Earth's axis. We must now consider another "north," the extremely important *magnetic north*. There are very few places on the Earth's surface at which a magnetic compass points precisely toward true north. Even that coincidence is temporary, since the magnetic north pole is gradually moving. The difference between the two poles is unimportant for general direction-finding, but is critical when using precision maps and measurements.

The angular difference between true and magnetic north is called *magnetic declination*. Declination varies from place to place. For example, in a diagram appearing in the very informative text *Map Use* (Muehrcke, 1978), there is no declination at all along a line from Michigan to Florida. However, declination increases to 6° across western Maryland and reaches 20° across Maine. Toward the west, declination is 12° along a line from North Dakota to New Mexico and is 22° across Washington state. Declination also changes over time; in central West Virginia, for example, it is currently changing at the rate of 7.5 minutes (1/8 degree) per year.

If you check the margin of a topographic quadrangle map, you will find a small diagram showing the declination for that map area as of the date of publication. Such a diagram is shown here, reproduced from the Cumberland, MD-PA-WV quadrangle.

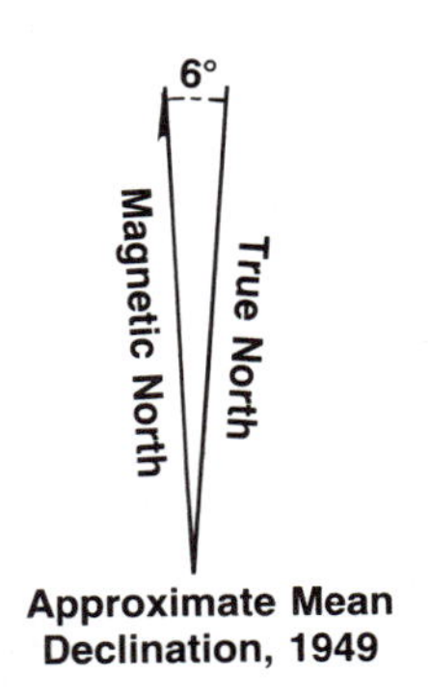

Since published quadrangles are oriented toward true north, the only time we need to be concerned with magnetic north while using these quadrangles is when we are dealing with bearings or azimuths which are measured by a magnetic compass in the field. To plot such readings on a quadrangle, we must first *convert* them to true-north values.

Figure A–8 shows a magnetic declination of 8° west. This means that *true* north has a *magnetic bearing* of N8°E, and a *magnetic azimuth* of 8°. Conversely, *magnetic* north has a *true bearing* of N8°W and a *true azimuth* of 352°. From reference ground point **X,** the *magnetic* bearing to point **A** is N50°W. The *true* bearing to **A** is 50° + 8° = N58°W. From **X** to **B,** the magnetic bearing is N50°E. True bearing from **X** to **B** is 50° − 8° = N42°E.

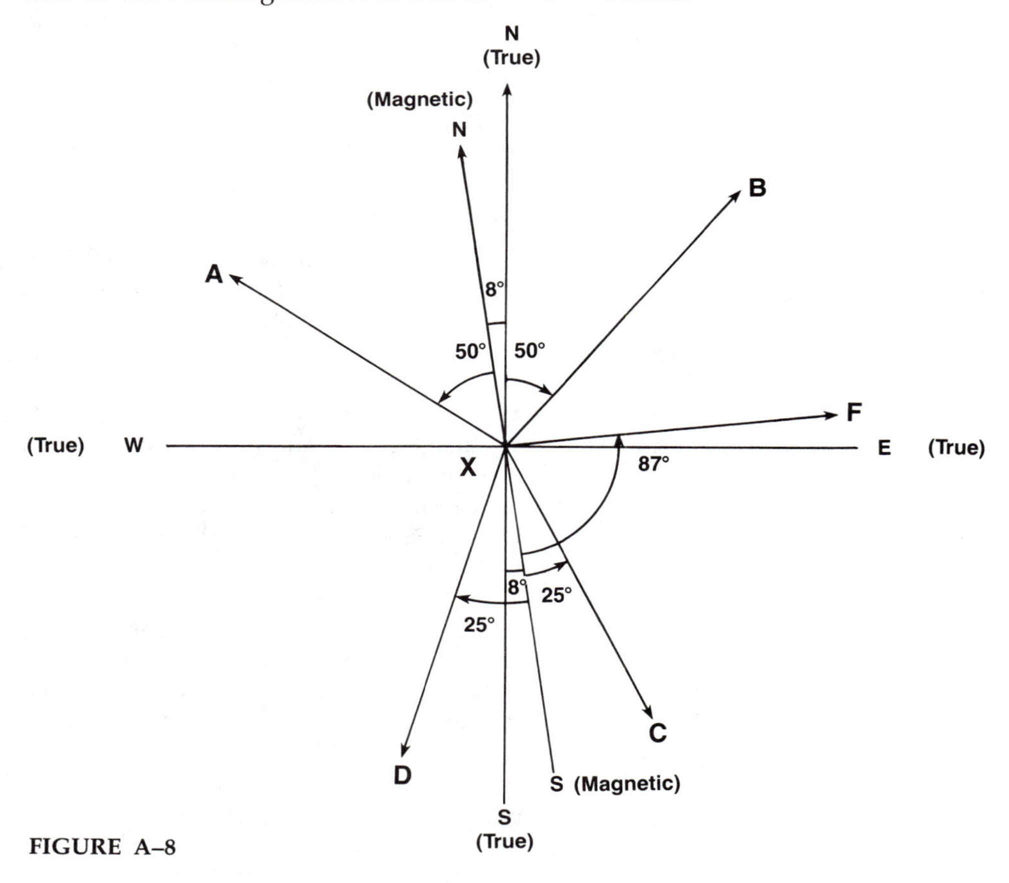

FIGURE A–8

Exercise. (Figure A–8.)

1. The magnetic bearing to **C** is S25°E. What is the true bearing?
2. The magnetic bearing to **D** is S25°W. What is the true bearing?
3. The magnetic bearing to **F** is S87°E. What is the true bearing?
4. In Table A–3, calculate and fill in the missing values.

TABLE A–3
Exercise—Calculate and Enter Missing Values.

	Magnetic Readings		True Readings	
	Bearing	Azimuth	Bearing	Azimuth
A	N50°W		N58°W	
B	N50°E		N42°E	
C	S25°E			
D	S25°W			
F	S87°E			
G		4°		
H		97°		
I		190°		
J		301°		

SCALE AND DISTANCE

We have learned how to determine the *direction* from **A** to **B**. Now, *how far* is it from **A** to **B**? Have you ever heard someone say, "We can easily make it to Phoenix in time for lunch; it is only a couple of inches away on the map"?

Scales

Every map worthy of the name indicates its scale. Scale is the relation between the size of a feature *on the ground* and the size of the same feature *on the map*. For example, suppose we have a map on which a ground distance of 2000 feet is shown as a line one inch long. How may we state or show the scale of this map? One way is just to say it: "one inch *equals* 2000 feet," or "one inch *to* 2000 feet." On a map of this scale, any linear inch represents 2000 feet on the ground. Thus a map length of 1.63 inches represents a ground distance of 3260 feet (1.63 × 2000). A mile contains 5280 feet, which would be represented by a line 2.64 inches long (5280 ÷ 2000). We might round these figures and state that the scale is "approximately 2½ inches to the mile." (This scale expressed in words is a *verbal scale*.)

One drawback to expressing scale as "inches per feet" or "inches per mile" is that it is virtually useless to someone using different units of length, such as centimeters or kilometers. We can avoid this kind of problem by converting the scale to a *unitless fraction*. For example, there are 24,000 inches in 2000 feet (12 × 2000 = 24,000). We may say that, if one inch equals 2000 feet, one inch also equals 24,000 inches. This can be expressed in fractional form:

$$\text{Scale} = \frac{\text{map distance}}{\text{ground distance}} = \frac{\text{1 inch}}{\text{24,000 inches}}$$

In this fraction we can cancel the "inch" and "inches," and end up with a unitless fraction:

$$\text{Scale} = \frac{1}{24{,}000}, \text{ or } 1/24{,}000, \text{ or } 1{:}24{,}000$$

The beauty of this kind of scale is that it applies equally to any unit of length. One centimeter on the map equals 24,000 centimeters on the ground. One millimeter on the map equals 24,000 millimeters on the ground. In fact, you may use the width of your thumbnail as a unit: one thumbnail width on the map equals 24,000 thumbnail widths on the ground!

How can we *convert* a scale of "one inch to the mile" to a fractional scale? All we need to know is that there are 5280 feet per mile, and 12 inches per foot. Multiply together: there are 63,360 inches in a mile.

$$\text{Scale} = \frac{\text{map distance}}{\text{ground distance}} = \frac{\text{1 inch}}{\text{63,360 inches}} = \frac{1}{63{,}360}$$

Since 63,360 is an extremely awkward number with which to work, for many years the federal government rounded that number to 62,500. Federal agencies have published maps at a scale of 1/62,500, which is stated as, "one inch equals *approximately* one mile." Similarly, the scale of 1/125,000 is "one inch equals approximately two miles," 1/250,000 is "one inch equals approximately four miles," and 1/31,250 is "one inch equals approximately one-half mile."

Exercises. Convert the following verbal scales to fractional scales:

1. One foot equals six miles.
2. One centimeter equals ten meters.
3. One inch equals 25 yards.
4. One inch equals one kilometer. Round your answer reasonably. (Note: one kilometer equals 0.62 miles.)

Now convert the following fractional scales to verbal scales. Vary the units for practice, using inches, feet, miles, centimeters, or whatever:

1. 1/48,000
2. 1/500,000
3. 1/200,000
4. 1/24
5. 1/100
6. 1/15,625
7. 1/15,840

Graphic Scale

By far the most convenient type of scale is the *graphic* scale, which is a segmented line or bar that actually displays the ground distance on the map. For example, here is a graphic scale for a map with a scale of one inch equals one mile:

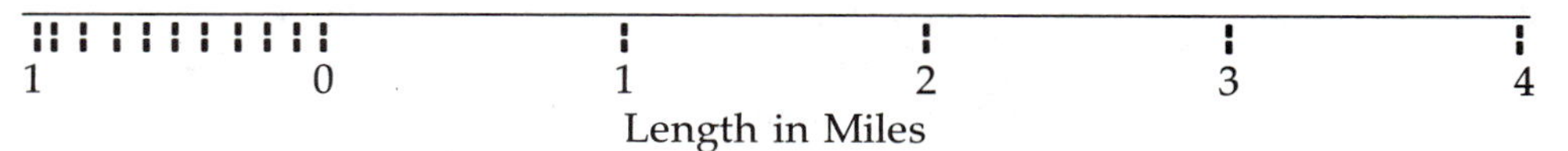

Note that the 0 lies one inch (one mile) in from the left end of the bar. This "end inch" (mile) is subdivided into tenths of an inch, which correspond to tenths of a mile.

Figure A–9 illustrates how we use this scale bar. To determine the ground distance between point **A** and point **B** on the map, place the edge of a sheet of paper along line **AB,** and mark points **a** and **b**. Then move the paper down to the bar scale. Place tick mark **a** at whichever inch mark will permit tick mark **b** to fall within the segmented left-hand inch. In this case, place **a** on the 2-

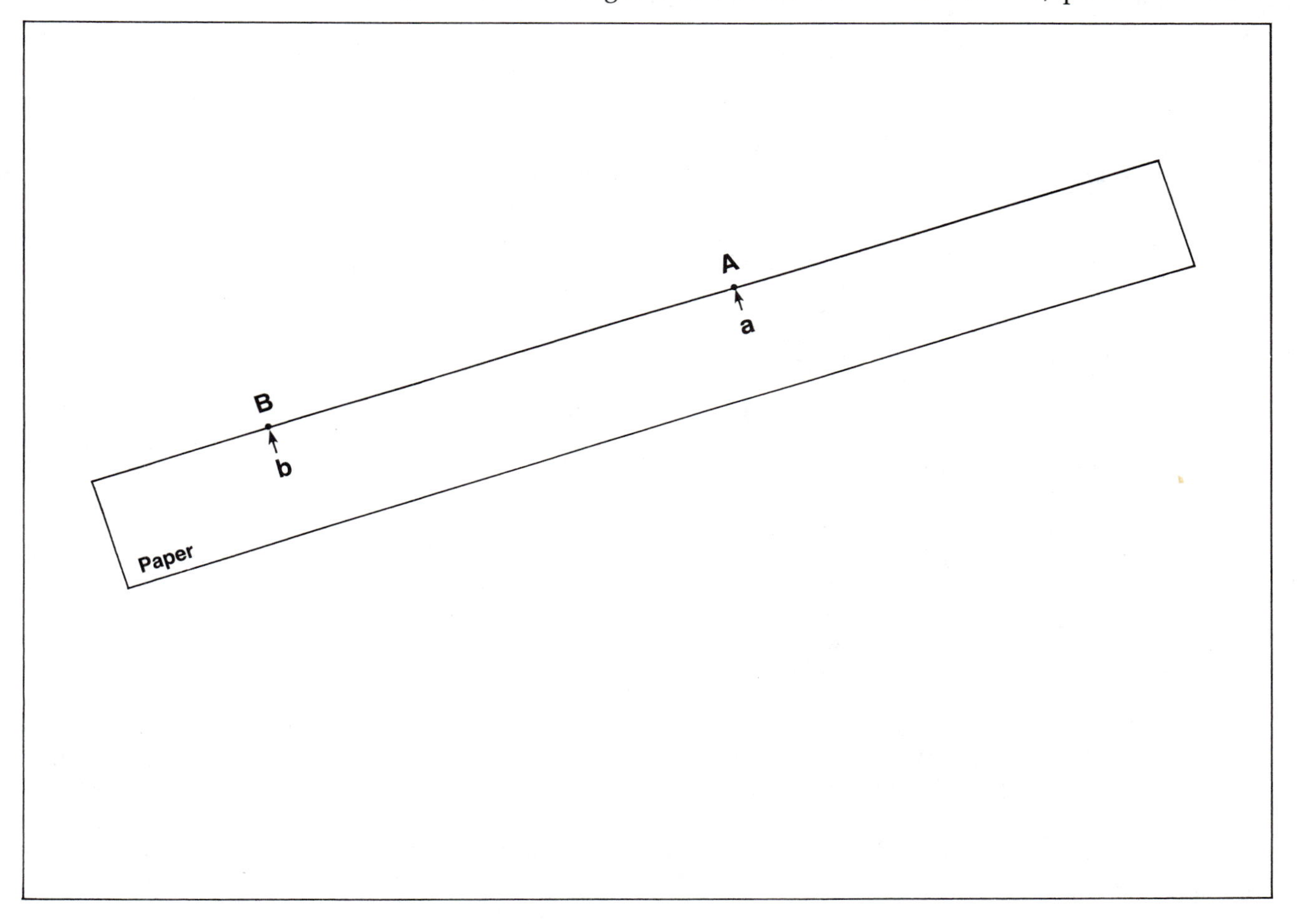

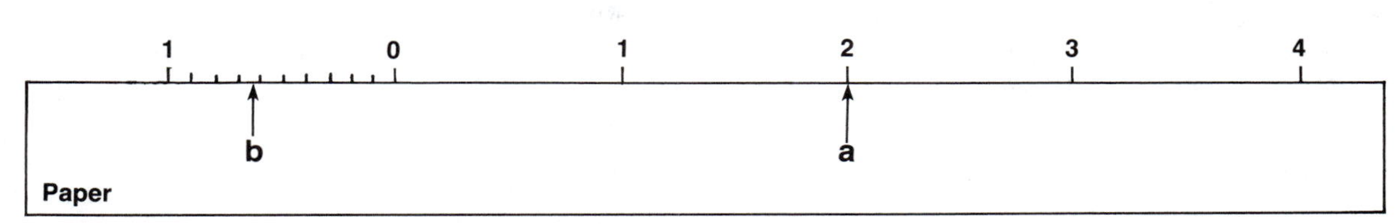

FIGURE A–9

mile mark. Then you can estimate quite accurately the position of **b,** which seems to be about 0.64 inch from 0. Ground distance **AB,** therefore, is almost exactly 2.64 miles in length. How many feet is that?

Whereas a map will have a single verbal scale and a single fractional scale, it usually has two or three graphic scales, one showing feet, another showing miles, and a third showing kilometers.

Exercises.

1. The scale of Figure A–2 is 1 in. = 500 ft. In feet, how long are ground distances:
   AB?
   AC?
   BC?
2. The scale of Figure A–6 is 1/48,000. In feet, how long are ground distances:
   AB?
   AC?
   CB?
3. In Figure A–7, ground distance **ED** is 5.2 miles.
   (a) What is the fractional scale of this map?
   (b) In feet, what is the ground distance **EF?**
4. On Figure A–2, plot point **D,** which is 1200 feet S50°W from point **B** (true bearing). Scale is 1 in. = 500 ft. What is the length of line **BD**?

The Area of a Map

How much ground *area* can be covered by a map? Two factors control this: *map dimension* and *map scale*. For a map of given dimensions, the ground area that can be covered depends entirely on map scale. For example, see Figure A–10, which shows three 20 in. × 30 in. maps of the same general area. Each map has a different scale:

- □ Map a, scale 1 in. = 1/2 mi. (or 2 in. = 1 mi.), covers 150 square miles.
- □ Map b, scale 1 in. = 1 mile, covers 600 square miles.
- □ Map c, scale 1 in. = 2 miles (or 1/2 in. = 1 mile), covers 2400 square miles.

These figures reveal that the area covered varies *inversely* with the *square* of the scale *difference*. For example, Map b has half the scale of Map a; one-half squared is one-fourth; one-fourth inverted is four; and sure enough, Map b covers four times the area (600 sq. mi) as Map a (150 sq. mi).

It is important to keep sight of this relation. Whenever you study a topographic map, you must *always* keep in mind the *size* of the area you are looking at, and the size of the individual features, such as stream valleys, slopes, and hills. Note, for example, that stream **2** (Map A–10b) is considerably longer than stream **1** (Map A–10a), but only about half as long as stream **3** (Map A–10c).

TOPOGRAPHY

A map that shows *where* an object is, and its *shape*, its *size*, and its *relation* to other objects, is a *planimetric* map. Probably the most widely used planimetric map is the highway map. Scale, distance, and direction are the principal char-

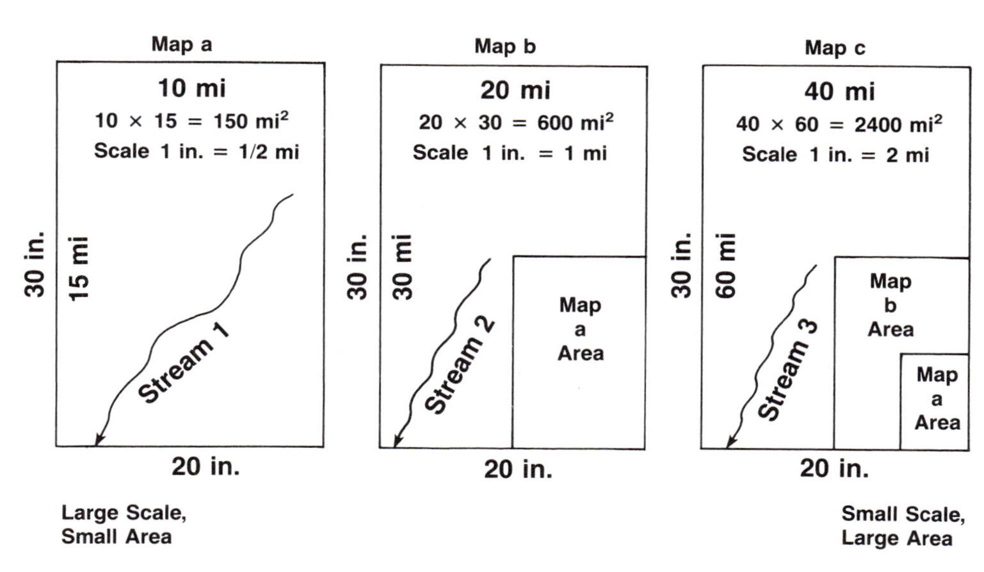

FIGURE A–10

acteristics of a planimetric map, as they are of any map, but a *topographic* map depicts more. A topographic map will show the same stream that is depicted on a simple planimetric map, and it will also tell whether that stream's valley is narrow or broad, canyon-like or flat-floored. It will tell us where the hills stand and how high they are, and whether their slopes are gentle or steep, uniform or step-like, concave or convex. Topographic maps literally add a third dimension to the area depicted—the *vertical* dimension.

A simple way to show the height of a hill compared to a low area is to indicate their elevations on the map (e.g. points **A** and **B,** Figure A–11). Hill **A** is 130 feet higher than point **B**. But that is *all* we know about these points—their elevations, and the difference in elevation.

We can make the picture better by adding more *spot elevations,* such as points **C** through **L**. What do these additional points tell us? They show that **A** is the highest point (hilltop) in the area. They also tell us that A is not surrounded by a uniformly low, flat area. The other spot elevations indicate that the area to the south and east of **A** (area **EDCL**) is lower than hilltop **A**. They also show that the area to the west and north (area **FGHIJ**) is **much lower**.

As you study the spot elevations, extracting these facts about the topography, note that your eyes must continually jump from one spot to another, back and forth, as you build a mental picture of the terrain. Not easy. And if your eyes leave the map, your mental picture disappears completely. When you turn back to the map, once again all you see is numbers and dots.

Contours

Instead of trying to construct and maintain a three-dimensional mental picture, we can use these same spot elevations to construct a graphic depiction of the terrain (Figure A–12). This is done by adding **contours**. A contour is a line connecting all points on the ground that have the same elevation. A **shoreline** is a natural contour line, because the edge of the water "connects" all points having the same elevation. In fact, the sea-level contour is used as the 0-foot level, or reference level, or "datum," for most topographic maps.

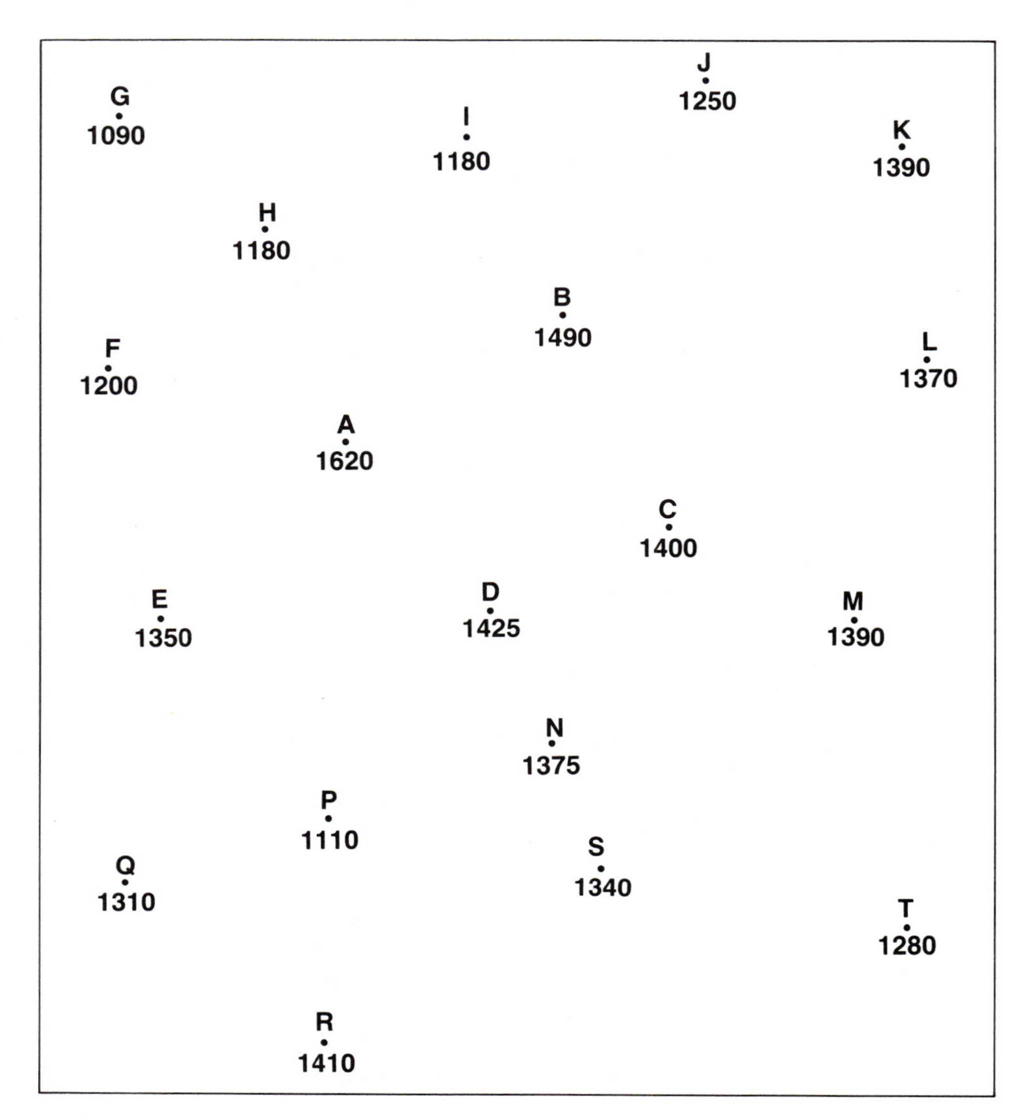

FIGURE A–11

Let us add contours to Figure A–12. Start with point **C,** which is a convenient 1400 feet (above mean sea level). We have to start the line in some direction, so let us choose west, toward **A** and **D**. But the land between **D** and **A** is all **above** 1400 feet (**D** = 1425 ft, **A** = 1620 ft). However, between **D** (1425 ft) and **N** (1375 ft) there must be a 1400-foot point. And there must be another between **D** (1425 ft) and **E** (1350 ft), and between **E** (1350 ft), and **A** (1620 ft). These three 1400-foot points we have added to the map as dots **1, 2,** and **3**.

We have to assume that the slope between any two adjacent ground points is a uniform slope. So we position point **1** midway between **N** and **D** (**N** is 25 feet below 1400 feet; **D** is 25 feet above 1400 feet). Accordingly, we position point **2** two-thirds of the way from **E**. (**E** is 50 feet below 1400 feet; **D** is 25 feet above 1400 feet.) Our assumption of uniform slopes explains the positioning of point **3** much closer to **E** than to **A**.

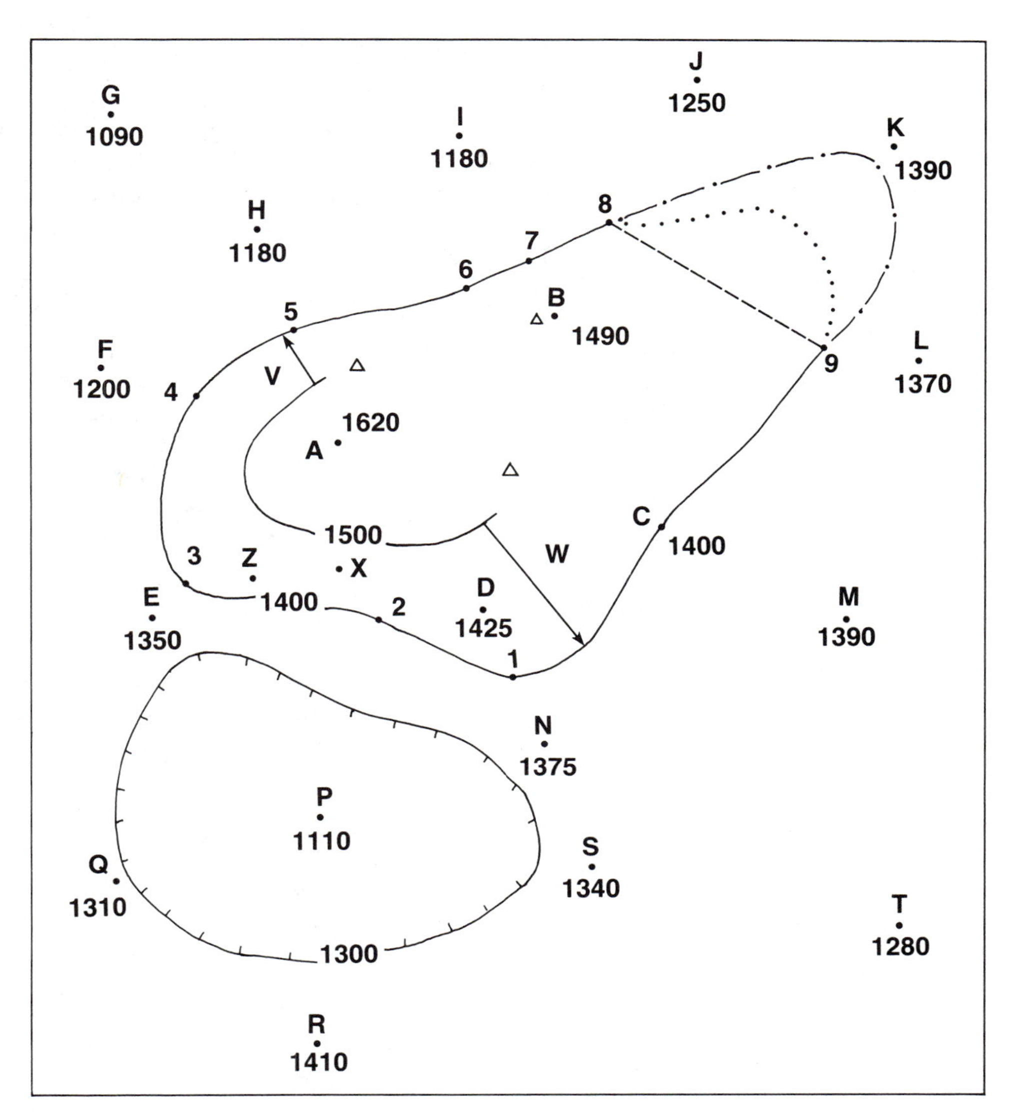

FIGURE A–12

We continue to find 1400-foot points: point **4** between **F** and **A, 5** between **H** and **A, 6** between **H** and **B, 7** between **I** and **B, 8** between **J** and **B,** and **9** between **L** and **B**.

We can connect these points with a solid line, signifying that we are sure of ourselves. Starting at **C,** we draw the line to **1, 2,** and so on to **8**. And we can connect **C** northeastward to **9**. Connecting **8** and **9** poses a problem. We can visualize a 1400-foot elevation between **K** and **B,** but we cannot assume a 1400-foot elevation between **J** and **L,** both of which are below that elevation. It would not be "wrong" to simply connect **8** and **9** with a dashed line, as we have, but as you can see, it *looks* wrong, and probably is.

Could lines **7–8** and **C–9** continue northeastward nearly to **K** (dot-dash line)? Or should the connection be the dotted line? The choice is a matter of judgment (though we can safely reject the straight dashed line).

The line we have drawn—and smoothed, rather than leaving it as a series of straight segments—is the 1400-foot *contour*. We have labeled it in the traditional manner, by inserting **1400** into the broken contour.

Since this contour completely encircles a central high area (all ground points within the circled area are *above* 1400 feet), we refer to this as a *closed contour*. In contrast to this contour which encircles a hill, note the 1300-foot contour in the south part of the map area. This contour encircles a *depression* (**P** is the lowest point in an area surrounded by higher points). This too is a closed contour. To signify that it is a depression, we have added small, perpendicular ticks to the 1300-foot contour. These ticks, which point toward lower elevation, are called *teeth* or *hachures*. This is a *depression contour*.

It is helpful to think of contours as "potential shorelines." If the land area in Figure A–12 were real, you could flood it, filling it to the 1400-foot contour level. The portion of the hill above the 1400-foot line would become an island. And when you drained the area, there would remain a lake, 190 feet deep at its center **P** (1300 ft − 1110 ft = 190 ft).

Contour Interval

A map with only one or two contours is of very limited value. Our next job is to draw in more contours. Since we may think of the 1400-foot contour as a potential shoreline, we might next add a line to show where the shoreline would be if "sea level" were to rise one foot, to 1401 feet. We *could* draw such a contour, but if we continued to add contours at *vertical intervals* of only one foot, we would end up with hundreds of contours and a plethora of lines so close together that we would be unable to distinguish one from another.

A far more reasonable approach would be to select a *contour interval* of something like 10, 25, 50, or 100 feet. (Contour interval is the vertical distance separating any two adjacent contours.) The interval to choose depends on controlling factors such as map scale and topographic relief. (*Relief* is the difference between the highest and lowest points in an area.) For example, an interval of 5 feet would be very reasonable in an area having a relief of 68 feet. But it would be absurd in an area having a relief of 4000 feet. Conversely, in a low-relief area, a contour interval of 50 feet would be pointless, resulting in two or three lines wandering across the map, signifying nothing.

Though in most cases a single contour interval (C.I.) is used for an entire map, there are times when two intervals must be used. For example, a steep-sloped mountainous area might be depicted by contours drawn at a large interval (C.I. = 100 feet). But an adjacent flat expanse might require the use of a much-smaller interval (C.I. = 5 feet). This is because its topographic irregularities would not be picked up by large-interval contours, but would fall between two contour planes.

Let us choose an interval of 100 feet for our map (Figure A–12). Consider where we may extend the 1500-foot contour, of which we have already drawn about half. To help you complete this contour, we have added several points (triangles) like our points **1** through **9**. Connect these points to form a smooth, rounded 1500-foot line similar to the 1400-foot contour.

Now that you have completed this contour, we can state a fact about the strip of land lying between the two contours: all points within that strip are more than 1400 feet, and less than 1500 feet, above sea level. Also, just as we assume that the slope between any two spot elevations is uniform, we also assume that the slope between any two contour lines (at any given point) is uniform. For example, we can assume that point **X,** which is midway between the contours, is at 1450 feet elevation. The elevation of point **Z,** which is one-fourth the distance from the lower to the higher contour, is 1425 feet. (Indicate where you would pinpoint a nearby ground elevation of 1480 feet.)

Some Do-Not-Forgets

The most important things to remember about these two contours: (1) Their horizontal separation *varies* from place to place—they are *not parallel,* and (2) they are by construction *uniformly separated vertically*. Note, for example, that at **W** their map or horizontal separation (arrow **W**) is about three times as great as at **V** (arrow **V**). This contrast tells us that at **W** you would have to travel horizontally three times as far to descend 100 feet, as you would have to travel at **V** to descend 100 feet. In other words, the slope at **V** is much steeper than the slope at **W**!

The ability to visualize this essential relation, between *slope magnitude* and *contour spacing,* is the most-important key to reading topographic maps, and ultimately to interpreting them. No one can ever hope to glance at a map and declare, "This is a slope of 16°." But we can always point to two slopes on any given map, and with absolute certainty state, "This slope is steeper than that slope."

Exercise. Our mapping of the topography in Figure A–12 is still incomplete. Before continuing to our next topic, complete the contouring that we have begun. This will include completing the 1500-foot contour, plus the other 100-foot-interval contours (1300 feet, 1200 feet, etc.). Be sure not to overlook the central part of depression **P** in the south. Point **P** is 190 feet below the 1300-foot contour. Remember to add hachures to your contours in the depression!

The Contour V

Examine Figure A–13, a block diagram of a partially submerged hill (now an island). Into its flanks a couple of streams have cut valleys (**A** and **B**). Pay particular attention to the way in which the sea extends into the valleys to form deep embayments. Each embayment comes to a distinct point at the very mouth of its entering stream. The shoreline is the 0-foot contour, and this contour forms a *V-shaped bend* at the mouth of each stream.

If sea-level were to rise 100 feet, the lower 100 feet of the present island would be submerged, and a new shoreline would be formed (dashed 100-foot contour line). Note on Figure A–14, a topographic map of this island, that this same contour also forms a V-shaped "embayment" up each valley. These are

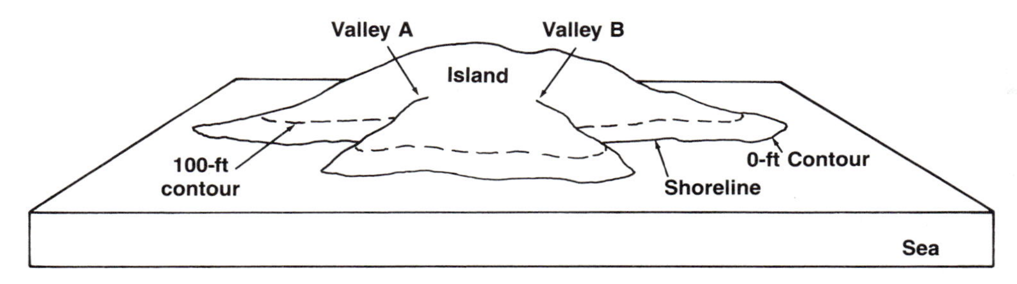

FIGURE A–13

examples of *contour V's*. Whenever a contour bends upstream to a point where it crosses the stream itself, it forms this telltale V. And remember: the *V always points upstream*.

Note the shoreline and contour configurations in the northern part of the island (area **CDE,** Figure A–14). The present shoreline forms a V which converges to point **E,** the mouth of small north-flowing stream **E.** Question: if there is a stream **E** here, why is it not shown by a distinct line, as streams **A** and **B** are shown?

Select at random almost any U.S. Geological Survey (USGS) topographic quadrangle. You will find that although some of the drainage lines are shown as solid or broken blue lines, many others are not. The more familiar you become with topographic maps, the more you will develop the habit of *recognizing streams by their contour V's,* and the less dependent you will become on searching for blue lines.

In Figure A–14, you can see the match between stream **A** and its contour V, and likewise for stream **B** and its contour V. Rest assured that contour V's **C, D,** and **E** indicate the presence of streams **C, D,** and **E.**

What we must now do is determine how streams **C, D,** and **E** relate to one another. Stream **C** flows northeasterly toward **E.** Stream **D** flows northwesterly toward **E.** Point **E** is the mouth of our north-flowing steam **E.** Stream **E** must be a short trunk stream into which streams **C** and **D** are tributaries. (Add streams **C, D,** and **E** to Figure A–14.)

With few exceptions, all maps in this book are devoid of drainage lines: nearly all their streams and valleys are indicated only by contour V's. As you look through the maps, you will note that in some (Figures 8–4, 8–10, 8–11, and 10–1 in the back of this book), the contour bends and V's are relatively widely spaced, denoting systems of drainage lines that are also widely spaced. In others, such as in Figures 6–2 and 11–6 (in back of book), note that the contour bends and V's are remarkably small-scaled and very closely packed together, indicating fine dissection by a plethora of small, closely spaced streams.

If you were to trace all the streams on one of the latter maps, you would not only soon become frustrated and bored by the endless task. You also would end up with a fine mesh of meaningless miniature lines.

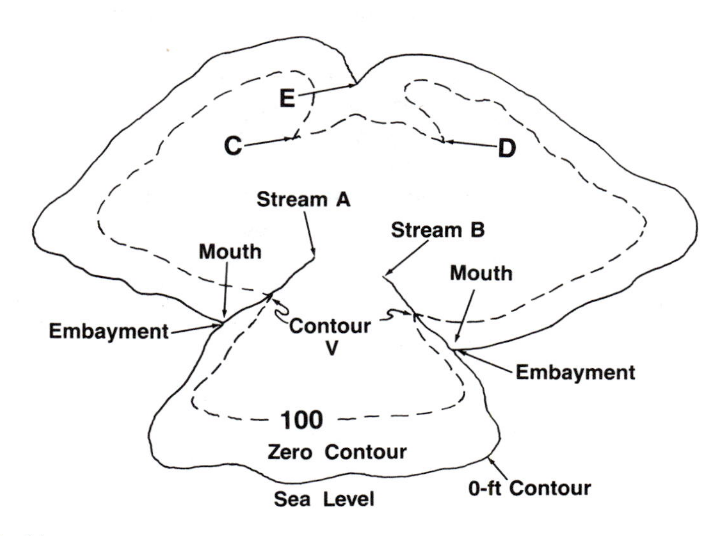

FIGURE A–14

Exercise. On the other hand, it would be easy and highly instructive for you to select a more coarsely textured area, such as the western half (or all) of Figure 1–10 (in back of book), and painstakingly trace all the streams you can identify by contour configuration. First overlay the map with a sheet of thin—e.g. 0.03 mm—frosted Mylar and trace the streams on it. This will protect the book, and it will permit you to make corrections, which is almost always necessary no matter how much you have worked with maps.

Profiles

First, here are three important definitions to help you understand the differences among *map view, cross section,* and *profile:*

> Imagine a seven-layer birthday cake. When you look down upon it from directly above and read "Happy Birthday," you are seeing the *map view* (also called plan view). If you slice the cake and view the layers from the side, you are seeing a *cross section*. If you don't slice the cake, and look at the outline of it from the side, you are seeing its *profile,* or outline.

On a topographic map, contours show slope directions, steepness, curvatures, and variations, from a vantage point directly overhead. We look at a map in the same way that we look down at the ground from an aircraft. Although contours permit us to *visualize* the ground's slopes, they do not permit us to *see* those slopes directly.

The only way to *see* a slope is to view it from the *side,* in *profile.* And the only way to do that is to *draw* a profile of the ground, and look at *that*. Here is how to construct a profile.

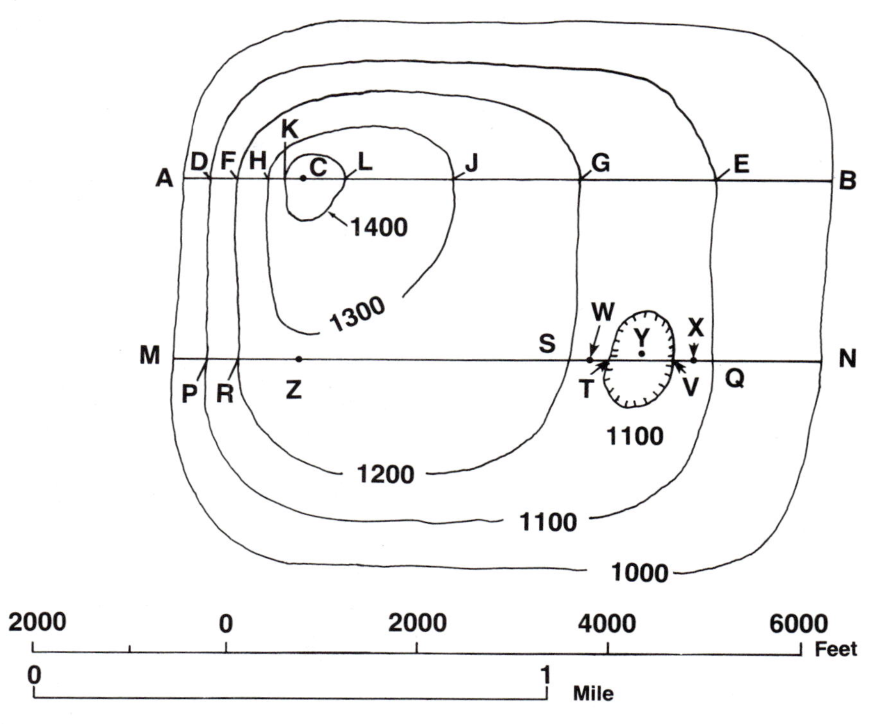

FIGURE A–15

First we must select the line on the ground that we want to view from the side. This is the line along which we will draw our profile. Figure A–15 is a map with a scale of 1/24,000 (one inch equals 2000 feet) and a contour interval of 100 feet. We will select line **AB** for profiling.

The lowest points on line **AB** are points **A** and **B,** both of which lie on the 1000-foot contour. We do not know the exact elevation of point **C,** the highest point on the line. We **do** know that it is above 1400 feet, and below 1500 feet, because:

1. The entire area encircled by contour 1400 is higher than that contour.
2. If point **C** were above 1500 feet, a 1500-foot contour would have been added surrounding the high point **C**.
3. Were **C** exactly 1500 feet high, it would be marked by a dot, a small ×, or some other symbol, and its "spot elevation" indicated by the number 1500.

When we construct our profile, what will be its vertical dimension? In Figure A–16, will it resemble Profile **I**, or will it resemble Profile **II**? We are free to draw it with any *vertical scale* we wish, perhaps to resemble either Profile **I** or **II,** or something in between, or more severe than either.

Since the hill is less than 500 feet high, if we draw the profile with a vertical scale equal to the horizontal scale (1/24,000, or one inch = 2000 feet), we will draw it as Profile **I**. It is only 1/4 inch high (since 1/4 inch represents 500 feet, if 1 inch represents 2000 feet). But Profile **I,** though accurate and "to scale," does not tell us very much.

If we wish to stress the vertical element of the topography along line **AB,** we must *exaggerate* the vertical scale. The peak of Profile **II** is 2 inches above its base, which means that the vertical scale of this profile is 2 inches equals 500 feet, or one inch equals 250 feet. To show a height of 2000 feet at this scale, we would have to draw a profile 8 inches high! This means that the vertical scale of Profile **II** is 8 times as great as the horizontal scale: its *vertical exaggeration* is × 8.

Although Profile **II** may be exaggerated (distorted), it forcefully brings out the contrast between steep slope **CA** and gentle slope **CB**. This contrast is detectable in Profile **I,** and with a protractor, you can actually measure the difference in its slopes. Nevertheless, visual study of this low profile is not very revealing or informative.

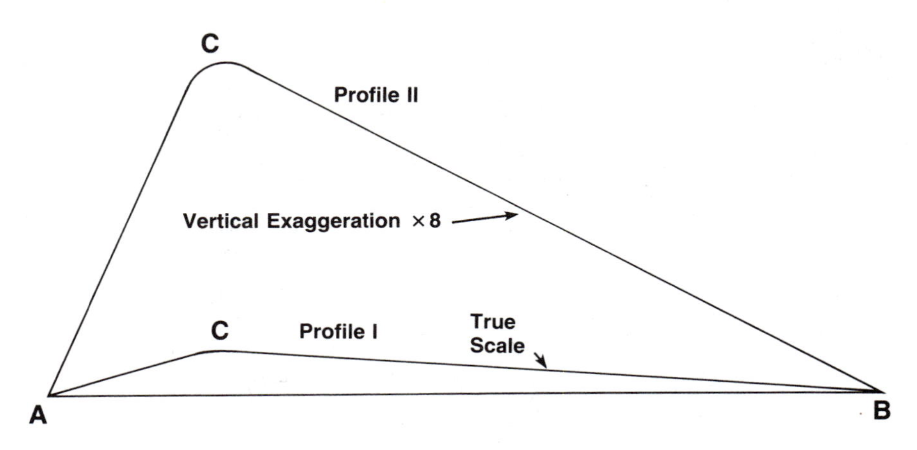

FIGURE A–16

So, the first thing to establish is the vertical scale (exaggeration) with which you plan to work. Let us construct a profile along line **AB** using a vertical scale of one inch to 250 feet (vertical exaggeration of × 8). (We must proceed slowly and painstakingly, because students experience difficulty and create problems for themselves by trying to rush a procedure they really have not mastered.)

A contour is a potential shoreline. What, then, is a shoreline? Think of a shoreline as the contact between (or intersection of) a *horizontal plane*—the surface of the sea—and a *sloping surface*—the ground (Figure A–13). In Figure A–15 we may think of the 1000-foot contour as the shoreline of a sea which is 1000 feet higher than present sea level. It is the intersection between the ground and a 1000-foot-high horizontal plane.

By the same token, the 1100-foot contour is the intersection between the sloping ground and an 1100-foot-high horizontal plane, an 1100-foot-high "sea level." And the 1200-, 1300-, and 1400-foot contours are lines of intersection between the sloping ground surface and 1200-, 1300-, and 1400-foot-high horizontal planes or "sea levels."

Points **D** and **E** are simply two points on the contour of intersection between the ground and the 1100-foot-high horizontal plane (the 1100-foot-high "shoreline"). The only thing special about these two points is that they also happen to lie on line **AB,** the line along which we are going to construct our profile.

The next step is to draw the sea-level planes as they would appear from the *side* (Figure A–17). The 1000-foot plane is a horizontal line that we label 1000. To add the 1100-foot plane, we draw another line, parallel to and 0.4 inch above the 1000-foot line. (Since 1 inch equals 250 feet on our chosen vertical scale, 0.4 inch must represent 100 feet, which is the vertical interval between the two horizontal planes.) Then, at 0.4-inch intervals, we add the lines that represent the 1200-, 1300-, and 1400-foot planes. We must also add the 1500-foot plane, to avoid plotting point **C** too high (it must lie below the 1500-foot level).

Next, note on Figure A–15 the points **D** though **L** at which line **AB** cuts the higher contours. These are the control points we will transfer to the horizontal lines we have just drawn in Figure A–17. When we connect these transferred control points we will have our profile.

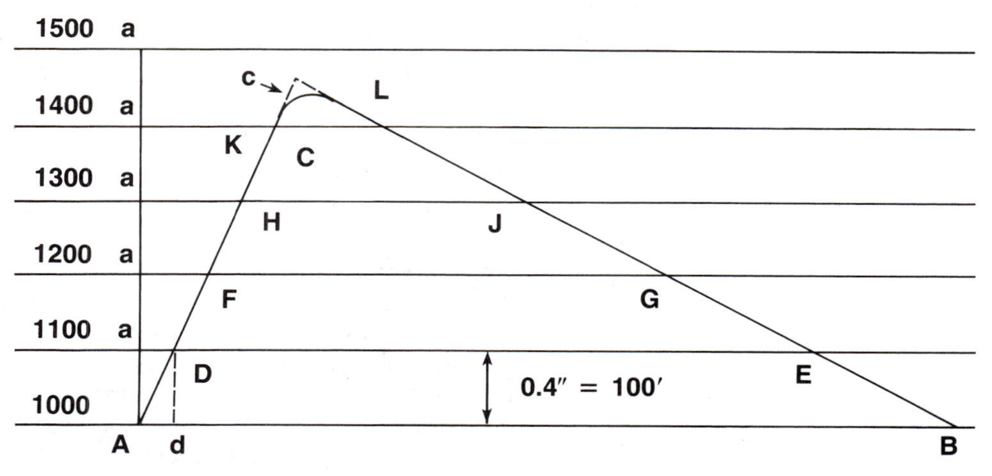

FIGURE A–17

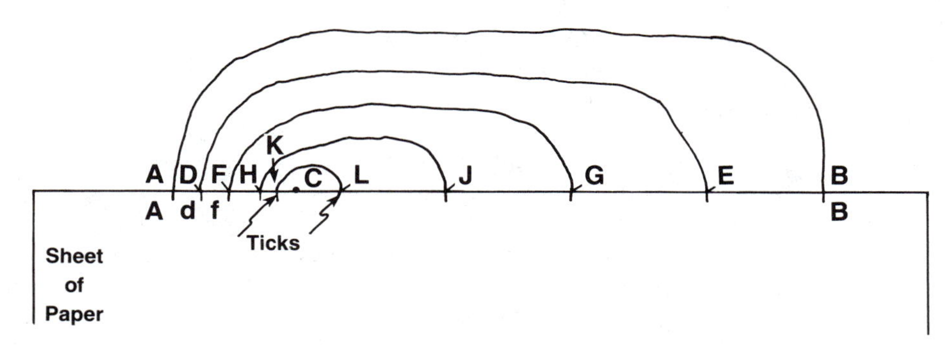

FIGURE A–18

Figure A–18 shows the next step. Lay a sheet of paper on Figure A–15, with the edge along line **AB**. On this edge, mark points **A** through **L,** including point **C**.

Then lay this paper edge along the 1000-foot line of Figure A–17. Mark points **A** and **B**. We can then "see" on Figure A–17 the *profile* view of these two points.

Points **D** and **E** are on the 1100-foot plane, so we will plot them on the 1100-foot profile line of Figure A–17. We must be extremely careful, since this is the most critical step in our procedure.

As you can see on Figures A–15 and A–18, point **D** lies not far to the right of point **A**. We can plot this point on Figure A–17 in either of two ways: (1) We can mark distance **Ad** on the 1000-foot level line of Figure A–17, to point **d,** and erect a perpendicular from **d** to **D** on the 1100-foot line. (2) Or, we may first erect a perpendicular from **A** to **a** on the 1100-foot line, and then mark off distance **aD** to locate **D**.

Since we will be plotting similar points on the higher-level lines, it is simplest to extend the perpendicular up from **A** to intersect them all. Then we can mark the other points to the right of this perpendicular. We may refer to all points on the perpendicular as points "**a**." Thus, point **F** is plotted at distance **aF** along the 1200-foot line, point **H** at distance **aH** along the 1300-foot line, and so on. (Leave point **C** for last.)

After plotting all points except **C,** connect **A** through **K,** and **B** through **L**. Use smooth lines, since from **A** to **K** and from **B** to **L** the contours are about equally spaced.

As for **C,** were we simply to extend these straight lines upward until they intersect (dotted lines up to point **c**), we would draw an unnaturally pointed hilltop. But if we round the top a bit (solid line), we will place **C** in a more-reasonable position, and likely more correct.

Exercise. Now, construct a profile on your own. Use Figure A–15, line **MN**. Use the same vertical scale of one inch equals 250 feet, and thus the same vertical exaggeration of × 8. Note that the highest point, **Z,** at about 1270 feet, is far lower than point **C** at over 1400 feet. Since distant point **S** is a mere 70 feet lower, the slope from **Z** down to **S** must be extremely gentle (much gentler than the slope from **C** down to **G** on profile line **AB**).

Also note that profile **MN** cuts across a closed depression (center at **Y**). Starting at **N** and proceeding westward (left), we rise to **Q** at 1100 feet, and then on to **X** (above 1100 feet, but below 1200 feet).

Since the contour around **Y** is a depression contour, along that line the slope is downward, toward **Y**. Thus the ground immediately surrounding that contour must be yet higher. There must therefore be another 1100-foot contour, since that must be the elevation of any contour just below **X**. We do not know the exact elevation of **Y** (we *do* know that it lies below 1100 feet and above 1000 feet).

Continuing to the west past **Y,** we cross the 1100-foot contour again at **T**. The next contour is the closed 1200-foot, which gives us the elevation not only of point **S,** but also of point **W**. **W** lies outside of (above) the 1100-foot depression contour, and outside of (below) the 1200-foot closed normal or regular contour.

It is important that you comprehend why points **X** and **W** lie at approximately the *same* elevation of 1150± feet. **X** falls between a closed 1100-foot contour and an 1100-foot depression contour. **W** falls between a closed 1200-foot contour and the *same* 1100-foot depression contour.

Drawing topographic profiles is not nearly as difficult as it may at first appear. On the other hand, it is not as easy as it appears, either. If you can draw a profile along line **MN,** and do it correctly, you are on firm ground.

LATITUDE AND LONGITUDE

Topographic Quadrangles

Each topographic map published by the USGS is bordered on the north and south by parallels of latitude, and on the east and west by meridians of longitude. *Latitudes* are parallel east-west lines (see Figure A–19). The *equator* is designated as 0°. The maximum value of *north latitude* is 90° at the North Pole, and the maximum value of *south latitude* is 90° at the South Pole.

Longitudes are north-south lines. They are numbered from the *prime meridian* (0°) located at Greenwich, England. Longitude lines east of the prime me-

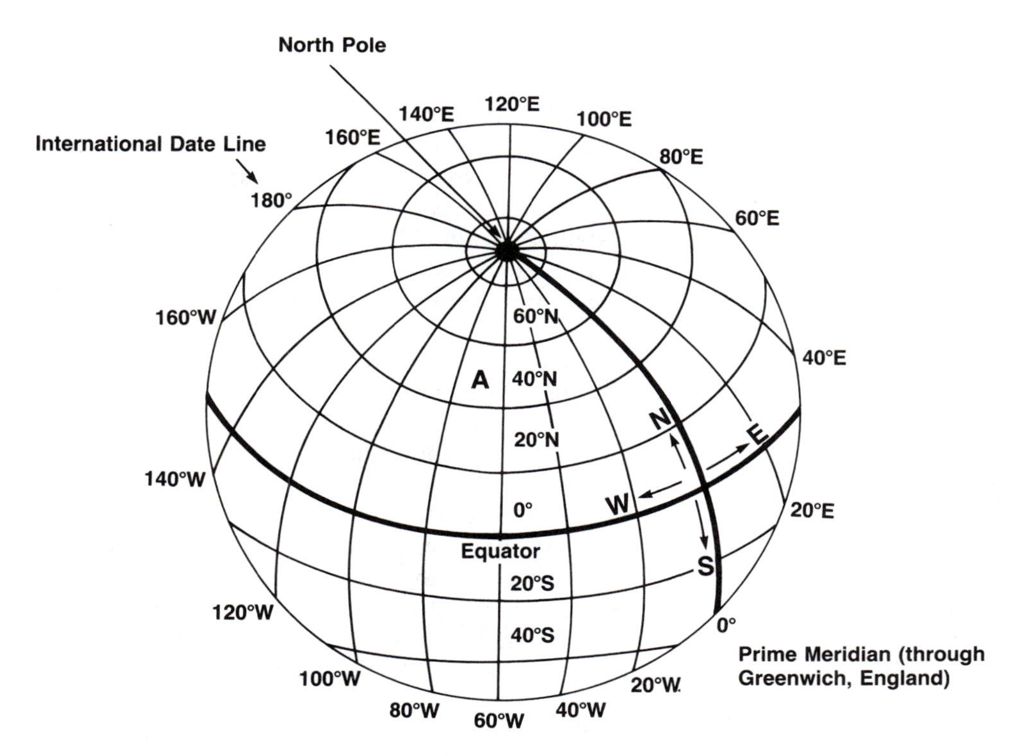

FIGURE A–19

ridian are designed as *east longitude*; those west of the prime meridian are *west longitude*. The 180° longitude line, located on the opposite side of the Earth from the prime meridian, is the *international date line*, where the "day" officially changes (e.g. from Monday to Tuesday) when you travel across it.

Latitude and longitude lines make a grid for locating points on the Earth. It is called the *geographic grid*. On the Earth's curved surface, latitude and longitude lines intersect at spherical right angles. But on a flat map, the angles cannot be 90°. And where longitude lines converge at the poles, the sections of the grid are triangular. In the Northern Hemisphere, converging longitude lines make the north border of each grid slightly shorter than the south border. The area covered by a map that uses latitude and longitude for boundaries is not a rectangle (four right angles), but a *quadrangle* (four angles).

The most-common large-scale USGS quadrangle scales are:

- □ 1/250,000, covering one degree of latitude and two degrees of longitude.
- □ 1/125,000, covering 30 minutes of latitude and longitude.
- □ 1/62,500, covering 15 minutes of latitude and longitude.
- □ 1/31,680, covering 7.5 minutes of latitude and longitude.
- □ 1/24,000, also covering 7.5 minutes of latitude and longitude.

Both the 1/31,680 and the 1/24,000 maps cover the same ground area, but the 1/24,000 map is printed on a larger sheet of paper.

Location of Points Using Latitude and Longitude

The location of any ground point shown on a topographic quandrangle can be precisely designated by latitude and longitude. Here is a very general example: Point **A** (Figure A–19) lies between 40° and 50° north latitude. Its approximate location is 43°N. This locates point **A** as being anywhere on a "ring" around the Earth, 43° north of the equator.

To determine specifically where point **A** is on that "ring"—i.e. how far east or west point **A** is from the prime meridian—the north-south longitude lines are used. On Figure A–19, point **A** is between 60° and 80° west longitude. Its approximate location is 78°W. Thus, the approximate location point **A** may be stated as "about 43N, 78W."

Each degree of either latitude or longitude is divided into 60 minutes (60′) and each minute into 60 seconds (60″). A more-exact location of point **A** might be 42° 48′ 16″ north and 78° 06′ 24″ west.

There are 360° in a circle. The circumference of the Earth is approximately 25,000 miles. If you divide the circumference by 360°, at the equator one degree of either latitude or longitude is about 69 miles wide. We round this figure for convenience, and our rule of thumb becomes "a degree covers a distance of about 70 miles." If we are studying a 15-minute quadrangle, we know that its north-south ground dimension is 70 miles ÷ 4, or about 17 miles.

We cannot make such a direct estimate of the map area's east-west dimension, because meridians converge poleward. Hence, the farther the area from the equator, the shorter the ground width of the area. However, since we can estimate the north-south dimension, all we need is to compare the "width" of the map to the "height." These two estimated dimensions give a fairly good idea of the ground area that the map depicts.

Although geographical coordinates (latitude and longitude) provide extremely accurate location designations, they have the disadvantage of being awkward and time-consuming to work with, since each location requires map measurements and calculations. For example, if a ground point is shown on a map to lie 3¾ in. to the north of the 35° north latitude line, and the 35° 15′ north latitude line lies 16 in. north of the 35° line, what is the latitude (in degrees, minutes, and seconds) of the ground point? Consider how long it takes you to answer this question, and imagine that you are required to give the precise locations of 150 such ground points!

Exercise. At 60° north and 60° south, one degree of longitude is one-half as wide as one degree of longitude at the equator. Question: 60° north latitude is two-thirds of the distance to the North Pole. 45° north latitude is one-half the distance to the Pole. Why is it not at 45°N that the length of the degree is one-half the length at the equator?

TOWNSHIP AND RANGE

In the United States, north and west of the Ohio River, there is an additional system of land division. This huge region has been divided into 32 survey areas. In each, the basic area unit is a six-mile square known as a *township* (Figure A–20).

In the latitude/longitude system, Earth's geographical grid is based on the Greenwich meridian (0° longitude) and the equator (0° latitude). In the *Township and Range System,* each of the 32 survey areas is based on its own 0° *principal meridian.* To the east and west of each principal meridian, six-mile-wide

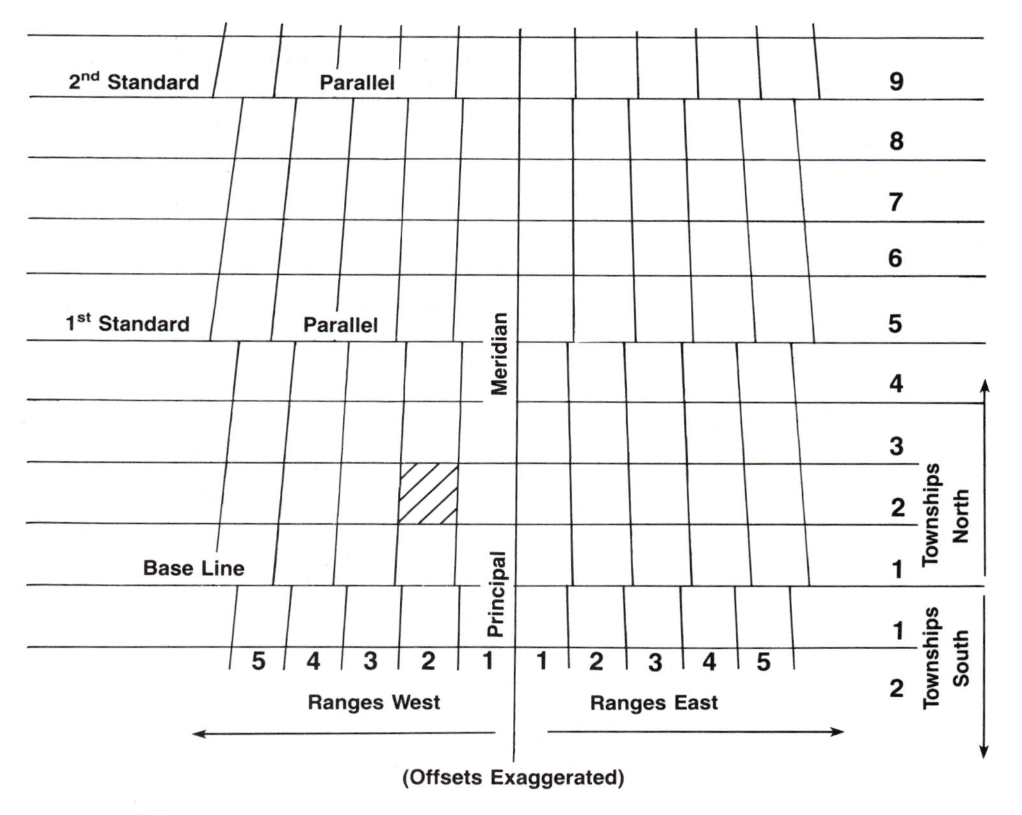

FIGURE A–20

rows of *ranges* are marked off. Each of the 32 areas has its own *base line,* to the north and south of which are six-mile-high tiers of *townships*.

Since they are meridians of longitude, the north-south lines that border the ranges converge toward the north. This causes severe narrowing of the ranges to the north of the base line, and corresponding severe widening of the ranges to the south of the base line. To avoid this problem, secondary base lines, called *standard parallels,* were surveyed at intervals of four townships. The small amount of convergence that occurs northward from one standard parallel to the next, a distance of twenty-four miles, is corrected for by minor offsets, as shown in Figure A–20.

Motorists unfamiliar with the U. S. Public Land Survey (official title of the Township and Range System) may be puzzled while driving straight roads that follow this survey's lines, when they encounter the offsets caused by survey adjustments along the standard parallels. And such offsets are not restricted to open country roads: residents of Phoenix, Arizona, who must cross Base Line Road, accept its bends and offsets as part of their daily lives (Figure A–21).

Townships alone do not describe precise locations. To state that a small area or a specific point happens to lie within a particular 36-square-mile township (e.g. in Figure A–20, highlighted **Township 1 North, Range 2 West**) falls short of pinpoint accuracy. To permit more specific location designation, each township is subdivided into one-square-mile units known as *sections* (Figure A–22).

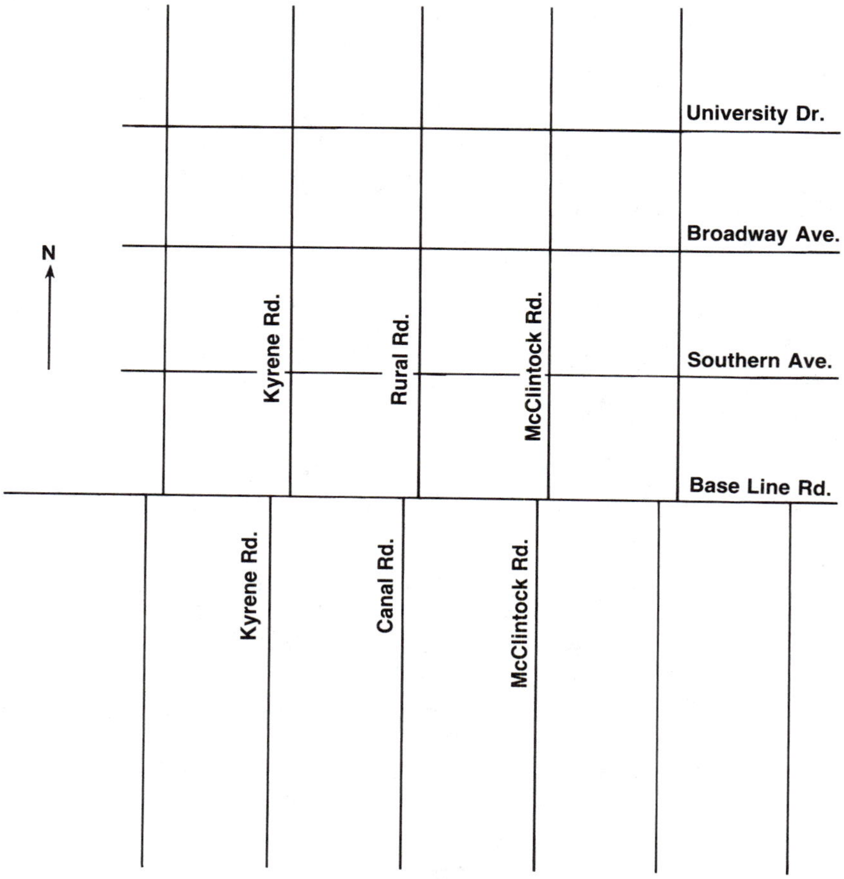

FIGURE A–21

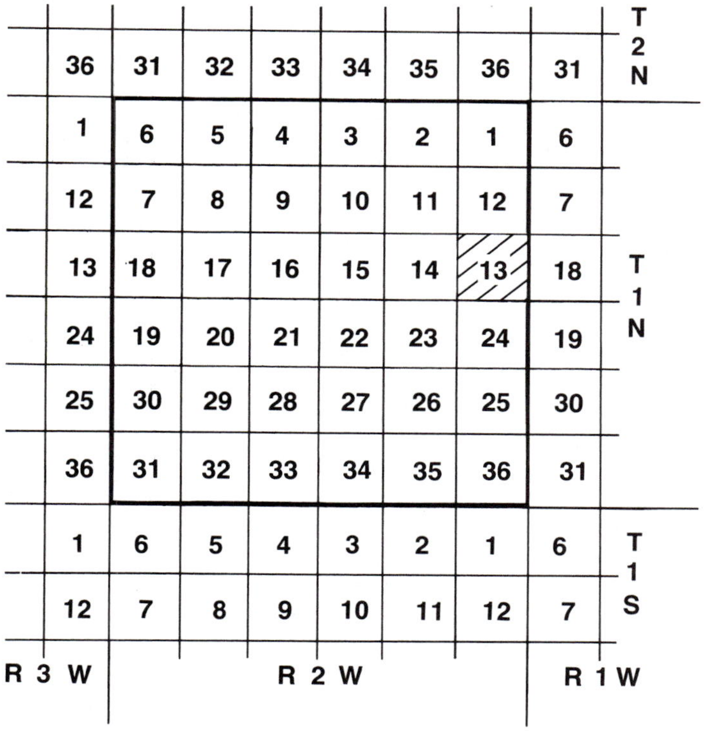

36	31	32	33	34	35	36	31	T 2 N
1	6	5	4	3	2	1	6	
12	7	8	9	10	11	12	7	
13	18	17	16	15	14	13	18	T 1 N
24	19	20	21	22	23	24	19	
25	30	29	28	27	26	25	30	
36	31	32	33	34	35	36	31	
1	6	5	4	3	2	1	6	T 1 S
12	7	8	9	10	11	12	7	
R 3 W			R 2 W				R 1 W	

FIGURE A–22

These are numbered **1** through **36,** in zigzag fashion. The highlighted section is identified by both section and township numbers: **Section 13, Township 1 North, Range 2 West**. This may also be written **Sec 13, T1N, R2W,** or **Sec 13-1N-2W**.

Figure A–23 shows how each section may in turn be further subdivided. It can be cut into halves (N 1/2, or simply N) and quarters (SE 1/4, or SE). Each quarter may be divided into halves and quarters, and so on, as far as one wants to carry the subdividing:

- □ Area **A** is thus N 1/2 Sec 13, or N Sec 13
- □ Area **B** is SW 1/4 Sec 13, or SW Sec 13.
- □ Area **C** is SW 1/4 SE 1/4 Sec 13, or SW SE Sec 13.
- □ Area **D** is SW 1/4 NE 1/4 SE 1/4 Sec 13, or SW NE SE Sec 13.
- □ Area **E** is S 1/2 NE 1/4 NE 1/4 SE 1/4 Sec 13, or S NE NE SE Sec 13.

Exercise. As a review, identify the following areas shown in Figure A–24. Also, how many acres of land are included in each area? (There are 640 acres in a section.)

1. Area **A:**
2. Area **B:**
3. Area **C:**
4. Area **D:**

Finally, here is a question for which you might have to sketch a simple map. In Mesa, Arizona, you are driving northward on Dobson Road. Imme-

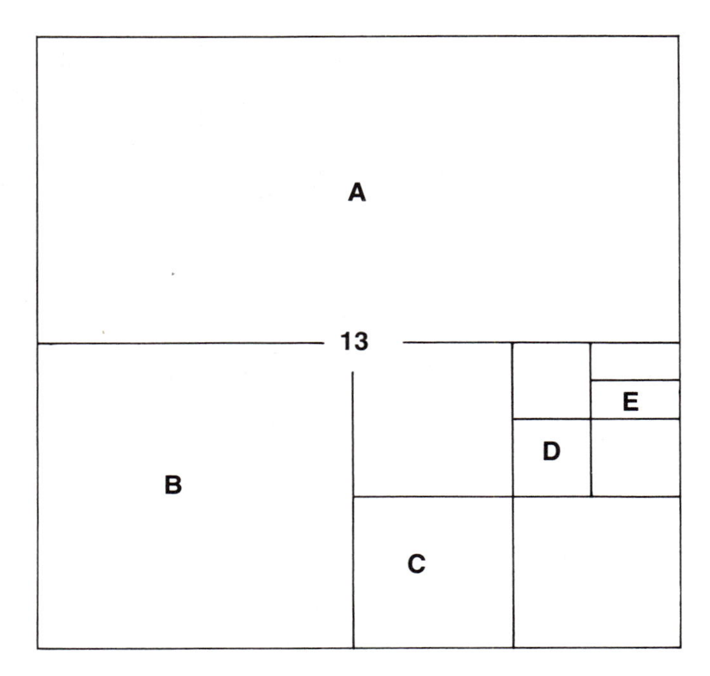

FIGURE A–23

diately after crossing Base Line Road, you must turn right, travel several hundred feet, and then turn left to continue due northward. Dobson Road is typical of all north-south roads that cross Base Line Road in the Phoenix area. Is the principal meridian of the Arizona Survey to the east or west of Dobson Road and its companions?

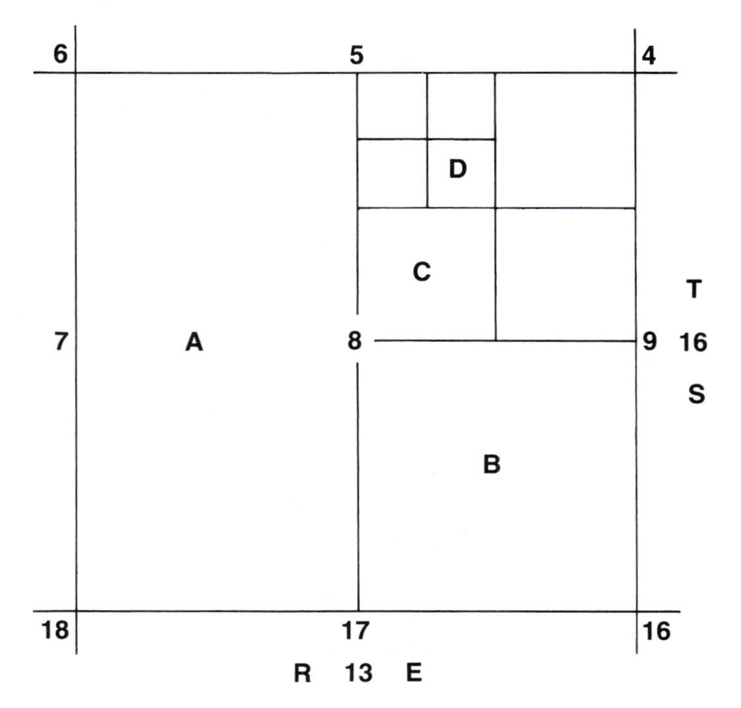

FIGURE A–24

GRADIENTS

Throughout this book we remind you that some slopes are steeper than others, and that the channels of some streams are steeper than others. Without exception, all such comparisons are qualitative rather than quantitative. This is because our main concern is *interpretation,* rather than measurement of topographic features and their relations.

However, there may be times when you must determine the actual inclination of a ground surface depicted on a map—an alluvial fan, a mountain slope, or the rate of descent of a river. There are several ways in which you can compute and express the inclination of such a slope. The inclination is called *gradient*.

1. The gradient may be expressed in *degrees,* measured down from the horizontal. A vertical cliff therefore has a gradient of 90°.
2. The gradient may be expressed as a *percentage* (%). A vertical slope has a gradient of 100%.
3. The gradient may be expressed *verbally*: "a gradient of 14 feet per mile." Note that the unit for the vertical difference (feet) is not the same as the horizontal distance (mile).
4. The gradient may be expressed as a *ratio:* "a gradient of 1 in 3," in which no particular unit of measurement is mentioned. (It is like a map scale.) The ratio "1 in 3" may be read, "over a horizontal distance of three kilometers, the ground descends one kilometer," or "over a horizontal distance of 3 yards, the ground descends one yard," or "there is one unit length difference in elevation for every 3 of the same unit lengths of distance."

To obtain from a topographic map a slope's gradient in either degrees or percentage, you must draw a profile of the slope and measure it with a protractor. On the other hand, verbal or ratio gradients are relatively easy to extract and compute from a topographic map.

Gradients in Feet Per Mile

In Figure A–15, let us express slope **LB** in feet per mile. Use the map's graphic scale to obtain the slope's horizontal length or distance (its *horizontal equivalent*) from **L** to **B**. This distance is 6336 feet, which, divided by 5,280 feet in one mile, equals 1.2 miles. The difference in elevation (*vertical interval*) between **L** and **B** is 400 feet (1400 − 1000 = 400). The gradient of this slope is 400 feet in 1.2 miles. Converted to a unit-mile basis (by dividing both values by 1.2), this is 333.3 feet in 1 mile, or *333.3 feet per mile*.

Gradients in Ratio Form

The same gradient equals 400 feet difference in 6336 feet. After dividing both sides by 400, we may state that the gradient equals 1 foot difference in elevation for every 15.84 feet of distance, which in turn may be expressed by the ratio *1 in 15.84*.

This means that, for every 15.84 units (any unit) of distance, there is a fall in height of 1 unit (same unit).

Stream Gradients

Figure A–25 depicts a stream which drops from 1200 feet at point **A** to 600 feet at point **C**. Though the stream makes many small bends and twists, its overall course from **A** to **B** is relatively straight, as is its course from **B** to **C**. When we measure the lengths of these two main stream segments, we obtain approximate horizontal equivalents of 1.25 and 2.5 miles, respectively, or a total length of 3.75 miles. We understand that the actual length of the stream from **A** to **C** is slightly greater than 3.75 miles, but for our purposes we may accept this figure as the *approximate* length.

The approximate average gradient of stream segment **AC** is thus 600 feet in 3.75 miles, or (divide both values by 3.75) *160 feet per mile.*

There are 19,800 feet in 3.75 miles (5280 × 3.75 = 19,800). The approximate average gradient is 600 feet difference over a distance of 19,800 feet, which reduces to a 1-foot difference in a distance of 33 feet. Expressed in ratio form this is: 1 in 33.

Note that the contours in the western half of segment **AB** are much closer together than those downstream. This indicates that the gradient of the upstream segment is considerably greater than the 160-feet-per mile (1 in 33) average gradient. However, downstream from bend **B,** the gradient is noticeably less than 160 feet per mile (perhaps 1 in 40).

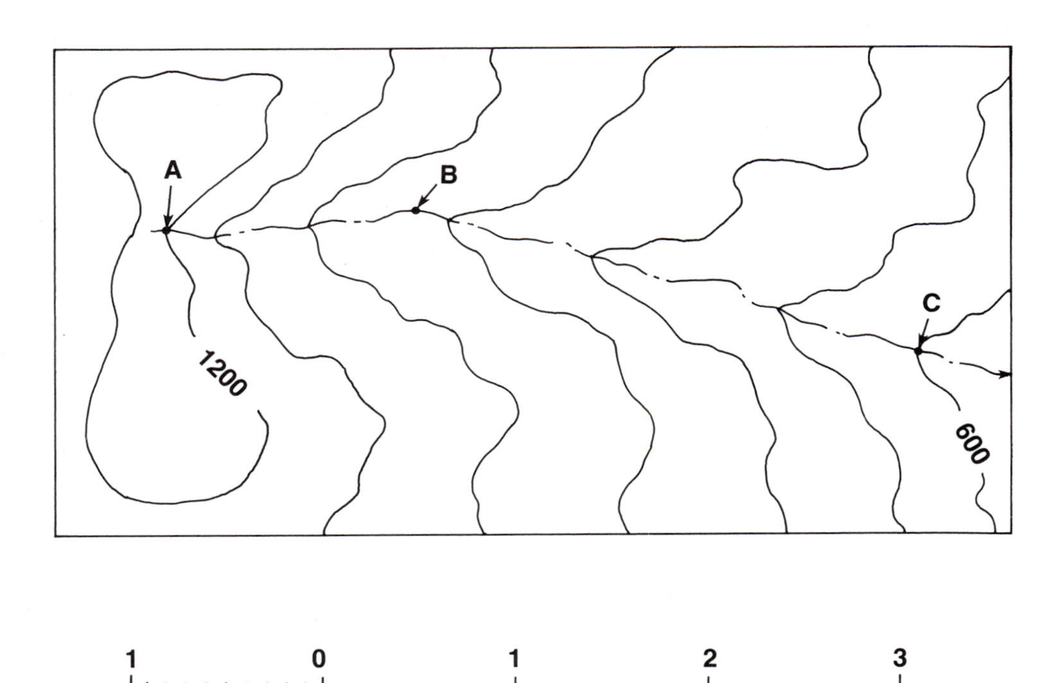

FIGURE A–25

References

Bates, Robert L., and Julia A. Jackson, eds. 1987. *Glossary of geology*. 3d ed. Alexandria, VA: American Geological Institute, 788 p.

Bennison, A. P., and J. M. Webb, comps. 1986. *Geological highway map of the mid-continent region*. Tulsa, OK: American Association of Petroleum Geologists.

Berg, T. M., and others, comps. 1980. *Geologic map of Pennsylvania*. Harrisburg, PA: Commonwealth of Pennsylvania, Department of Environmental Resources, Bureau of Topographic and Geologic Survey. Scale 1/250,000.

Calver, J. L., and C. R. B. Hobbs, Jr., eds. 1963. *Geologic map of Virginia*. Richmond: Virginia Division of Mineral Resources. Scale 1/500,000.

Cooper, John R. 1973. *Geological map of the Twin Buttes quadrangle, southwest of Tuscon, Pima County, Arizona*. Washington: U.S. Geological Survey Miscellaneous Geological Investigations Map I-745.

Davis, William Morris. 1909. *Geographical essays*. Boston: Ginn & Co. Reprint. New York: Dover Publications, 1954, 777 p.

Hernon, R. M., W. R. Jones, and S. L. Moore. 1964. *The geology of the Santa Rita quadrangle, New Mexico*. Washington: U.S. Geological Survey Geologic Quadrangle GQ-306.

Kalaswad, Sanjeev. 1983. Dynamic disequilibrium in the Utukok River–Lookout Ridge area, Alaska: A radar and map investigation. Master's thesis, Indiana State University, 78 p.

Klemic, Harry. 1966. *Geological quadrangle map of the Hammacksville quadrangle, Kentucky-Tennessee*. Washington: U.S. Geological Survey Geologic Quadrangle GQ-540.

Lewis, J. Volney, and Henry B. Kummel. 1910–1912. *Geologic map of New Jersey*. Trenton: State of New Jersey, Department of Conservation and Economic Development. Atlas sheet no. 40, scale 1/250,000.

Love, J. D., and A. C. Christiansen. 1985. *Geologic map of Wyoming*. Washington: U.S. Geological Survey. Scale 1/500,000.

Malde, Harold E., Howard A. Powers, and Charles H. Marshall. 1963. *Reconnaissance and geologic map of west-central Snake River Plain, Idaho*. Washington: U.S. Geological Survey Miscellaneous Geological Investigations Map I-373. Scale 1/125,000.

Miser, Hugh D., and others. 1954. *Geologic map of Oklahoma*. Washington: U.S. Geological Survey. Scale 1/500,000.

Muehrcke, Phillip C. 1978. *Map use—Reading, analysis, and interpretation*. Madison, WI: JP Publications, 474 p.

Oetking, Philip, Dan E. Feray, and H. B. Renfro, comps. 1967. *Geological highway map of the southern Rocky Mountain region*. Tulsa, OK: American Association of Petroleum Geologists.

Raisz, Erwin Josephus. 1952. *Map of the landforms of the United States*. Washington: Army Map Service.

Renfro, H. B., and Dan E. Feray, comps. 1968. *Geological highway map of the mid-Atlantic region*. Tulsa, OK: American Association of Petroleum Geologists.

Stose, George W. 1936. *Delaware water gap, PA-NJ*. Washington: U.S. Geological Survey. Topographic quadrangle map (scale 1/62,500) on one side; other side includes description of geology, with photos, geologic cross section, and geologic map (scale about 1/250,000).

Stose, George W., and O. A. Ljungstedt. 1932. *Geologic map of the United States*. Washington: U.S. Geological Survey. Scale 1/2,500,000.

Swadley, W. C. 1963. *Geology of the Flaherty quadrangle, Kentucky*. Washington: U.S. Geological Survey Geologic Quadrangle GQ-229.

U.S. Geological Survey. 1969. *Geologic map of Arizona*. Washington. Scale 1/500,000.

Vitaliano, Dorothy B. 1973. *Legends of the Earth—Their geologic origins*. Bloomington: Indiana University Press.

Glossary

Most of the definitions in this glossary were taken from or based on definitions in the *Glossary of Geology*, 3rd ed., edited by Robert L. Bates and Julia A. Jackson (1987) and were reprinted by permission of the American Geological Institute.

ablation (glacial) All processes by which snow and ice are lost from a glacier.

aggradation Long-term building up of a stream channel by deposition. (Cf.* *degradation*.)

alluvial fan Low, fan-shaped deposit of alluvium, deposited by streams where the gradient changes abruptly. In semiarid regions alluvial fans often occur at the mouths of canyons.

alluvial terrace Terrace cut in alluvium by renewed downcutting of a stream.

alluvium Unconsolidated sediment deposited by running water.

angular stream pattern See *stream patterns*.

angular unconformity Unconformity in which younger strata lie across the upturned, truncated edges of more steeply dipping older strata.

annular stream pattern See *stream patterns*.

anomaly Geomorphic feature or relation that does not seem to "fit" into its surrounding terrain.

antecedent stream Transverse stream course that predates the structure it crosses. Calling a stream "antecedent" implies that it developed and has maintained its original course by incision at a rate equal to or exceeding uplift of the land, thereby cutting through what seem to be structural barriers. Figure 9–2 is a striking example.

anticlinal lowland Topographic lowland produced by erosion of rocks at the anticlinal axis. (Cf. *synclinal lowland*.)

anticlinal nose Gently descending ridge developed at the axis of a plunging anticline. (Cf. *synclinal nose*.)

anticlinal ridge Ridge or mountain developed on resistant strata at the axis of an anticline. (Cf. *synclinal ridge*.)

anticlinal valley Valley produced by erosion along an anticlinal axis. (Cf. *synclinal valley*.)

anticline Fold in which the core contains stratigraphically older rocks; generally convex upward. (Cf. *syncline*.)

---

*Cf means *confer*, or *compare*.

apparent dip Angle that a structural surface makes with the horizontal, measured in any random, vertical section rather than perpendicular to strike.

arête Narrow, serrate mountain crest sculptured by glaciers and resulting from continued backward growth of walls of adjoining cirques.

asymmetric glaciation More intense alpine glaciation of northerly and easterly slopes than of southerly and westerly slopes (in the northern hemisphere).

available load Material that can be transported by a natural transporting agent such as streams, glaciers, wind, and waves. In streams it may be *dissolved load* (carried in solution), *suspended load* (sand, silt, etc., carried by turbulent flow), or *bed load* (sand, pebbles, and occasionally boulders) carried along the bottom.

azimuth Direction of a horizontal line as measured on an imaginary horizontal circle.

backwasting Retreat of a scarp or slope as a result of removal of earth material by weathering and erosion.

bajada See *piedmont slope*.

barbed tributary Tributary stream that joins the master stream at an acute angle that points in the master stream's upstream direction.

barchan Crescent-shaped sand dune, where the gentle convex side faces the wind and the steep slope on the concave side is between the horns.

barrier beach Long, elongate sand ridge slightly above high-tide level. Generally parallel to the shore, but separated by a lagoon.

beach drift Drift of sand along a beach by swash and backwash waves that strike the shore obliquely.

beach ridge Low, essentially continuous sand ridge, created by waves and currents on backshore of beach beyond present limit of storm waves or ordinary tides. Occurs individually or as series of approximately parallel deposits. Ridges that are essentially parallel to the shoreline represent successive positions of shoreline advance.

bearing Angular direction of any place or object at one fixed point to another, especially the horizontal direction of a line on the Earth's surface with reference to the cardinal points of the compass.

bedding plane Plane, corresponding to the original surface of a layer of sediment, that separates layers of stratified rocks from the preceding or following layer.

beheaded stream Stream that has lost its upstream portion due to diversion by capture or piracy.

block faulting Type of normal fault in which the crust is divided into structural or fault blocks of different elevations and orientations.

bluff Cliff with a steep, broad base.

braided stream Shallow stream characterized by numerous branching and reuniting channels, separated from each other by branch islands. Produces a pattern resembling braided hair.

breached anticline Eroded anticline in which the older strata are exposed along the axial beds. Often characterized by an anticlinal valley flanked by in-facing scarps developed in resistant strata.

caliche Gravel, rock, soil, or alluvium cemented with calcium carbonate in arid and semiarid climates. Also called *hardpan, duricrust,* or *calcrete* in some localities.

capacity Ability of a current of water or wind to transport detritus, as shown by the amount measured at a given point per unit of time. (Cf. *competence.*)

cap rock Resistant rock unit that forms steep-sided cap on a mesa, butte, or plateau front. Underlain by less resistant rock.

captor stream In stream capture or piracy, the stream that captures or beheads the victim stream.

capture (stream) Diversion of the headwaters of one stream into the channel of another, usually when one stream erodes headward into the flank of another.

centripetal stream pattern See *stream patterns.*

cirque Amphitheatre-like, steep-walled depression gouged out at the head of a mountain or alpine glacier.

clastic Rock or sediment composed of broken fragments derived from preexisting rocks or minerals that have been transported as solid particles from their place of origin.

clastic dike Dike consisting of clastic sediment that has filled a space in bedrock, either from above or below.

cliffed headland Marine cliff created by wave attack against a headland.

closed contour Contour that forms a closed loop.

cockscomb ridge Sawtoothlike line of sharp ridge segments produced when a steeply dipping hogback is segmented by a series of closely spaced water gaps.

competence Ability of a current of water or wind to transport detritus in terms of *particle size* rather than amount. Measured as the diameter of the largest particle transported. (Cf. *capacity.*)

component dip See *apparent dip.*

conformity Normal stratigraphic contact; a plane separating two beds that do not have an erosion interval between them. (Cf. *unconformity.*)

consequent stream See *stream genesis.*

contact Surface separating two bedrock units or formations.

contour On a topographic map, a line connecting points of equal elevation.

contour interval Difference in elevation between adjacent contour lines.

contour V A notch along the contour line that points toward the next-higher contour level on the topographic map. The V depicts the presence of a stream bed. Stream flow is in the opposite direction from the point of the V.

corrasion Process of erosion whereby rocks and soil are mechanically removed or worn away by abrasive action of solid particles moved along by wind, waves, running water, glaciers, or gravity.

cuesta As used in this book, an extremely asymmetric ridge having one long, gentle slope that coincides with dip direction.

cutbank Steep or overhanging slope on the outside of a meander curve, opposite the slip-off slope.

deflation Sorting, lifting, and removal of loose, dry, fine-grained particles by wind action.

degradation Wearing down or reduction of the Earth's surface by the processes of weathering and erosion; e. g., the long-term deepening by a stream of its channel (Cf. *aggradation*).

dendritic stream pattern See *stream patterns.*

depression contour Closed contour surrounding an area that is geographically lower than itself. Opposite to a normal closed contour, which surrounds an area higher than itself.

detritus Collective term for loose rock and mineral grains that have been worn off or removed by mechanical means such as disintegration or abrasion. Especially fragmental material, such as sand, silt, and clay, derived from older rocks and moved from its place of origin.

diastrophism General term for all Earth crustal movement produced by tectonic processes, including formation of ocean basins, continents, plateaus, and mountain ranges.

differential erosion Erosion of different rock units at different rates; a topographic etching process.

differential weathering Weathering of different rock units at different rates.

dike Tabular igneous rock body that cuts across the bedding or foliation of the country rock.

dip Angle of inclination of a planar rock body or structure (e. g., sedimentary unit or fault plane) measured down from the horizontal. Measured perpendicular to the strike of the structure and the vertical plane.

dip slope Topographic slope that coincides with the dip of the underlying tabular or planar bedrock.

discharge Rate of flow at a given moment, expressed as volume per unit of time (e.g., cubic feet per minute).

dissection See *stages in regional dissection*.

divide Line of separation (or ridge, summit, or narrow track of high ground) marking the boundary between two adjacent drainage basins or dividing surface waters that flow naturally in opposite directions.

divide migration Shifting of a divide as a result of more rapid erosion on one flank than on the other.

downdip In the direction of dip; opposite to updip.

downsection In a sequence of sedimentary strata, from youngest down to oldest (down in the stratigraphic section). (Cf. *upsection*.)

downwasting Topographic lowering by the processes of weathering and erosion.

drainage density Ratio between the total length of streams in a defined area and the size of the area.

drainage texture Qualitative term pertaining to drainage density (e.g., fine texture indicates high drainage density).

drumlin Streamlined hill, mound, or ridge, composed of glacial till or drift, oriented parallel to the flow direction of a former glacier.

elbow of capture Sharp, angular bend at the point where a stream has been diverted from its original course into that of the stream that has beheaded it.

elevation Height of the ground above a selected datum, usually mean sea level.

embayment Formation of a bay, as by the sea overflowing a land depression near the mouth of a river.

en echelon Features in an overlapping or staggered arrangement. Each is relatively short, but collectively they form a linear zone, in which the strike of the individual features is oblique to the zone as a whole.

entrenched meander Incised meander carved downward into the surface of the valley in which the meander originally formed.

ephemeral stream Stream that flows on rare occasions for short periods of time and whose channel is always above the water table. An ephemeral stream is dry more than an intermittent stream.

equilibrium (stream) Balanced state in which capacity and load are equal, so that transportation occurs rather than erosion or deposition.

esker Long, narrow, sinuous ridge, composed of gravel deposited by a subglacial or englacial stream flowing between ice walls or in an ice tunnel of a stagnant or retreating glacier, and left behind when the ice melted.

estuarine delta Delta that has filled, or is filling, an estuary.

estuary Seaward end of a river valley, where sea water and fresh water meet.

eustatic change Worldwide change in sea level that affects all the oceans.

exhumation Uncovering or exposure by erosion of a preexisting surface, landscape, or feature that had been buried.

fault-associated scarp Scarp that follows the trace of a fault. Its origin is unknown; it may be either a fault scarp or a fault-line scarp.

fault-line scarp Steep slope or cliff formed by differential erosion along a fault line (see Figure 1–5). (Cf. *fault scarp*.)

fault plane Plane along which fault slippage has occurred.

fault scarp Steep slope or cliff formed by direct movement along a fault (see Figure 1–4). (Cf. *fault-line scarp*.)

fault trace Fault line.

fjord Seaward end of a deeply excavated, partly submerged glacial trough.

flatiron One of a series of short, triangular hogbacks forming a spur or ridge on the flank of a mountain, having a narrow apex and a broad base, resembling a huge flatiron. Usually consists of a plate of steeply inclined resistant rock on the dip slope (see Figure 2–5).

floodplain Strip of relatively smooth land adjacent to a river channel, constructed by the present river in its existing regime. It becomes covered with water when the river overflows its banks.

genetic stream classification The relation of a stream to original slope, underlying bedrock, and structure.

geomorphology Science that treats the general configuration of the Earth's surface. Study of the classification, description, nature, origin, and development of present landforms and their relationships to underlying structures, and the history of geologic changes as recorded by these surface features.

graben Elongate, downdropped block, bounded by faults on its long sides.

graded stream Stream in which transporting capacity and load supplied to it are balanced. When averaged, neither degradation nor aggradation of the stream channel takes place.

gradient Steepness of a slope.

hachures Short, straight, evenly spaced lines used on a topographic map for indicating surfaces in relief. Drawn perpendicular to the contour lines. Also, inward-pointing ticks trending downslope for a depression contour.

headland Irregularity of land, especially of considerable height with a steep cliff face, jutting out from the coast into a large body of water; a bold promontory.

hillock Small, low hill.

hogback Any ridge with a sharp summit and steep slopes of nearly equal inclination on both flanks.

homoclinal ridge Ridge on either side of a fold axis.

homoclinal valley Valley between homoclinal ridges.

homocline General term for a series of rock strata having the same dip, e.g., one limb of a fold.

horst Elongate, relatively uplifted crustal block that is bounded by faults on its long sides.

hummock Rounded or conchoidal knoll, mound, hillock, or other small elevation. Also a slight rise of ground above a level surface.

imbricated Overlapping, as tiles on a roof or scales on a bud.

incised meander Generic term for an old stream meander that has become deepened by rejuvenation and that is more or less closely bordered by valley walls.

ingrown meander Continually growing or expanding incised meander, formed during a single cycle of erosion by enlargement or accentuation of an initial minor curve while the stream was actively downcutting. Meander that "grows in place."

ingrown stream Stream that has enlarged its original course by undercutting the outer banks of its curve.

inlier Area or group of rocks surrounded by rocks of younger age. (Cf. *outlier*.)

inselberg Prominent, isolated, residual knob or hill, rising abruptly from a lowland erosion surface in a desert region.

insequent stream See *stream genesis*.

intermittent stream Stream that flows only at certain times (e.g., during snow melt).

karst Type of topography formed on limestone, gypsum, and other rocks by dissolution, and characterized by sinkholes, caves, and underground drainage.

kettle Depression that is steep-sided and usually basin- or bowl-shaped, commonly without surface drainage. Occurs in glacial-drift deposits. Often contains a lake or swamp.

knob Rounded eminence, as a knoll, hillock, or small hill or mountain.

lacustrine Pertaining to, produced by, or formed in, a lake or lakes.

lagoon Shallow stretch of sea water, such as a sound, channel bay, or saltwater lake, near or communicating with the sea but partly or completely separated from it by a low, narrow, elongate strip of land.

Landsat Unmanned, Earth-orbiting NASA satellite that transmits multispectral images to Earth receiving stations.

lee Side of a hill (or prominent object) that is sheltered or turned away from the prevailing wind.

littoral current Ocean current caused by the approach of waves to a coast at an angle.

load Material that is moved or carried by a natural transporting agent, such as a stream, glacier, wind, or waves.

longitudinal profile Profile of a stream or valley, drawn along its length from the source to the mouth.

longshore drift Material (such as shingle, gravel, sand, and shell fragments) that is moved along the shore by a littoral current.

magnetic north Uncorrected direction indicated by the north-seeking end of a magnetic compass needle.

map interpretation *Explaining* the topography as it is depicted on a topographic map. (Cf. *map reading*.)
map reading *Recognition* of topography as it is depicted on a topographic map. (Cf. *map interpretation*.)
maturity (dissection) See *stages in regional dissection*.
maturity (streams) See *stream stages*.
microclimate Climatic structure close to the Earth's surface, affected by the character of surface materials.
misfit stream Stream that is either too large or too small to have eroded the valley in which it flows.
monadnock Upstanding rock, hill, or mountain of circumdenudation rising conspicuously above the general level of a peneplain in a temperate climate, representing an isolated remnant of a former erosion cycle in a mountain region that has been largely beveled to its base level.
monocline Local steepening in an otherwise uniform gentle dip.
moraine (glacial) Mound, ridge, or other distinct accumulation of unsorted, unstratified glacial drift, predominantly till. Deposited chiefly by direct action of glacier ice, in a variety of topographic landforms that are independent of control by the surface on which the drift lies.

nickpoint Any interruption or break in slope, especially in the longitudinal profile of a stream or its valley.

obsequent fault-line scarp Fault-line scarp that faces the opposite direction from the original scarp, or in which the structurally downthrown block is higher than the upthrown block (see Figure 1–5). (Cf. *resequent fault-line scarp*.)
obsequent slope Slope that runs opposite to dip direction. (Cf. *resequent slope*.)
obsequent stream See *stream genesis*.
obsequent topography In this book, general or regional topographic/structural discordance, such as is found in an area characterized by synclinal mountains and anticlinal valleys and lowlands, or by topographically low horsts and topographically high grabens.
outcrop V Intersection between valley or water gap slopes and the plane of a dipping bed (see Figure 1–7).
outlier Area or group of rocks surrounded by rocks of older age. (Cf. *inlier*.)
outwash (glacial) Stratified detritus removed from a glacier by meltwater streams and deposited in front of or beyond the end moraine or the margin of an active glacier.
outwash plain Broad, gently sloping sheet of outwash deposited by meltwater streams in front of the end moraine of the glacier. Coalescing outwash fans form an outwash plain.

parallel stream pattern See *stream patterns*.
pass Natural passageway through high, difficult terrain. E.g., a break, depression, or other relatively low place in a mountain range, affording a passage across or an opening in a ridge between two peaks, usually approached by a steep valley.
pediment Broad, gently sloping erosional surface or plain of low relief, typically developed by running water, in an arid or semiarid region at the base of an abrupt or receding mountain front. Underlain by bedrock that

may be bare, but is more often mantled with a thin, discontinuous veneer of alluvium derived from the upland masses and in transit across the surface.

peneplain Low, nearly featureless, gently undulating land surface of considerable area, produced by reduction of an area to base level by agents of erosion. Also, such a surface uplifted to form a plateau and subject to dissection.

periglacial Said of processes, conditions, areas, climates, and topographic features at the immediate margins of former and existing glaciers and ice sheets, and influenced by the cold temperature of the ice.

photogrammetry Art and science of obtaining reliable measurements from photographic images.

piedmont Lying or formed at the base of a mountain range.

piedmont slope Gentle slope at the base of a mountain in a semiarid or desert region, composed of a pediment (upper surface of eroded bedrock) and bajada (lower surface of aggradational origin).

piracy (stream) Capture of one stream by another (see *capture*).

playa Southwestern U.S. term for a dry, vegetation-free, flat area at the lowest part of an undrained desert basin underlain by clay, silt, or sand, and commonly by soluble salts. May be marked by an ephemeral lake.

plug (volcanic) A vertical, pipelike body of magma that represents the conduit to a former volcanic vent.

plunge (structural geology) Inclination of a linear structure, measured in the vertical plane. Mainly used to describe geometry of folds.

pluton Intrusion of igneous rock.

plutonic activity (plutonism) General term for phenomena associated with the formation of plutons.

radar Electronic detection system for locating or tracking a distant object by measuring elapsed time of travel of ultrahigh-frequency radio waves emitted from a transmitter and reflected back by the object. It determines range, bearing, elevation, and other characteristics of the object.

radial stream pattern See *stream patterns*.

rectangular stream pattern See *stream patterns*.

rejuvenation Stimulating a stream to renewed erosional activity, as by uplift or a drop in sea level.

relief Elevations or differences in elevation, considered collectively, of a land surface.

resequent fault-line scarp Fault-line scarp that faces in the same direction as the original fault scarp (see Figure 1–5). (Cf. *obsequent fault-line scarp*.)

resequent slope Slope paralleling the strata in the dip direction and eroded below the topmost strata. (Cf. *obsequent slope*.)

resequent stream See *stream genesis*.

resurrected Said of a surface, landscape, or feature that has been restored by exhumation to its present status in the existing relief.

roche moutonnée Small, elongate, protruding knob or hillock of bedrock, so sculptured by a large glacier as to have its long axis oriented in the direction of ice movement, an upstream (stoss) side that is gently inclined, rounded, and striated, and a downstream (lee) side that is steep, rough, and hackly.

saddle Low point in a ridge crest, commonly on a divide between the heads of streams that flow in opposite directions.

scarp slope Relatively steeper face of a cuesta, facing a direction opposite to the dip of the strata.

sinkhole Circular depression in a karst area (see *karst*).

slip-off slope Long, low, relatively gentle slope on the inside of a stream meander, opposite to the cutbank.

slope Inclined surface of any part of the Earth's surface.

stages in regional dissection (defined by W. M. Davis, 1909):

- **youth** Surface partially scored by valleys or canyons, but maintains intact most of its simple, fairly smooth upland.
- **maturity** Original upland has been destroyed as a surface, but that original upland is still recognizable by the accordance of the divides.

stillstand Stability of an area of land, such as a continent or island, with reference to the Earth's interior or mean sea level, as might be reflected by a relatively unvarying base level of erosion between periods of crustal movement.

stoss Side of a hill or knob that faces the direction from which a glacier or ice sheet advanced.

strath terrace Extensive remnant of a strath (a flat valley bottom) that belonged to a former erosion cycle, and that has undergone dissection by a rejuvenated stream following uplift.

striations Multiple scratches or minute lines, generally parallel, inscribed on a rock surface by a geologic agent (i.e., glaciers, streams, or faulting).

strike Direction or trend taken by a structural surface (e.g., bedding or fault planes) as it intersects the horizon.

stream genesis See following and Figure 1–3:

- **consequent stream** Stream developed on an original surface slope.
- **insequent stream** Stream following a course that is apparently not controlled by original slope, structure, or rock type.
- **obsequent stream** Short stream flowing down the scarp slope of a cuesta, or a stream flowing in a direction opposite to that of the dip of the local strata or the tilt of the land surface.
- **resequent stream** Stream that flows down the dip of underlying strata in the same direction as the original consequent stream, but developed later at a lower level than the initial slope, and is generally tributary to a subsequent stream.
- **subsequent stream** Stream developed independently of, and subsequent to, the original relief of a land area. Such streams are usually adjusted to rock structure.

stream patterns

- **angular** Variation of rectangular drainage pattern. Usually develops along intersecting fault or joint systems. Angle of intersection is other than a right angle.
- **annular** Ringlike pattern of streams developed over an eroded structure dome or basin.
- **centripetal** Pattern formed by several streams that flow inward toward a central area or point.
- **dendritic** Arrangement of streams in a pattern resembling the branching of a tree or root system (see Figure 1–2).
- **parallel** Arrangement of streams in a pattern consisting of parallel master and tributary streams.
- **radial** Pattern in which streams radiate outward from the central part of a circular (or oval) topographically high area, and flow down the flanks in all directions.

rectangular Drainage pattern in which the main streams and their tributaries display many right-angle bends and exhibit sections of approximately the same length.

trellis Drainage pattern characterized by parallel main streams intersected at right angles (or nearly right angles) by their tributaries. The tributaries in turn are fed by elongated secondary tributaries, parallel to the main streams. Resembles in plan the stems of a vine or a trellis.

stream piracy See *capture*.

stream stages

youth Major observable activity is downcutting; valley profile is V-shaped; waterfalls and rapids persist.

early maturity Partial development of a flood plain.

maturity Floodplain is noticeably wider than the meander belt of the stream.

structural terrace Terrace formed by differential erosion of horizontal stratified rocks.

structure contour Contour that portrays a structural surface such as a formation boundary or a fault (synonym: *subsurface contour*).

subsequent stream See *stream genesis*.

superposed stream Stream that was established on a new surface, and maintained its course despite different lithologies and structures encountered as it eroded downward into the underlying rocks.

superposition The order in which rocks are placed or accumulated in beds one above the other, the highest bed being the youngest.

synclinal lowland Wide valley following a synclinal axis. (Cf. *anticlinal lowland*.)

synclinal nose Plunging end of a syncline. (Cf. *anticlinal nose*.)

synclinal ridge Ridge following the axis of a syncline. (Cf. *anticlinal ridge*.)

synclinal valley Valley following the axis of a syncline. (Cf. *anticlinal valley*.)

syncline Fold in which the core contains stratigraphically younger rocks; generally concave upward. (Cf. *anticline*.)

tarn Cirque lake.

tectonism See *diastrophism*.

terminal moraine Outermost moraine of a glacier or ice sheet.

terrace Large bench or steplike ledge that breaks the continuity of a slope.

tidal meanders Meanders in a tidal channel.

tombolo Sand or gravel bar or barrier that connects an island with the mainland or with another island.

topographic grain Alignment and direction of the topographic-relief features of a region.

topographic relief Relief (differences in elevation) of topographic features such as mountains, valleys.

topography General configuration of a land surface or any part of the Earth's surface, including its relief and the position of its natural and human-made features.

trellis stream pattern See *stream patterns*.

triangular facet Physiographic feature having a broad base and apex pointing upward.

turbulence (stream) Water flow in which the flow lines are confused and heterogeneously mixed.

unconformity A plane of contact in which a younger stratum overlies older strata on a surface produced by erosion.

uncomformity superposition Stream valley initiated on an uncomformable cover of younger strata eroding through the cover into the underlying structure. This often results in a stream course that is discordant to the structure.

undercut slope See *cutbank.*

underfit stream A misfit stream that appears to be too small to have eroded the valley in which it flows.

updip Direction that is upwards and parallel to the dip of a structure or surface; opposite to downdip.

upsection In a sequence of sedimentary strata, from oldest up to youngest (up in the stratigraphic section). (Cf. *downsection.*)

valley Any low-lying land bordered by higher ground.

volcanism Processes by which magma (molten rock) and its associated gases rise into the Earth's crust and are extruded onto the surface and into the atmosphere.

water gap Deep pass in a mountain ridge, through which a stream flows.

wave-cut bench Level to gently sloping narrow surface produced by wave erosion, extending outward from a wave-cut cliff.

wave planation Leveling of a land surface by the lateral erosion of waves.

weathering Destructive process by which atmospheric agents change soil and rock in color, texture, composition, firmness, or form, with little or no transport of the loosened or altered material.

wind gap A former water gap, now abandoned by the stream that formed it.

youth (dissection) See *stages in regional dissection.*

youth (streams) See *stream stages.*

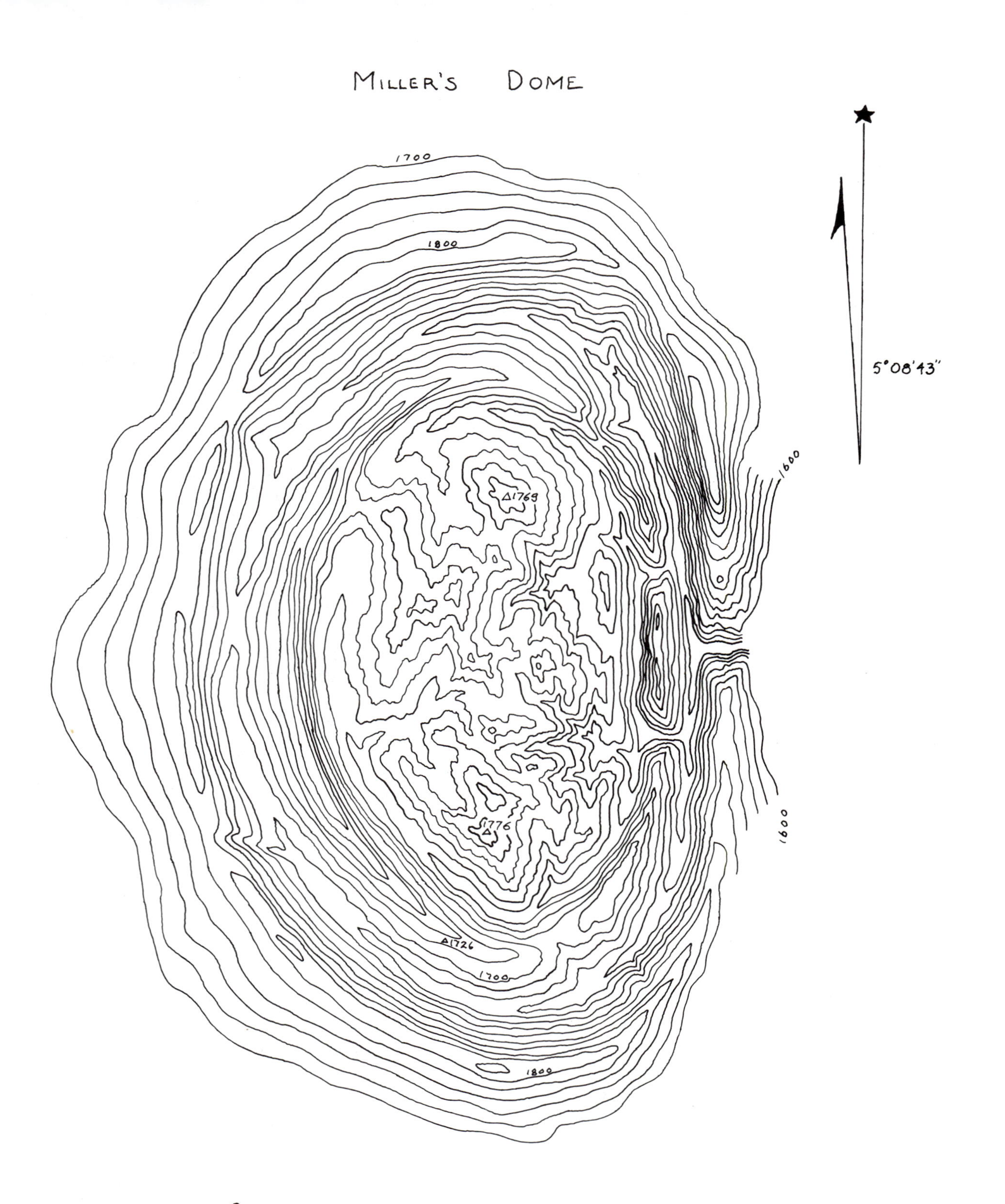

Idealized contour map of a deeply eroded dome drawn by Victor C. Miller for a class in map interpretation at Columbia University (1946).

Answers to Questions in Text

Questions and answers are keyed to the text pages on which they appear. The questions are repeated here as they appear in the text.

CHAPTER 8

Page 80

Question: Why are tributary valleys **A, B,** and **J** so linear?

Answer: Apparently they are subsequent streams, adjusted along joint or fault traces.

Page 89

Question 1: Why is the drainage density of Section **3** so much higher than in Section **2**?

Answer: There is more available relief, and therefore more incision.

Question 2: Is stream **N** destined to be captured by the tributaries of stream **Q**? If so, at what point within the map area will this capture occur? How would such a capture affect the continuing destruction of Section **2**?

Answer: Stream **N** will be captured by a westward-growing tributary of stream **Q**. The bend near **N** is close, and is a likely point of capture. The steeper gradient of the newly captured stream should hasten the destruction of **2**.

Question 3: Of Section **1**?

Answer: Section **1** is being destroyed at its border with **2**. If **2** is destroyed more rapidly, the destruction of **1** will be hastened.

CHAPTER 10

Page 116

Question: What is dip direction at **B**? At **D**? How can you tell?

Answer: The dip at **B** is to the south. At **D** it is to the west-southwest. Point **B** is the nose of a south-plunging syncline. Asymmetry discloses the dip at **F, A,** and **K,** which are on the flanks of this syncline. We cannot see the dip at **B** or **D**. But we know it because in each case it fits the overall structure defined by the many dips we *can* see.

Page 119

Question: What is the *structure* of ridge **GIK**?

Answer: Homoclinal ridge.

Question: What is the *structure* of valley **H**?

Answer: East-northeast-plunging anticline.

Question: What is the dip at **F**?

Answer: North-northwest, just as it is at nearby **G**.

Page 123

Question: Finally, note wind gap **H**. Could capture have done this? If so, what captured what? When? And Why?

Answer: A small east-flowing stream may have once flowed through gap **H,** just as one now flows through gap **T,** about four miles to the south. A younger, shorter, subsequent tributary **F** to the main stream **MP** could easily have extended itself headward along the nonresistant outcrop it now follows, to behead stream "**H**" and leave its gap abandoned. Since the capture, stream **F** has downcut about 100 feet, so we may say that the capture took place about "100 feet ago."

CHAPTER 12

Page 141

Question: How do we know the dip here (at **Z**) is to the east? How do we know the dip is northeast at broken arrow **K?**

Answer: The dip at **Y** tells us the relative ages of the strata on each side of this resistant unit. Since dip at **Y** is to southeast (toward the younger units), the dip at **Z** must be to the east, and that at **K** must be to the northeast. Also, these dips at **Z** and **K** coincide with adjacent dips **W** and **J.**

Page 149

Question: Pediment **L** is, apparently, also younger than pediment **fhPQ.** Why?

Answer: The lowest (southeastern) part of slope **L** lies about 100 feet below the upper (western) edge of slope **Q,** which once must have extended westward and northwestward across and above what is now slope **L.** Slope **L,** therefore, must have been developed following the upper part of slope **Q.**

Page 156

Question: That is, if the fault-associated scarps in this area of Tertiary and Quaternary rocks are fault scarps, and not fault-line scarps. What do you think they are? Why?

Answer: To answer this question, you need to know: (1) the difference between a fault scarp and a fault-line scarp, and (2) the position of scarp.

A *fault scarp* is produced directly by displacement along a fault. It does not have to retain the appearance of a newly exposed, clean-cut fault plane. A *fault-line scarp* is *not* the product of initial fault displacement, but of prolonged differential erosion. For the fault-line scarp to develop, the original fault scarp must first be completely destroyed.

We have indicated on the map (Figure 12–9) which side was downdropped, using bar-and-ball symbols. You can also determine the up and down sides of the fault by looking at the ages of the rocks along it. In areas of relatively simple structure, older rocks are faulted up against younger rocks. The text indicates that the faults in this area are *normal faults.* This means that the downdropped side is the hanging wall, and the side that is faulted up forms the fault scarp.

If you place the tracing you made of the drainage patterns over the topographic map, you can locate the position of the faults. The long fault that runs from the northwest to the southeast is easy to locate. So is the fault at **G.** The scarp of the long diagonal fault faces southwest. This is the position that the upthrown side would have from direct displacement. The scarp at **G** faces southeast, also the position of direct displacement. Therefore, these scarps are fault scarps.

Question: We have indicated many other interesting scarps, such as **M, L, B, K, S, U,** and **V.** According to the geologic map, these are not fault-associated. What do you think?

Answer: If you mark these scarps with a colored pencil and then overlay the transparency you made (with the structure) onto the topographic map, the trend of these scarps (in Tertiary rock) is consistent with the pattern of indicated faults. It is hard to imagine that these scarps are not structurally controlled. Note that some of these scarps are along stream beds which appear to be structurally controlled.

CHAPTER 13

Page 161

Question: Why is there but one ridge at **I**?

Answer: We do not know, but there are at least five apparent possibilities: (1) resistant unit **G** may be duplicated by faulting to form ridge **H** as a local feature; (2) ridge **H** may be cut off to the southeast by an oblique fault; (3) ridge-forming unit **H** may have been eroded off near **I** because dip there is locally very steep; (4) the sandstone at **H** may pinch out to the southeast; or (5) the sandstone at **H** may change facies to a less-resistant lithology (e.g., shale) to the southeast.

Question: Can you explain the "broken" contours in the areas marked by small **x**'s?

Answer: These areas have been excavated, apparently in some kind of mining or quarrying operation.

Question: Why are these odd topographic features restricted to ridge **I**?

Answer: Apparently the type rock which is being mined or quarried at **G** does not occur in association with the other ridge-forming units.

Page 170

Question: Are these structure-controlled surfaces? Or do you have some other explanation?

Answer: The topography of the lowlands in which these slopes lie is erosional. It is characterized by well-integrated drainage systems and gentle, rounded slopes. Since slopes **U** and **N** descend away from their respective border scarps, we consider them to be erosional landforms (pediments) and not structure-controlled.

Question: What is the significance, if any, of depression **K,** in the east?

Answer: This is a relatively arid area in which carbonate rocks are generally resistant. However, depression **K** indicates that there is sufficient precipitation to promote at least some local solution.

Page 172

Question: Could they (streams **B** and **O**) merely have been superposed from a concordant sequence of overlying nonresistant strata?

Answer: Such superposition could have produced the present relationships among these streams and their valleys, the geologic structure, the several exposed lithologies, and the overall topography.

Question 1: What stratigraphic relation exists between cuesta divide **Q** and divide **N**?

Answer: Divides **Q** and **N** are developed on the same resistant sandstone.

Question 2: Is **P** a dip slope? A resequent slope? Both?

Answer: **P** is both a resequent and a dip slope.

Question 3: What is the dip direction at **H**?

Answer: South-southeast.

Question 4: Is the rock at **H** older than, younger than, or the same age as that at **N** (assume no strike faults)?

Answer: Since rock unit **H** underlies unit **N,** it is older than unit **N**.

Question 5: Is the resistant unit at **f** older than, younger than, or the same age as that at **d** (assume no strike faults)?

Answer: The unit at **f** is older than that at **d,** which it underlies.

Question 6: Is the resistant unit at **f** older than, younger than, or the same age as that at **H**?

Answer: The unit at **f** appears to be the same age as that at **H**.

Question 7: How many resistant formations can you identify in this map area? Label them chronologically, starting with 1 for the oldest.

Answer: There are two resistant units: Unit 1 at **f** and **H;** Unit 2 at **N, Q, W,** and **d**.

Page 174

Question: Might stream **D** once have flowed southward through this gap to join stream **V** at or near **R**?

Answer: Stream **D** might once have flowed southward through what is now wind gap **Q**.

Question: If so, why does it now turn eastward to flow through water gap **S**?

Answer: Stream **D**'s abandonment of gap **Q** is best explained by **D**'s being captured by an east-flowing tributary to the stream that cuts water gap **S** through ridge **PO**.

Page 177

Question: How would you explain wind gap **O**?

Answer: Wind gap **O,** like wind gap **Q** in Figure 13–9, is readily explained by the capture of the stream that once flowed through it. It is also possible that this gap was never occupied by a through-flowing stream. It may be the result of preferential erosion along the trace of a cross fault or fault zone.

APPENDIX

Page 191

Question: Determine the bearings and azimuths of directions **BC** and **CB**. Use Figure A–2.

Answers: BC's bearing is S11°W; azimuth is 191°. CB's bearing is N11°E; azimuth is 11°.

Page 192

Question: Determine the bearing and azimuth of each of the directions in Figure A–7, and enter results in Table A–2.

Answers:

DE	S78°W	258°.
ED	N78°E	78°.
DF	N54°W	306°.
FD	S54°E	126°.
EF	N30°E	30°.
FE	S30°W	210°.

Page 195

Question 1: The magnetic bearing to **C** is S25°E. What is the true bearing?

Answer: S33°E.

Question 2: The magnetic bearing to **D** is S25°W. What is the true bearing?

Answer: S17°W.

Question 3: The magnetic bearing to **F** is S87°E. What is the true bearing?
Answer: N85°E.

Question 4: Calculate and enter the missing values in Table A–3. (Missing values are shown in **bold** type.)

TABLE A–3 (with answers)

	Magnetic Readings		True Readings	
	Bearing	Azimuth	Bearing	Azimuth
A	N50°W	**310°**	N58°W	**302°**
B	N50°E	**50°**	N42°E	**42°**
C	S25°E	**155°**	**S33°E**	**147°**
D	S25°W	**205°**	**S17°W**	**197°**
F	S87°E	**93°**	**N85°E**	**85°**
G	**N4°E**	4°	**N4°W**	**356°**
H	**S83°E**	97°	**N89°E**	**89°**
I	**S10°W**	190°	**S2°W**	**182°**
J	**N59°W**	301°	**N67°W**	**293°**

Page 196

Question: Convert the following verbal scales to fractional scales:
Answers:

1. One foot equals six miles: 1/31,680.
2. One centimeter equals ten meters: 1/1000.
3. One inch equals 25 yards: 1/900.
4. One inch equals one kilometer: 1/39,283 (rounds to 1/40,000).

Question: Convert the following fractional scales to verbal scales. Vary the units for practice, using inches, feet, miles, centimeters, or whatever:

Answers:

1. 1/48,000 — One inch equals 4000 feet.
2. 1/500,000 — One millimeter equals 500 meters; one centimeter equals 5000 meters; 1 inch equals about 8 miles.
3. 1/200,000 — 1 centimeter equals 2000 meters.
4. 1/24 — 1 inch equals 2 feet.
5. 1/100 — 1 centimeter equals 1 meter.
6. 1/15,625 — 1 inch equals 1/4 mile; 4 inches equals 1 mile.
7. 1/15,840 — 1 inch equals 1/4 mile; 4 inches equals 1 mile.

Page 198

Question: Ground distance **AB** is, therefore, almost exactly 2.64 miles in length. How many feet is that?

Answer: 13,939 feet.

Page 198

Question 1: The scale of Figure A–2 is 1 inch = 500 feet. In feet, how long are the ground distances shown?

Answer:

- ☐ **AB**? 1250 ft.
- ☐ **AC**? 1375 ft.
- ☐ **BC**? 1050 ft.

Question 2: The scale of Figure A–6 is 1/48,000. In feet, how long are the ground distances shown?

Answers:

- ☐ **AB**? 10,000 ft.
- ☐ **AC**? 11,000 ft.
- ☐ **CB**? 8400 ft.

Question 3: In Figure A–7, ground distance **ED** is 5.2 miles.
(a) What is the fractional scale of this map? 1/125,000.
(b) In feet, what is the ground distance **EF**? 22,440 ft.

Question 4: On Figure A–2, plot point **D**, which is 1200 feet S50°W from point **B** (true bearing). Scale is 1 in. = 500 ft. What is the length of line **BD**?

Answer: **BD** = 2.4 in.

Page 211

Question: 60° north latitude is two-thirds of the distance to the North Pole. 45° north latitude is one-half the distance to the North Pole. Why is it not at 45°N that the length of the degree is one-half the length at the equator?

Answer: In a 30°-60°-90° triangle, the short leg is one-half the length of the hypotenuse. A 30°-60°-90° triangle is formed by the Earth's axis, the line from the center of the Earth to the 60° north parallel, and the radius of the 60° north latitude plane. The radius of the 60° north parallel is the short leg. The line from the Earth's center to the 60° parallel is the hypotenuse. The radius at the equator is twice the radius length of the 60° parallel.

Page 213

Question: As a review, identify the following areas shown in Figure A–24. Also, how many acres of land are included in each area?

Answers:

1. Area **A**: W 1/2, Sec 8, T16S, R13E (W, Sec 8); 320 acres.
2. Area **B**: SE 1/4, Sec 8, T16S, R13E (SE, Sec 8); 160 acres.
3. Area **C**: SW 1/4, NE 1/4, Sec 8 (SW, NE, Sec 8); 40 acres.
4. Area **D**: SE 1/4, NW 1/4, NE 1/4, Sec 8 (SE, NW, NE, Sec 8); 10 acres.

Page 214

Question: Is the principal meridian of the Arizona Survey to the east or west of Dobson Road and its companions?

Answer: West.

Topographic Maps

Most topographic maps reproduced in this book are parts of one or more USGS quadrangles for which we obtained reproducible transparencies showing only the contours. Many of these are reproduced at original scale, but several have been enlarged. A few were reduced to fit the page. The remaining topographic maps are tracings of published quadrangles (transparencies were not available). The contour interval (C.I.) for each map is indicated.

FIGURE 1–9 Part of the Bolivar, Missouri topographic quadrangle, 1/125,000, C.I. 50 feet (1884). Small rectangle defines the area covered by Figure 1–10.

FIGURE 1–10 Part of the Bolivar, Missouri topographic quadrangle, 1/24,000, C.I. 10 feet (1959). Covers the area outlined on Figure 1–9.

FIGURE 2–1 Part of the Fremont Point, Oregon topographic quadrangle, 1/24,000, C.I. 20 feet.

FIGURE 2–2 Part of the Upper Slide Lake, Wyoming topographic quadrangle, 1/24,000, C.I. 40 feet.

FIGURE 3–1 Part of the Palmyra, New York topographic quadrangle, 1/62,500, C.I. 20 feet (enlarged × 1.4).

FIGURE 3–2 Part of the Hyannis, Massachusetts topographic quadrangle, 1/24,000, C.I. 10 feet.

FIGURE 3–3 Part of the Greenfield, Massachusetts topographic quadrangle, 1/24,000, C.I. 10 feet.

FIGURE 3–4 Part of the Pound Ridge, Norwalk North, Norwalk South, and Stamford, Connecticut topographic quadrangles, 1/24,000, C.I. 10 feet.

FIGURE 3–5 Part of the Marmot Mountain, Montana topographic quadrangle, 1/24,000, C.I. 40 feet.

FIGURE 4–1 Part of the Larimore and Emerado, North Dakota topographic quadrangles, 1/62,500, C.I. 10 feet.

FIGURE 4–2 Part of the Redondo Beach, California topographic quadrangle, 1/24,000, C.I. 25 feet.

FIGURE 4–3 *Orienting Block Diagrams with the Corresponding Map.* In Figure 4–3 you are looking from SE to northwest. The arrow on the block diagram points to the northeast. (Use map north arrow and diagram northeast arrow to see this relationship.) Part of the Wilton, Virginia topographic quadrangle, 1/24,000, C.I. 10 feet.

FIGURE 4–4 Part of the Rehoboth, Delaware topographic quadrangle, 1/62,500, C.I. 10 feet.

FIGURE 4–5 Part of the Mathews, Virginia topographic quadrangle, 1/24,000, C.I. 5 feet.

FIGURE 4–6 Part of the Honōkane, Hawaii topographic quadrangle, 1/24,000, C.I. 40 feet.

FIGURE 4–7 Part of the Pemaquid Point, Maine topographic quadrangle, 1/24,000, C.I. 10 feet (also depicts 30-, 60-, and 120-foot submarine contours).

FIGURE 5–1 Parts of the Medicine Lake and Timber Mountain, California topographic quadrangles, 1/62,500, C.I. 40 feet (enlarged × 1.4).

FIGURE 5–2 Part of the S P Mountain, Arizona topographic quadrangle, 1/62,500, C.I. 40 feet (enlarged × 1.4).

FIGURE 5–3 Part of the Grand Canyon, Arizona topographic quadrangle, 1/250,000, C.I. 200 feet (enlarged × 1.4).

FIGURE 5–4 Part of the Ship Rock, New Mexico topographic quadrangle, 1/62,500, C.I. 20 feet (enlarged × 1.4).

FIGURE 5–5 Part of the Spanish Peaks and Walsenburg, Colorado topographic quadrangles, 1/125,000, C.I. 250 and 125 feet (125-foot contours are dashed).

FIGURE 6–1 Parts of the Manly Peak and Trona, California topographic quadrangles, 1/62,500, C.I. 40 feet.

FIGURE 6–2 Part of the Avawatz Pass, California topographic quadrangle, 1/62,500, C.I. 40 feet.

FIGURE 6–3 Part of the Wellington, Utah topographic quadrangle, 1/62,500, C.I. 250 and 125 feet.

FIGURE 6–4 Part of the Twin Buttes, Arizona topographic quadrangle, 1/62,500, C.I. 50 feet (enlarged × 1.4).

FIGURE 6–5 Part of the Sacaton, Arizona topographic quadrangle, 1/24,000, C.I. 10 feet (1963).

FIGURE 6–6 Part of the Sacaton, Arizona topographic quadrangle, 1/62,500, C.I. 50 feet (1904–06) (enlarged × 1.7).

FIGURE 7–1 Part of the Anton, Texas topographic quadrangle, 1/62,500, C.I. 10 feet.

FIGURE 7–2 Part of the Mammoth Cave, Kentucky topographic quadrangle, 1/62,500, C.I. 20 feet (enlarged × 1.7).

FIGURE 7–3 Part of the Monticello, Kentucky topographic quadrangle, 1/24,000, C.I. 20 feet (reduced × 0.6).

FIGURE 7–4 Part of the Trenton, Kentucky topographic quadrangle, 1/24,000, C.I. 10 feet.

FIGURE 8–1 Part of the Bluemont, Virginia topographic quadrangle, 1/24,000, C.I. 10 feet.

FIGURE 8–2 Part of the Orlean, Virginia topographic quadrangle, 1/24,000, C.I. 20 feet.

FIGURE 8–3 Part of the Eastatoe Gap, North and South Carolina topographic quadrangle, 1/24,000, C.I. 40 feet.

FIGURE 8–4 Part of the Santa Rita, New Mexico topographic quadrangle, 1/24,000, C.I. 25 feet (enlarged × 1.4).

FIGURE 8–4b Part of USGS Geologic Quadrangle Map 306, *The Geology of the Santa Rita Quadrangle, New Mexico,* 1/24,000 (enlarged × 1.4). Covers the area depicted in Figure 8–4.

FIGURE 8–5 Part of the Kayjay, Kentucky topographic quadrangle, 1/24,000, C.I. 20 feet.

FIGURE 8–6 Part of the Paxton, Nebraska topographic quadrangle, 1/125,000, C.I. 20 feet (enlarged × 1.6).

FIGURE 8–7 Part of the Frankfort East and Frankfort West, Kentucky topographic quadrangles, 1/24,000, C.I. 10 feet.

FIGURE 8–8 Part of the Gratz, Kentucky topographic quadrangle, 1/24,000, C.I. 20 feet.

FIGURE 8–9 Part of the Ashland City, Tennessee topographic quadrangle, 1/24,000, C.I. 20 feet.

FIGURE 8–10 Part of the Kaaterskill, New York topographic quadrangle, 1/62,500, C.I. 20 feet.

FIGURE 8–11 Part of the Zionville, North Carolina/Tennessee topographic quadrangle, 1/24,000, C.I. 40 feet.

FIGURE 8–12 Part of the Lubbock, Texas topographic quadrangle, 1/250,000, C.I. 100 feet.

FIGURE 8–13 Part of the Lookout Ridge, Alaska topographic quadrangle, 1/250,000, C.I. 100 feet.

FIGURE 9–1 Part of the Cumberland, Maryland topographic quadrangle, 1/24,000, C.I. 20 feet.

FIGURE 9–2 Part of the Yakima East, Washington topographic quadrangle, 1/24,000, C.I. 20 feet.

FIGURE 9–3 Parts of the Bluemont, Virginia and Round Hill, West Virginia topographic quadrangles, 1/24,000, C.I. 10 and 20 feet.

FIGURE 9–4 Part of the Delaware Water Gap, Pennsylvania/New Jersey topographic quadrangle, 1/62,500, C.I. 20 feet (enlarged × 1.4).

FIGURE 9–5 Part of the Independence Rock, Wyoming topographic quadrangle, 1/24,000, C.I. 20 feet.

FIGURE 9–6 Part of the Lake Killarney, Missouri topographic quadrangle, 1/24,000, C.I. 20 feet.

FIGURE 10–1 Part of the Cambridge NE, Kansas topographic quadrangle, 1/24,000, C.I. 10 feet.

FIGURE 10–2 Part of the Marmot Mountain, Montana topographic quadrangle, 1/24,000, C.I. 40 feet.

FIGURE 10–3 Part of the Booneville, Arkansas topographic quadrangle, 1/62,500, C.I. 20 feet (enlarged × 1.7).

FIGURE 10–4 Part of the Jacob Lake, Arizona topographic quadrangle, 1/62,500, C.I. 50 feet.

FIGURE 10–5 Part of the New Enterprise, Pennsylvania topographic quadrangle, 1/24,000, C.I. 20 feet.

FIGURE 10–6 Part of the Nashville, Indiana topographic quadrangle, 1/24,000, C.I. 10 feet (enlarged × 1.7).

FIGURE 10–7 Part of the Everett West, Pennsylvania topographic quadrangle, 1/24,000, C.I. 20 feet (reduced × 0.67).

FIGURE 10–8 Part of the Millheim, Pennsylvania topographic quadrangle, 1/62,500, C.I. 20 feet (enlarged × 1.7).

FIGURE 10–9 Part of the Cato, Arkansas topographic quadrangle, 1/24,000, C.I. 10 feet.

FIGURE 10–10 Part of the Strasburg, Virginia topographic quadrangle, 1/62,500, C.I. 40 feet.

FIGURE 10–11 Part of the Greenland Gap, West Virginia topographic quadrangle, 1/62,500, C.I. 50 feet (enlarged × 1.7).

FIGURE 10–12 Parts of the Karbers Ridge and Herod, Illinois topographic quadrangles, 1/24,000, C.I. 20 feet.

FIGURE 10–13 Part of the Booneville, Arkansas topographic quadrangle, 1/62,500, C.I. 20 feet (enlarged × 1.7).

FIGURE 10–14 Part of the Mendota, Virginia topographic quadrangle, 1/24,000, C.I. 20 feet.

FIGURE 11–1 Part of the Kanab Point, Arizona topographic quadrangle, 1/62,500, C.I. 80 feet.

FIGURE 11–2 Part of the Calf Creek, Utah topographic quadrangle, 1/24,000, C.I. 40 feet.

FIGURE 11–3 Part of the Tulelake, California topographic quadrangle, 1/62,500, C.I. 20 feet (enlarged × 1.7).

FIGURE 11–4 Part of the Springer, Oklahoma topographic quadrangle, 1/24,000, C.I. 10 feet.

FIGURE 11–5 Part of the Silver Bay, New York topographic quadrangle, 1/24,000, C.I. 20 feet.

FIGURE 11–6 Part of the Andersonville, Virginia topographic quadrangle, 1/24,000, C.I. 10 feet.

FIGURE 12–1 Part of the Flaherty, Kentucky topographic quadrangle, 1/24,000, C.I. 20 feet.

FIGURE 12–2 Part of the Rawlins, Wyoming topographic quadrangle, 1/24,000, C.I. 20 feet.

FIGURE 12–3 Part of the Quanah Mountain, Oklahoma topographic quadrangle, 1/24,000, C.I. 10 feet.

FIGURE 12–4 Part of the Golden, Colorado topographic quadrangle, 1/24,000, C.I. 10 feet.

FIGURE 12–5 Part of the Wind River, Wyoming topographic quadrangle, 1/24,000, C.I. 20 feet.

FIGURE 12–6 Part of the Ash Mountain, New Mexico topographic quadrangle, 1/62,500, C.I. 40 feet (enlarged × 1.4).

FIGURE 12–7 Part of the Buffalo Gap, South Dakota topographic quadrangle, 1/24,000, C.I. 10 feet.

FIGURE 12–8 Part of the King Hill, Idaho topographic quadrangle, 1/62,500, C.I. 40 feet.

FIGURE 12–9 Part of the *Reconnaissance and Geologic Map of the West-Central Snake River Plain, Idaho* (USGS Miscellaneous Geologic Investigations Map I-373), 1/125,000 (enlarged × 1.83).

FIGURE 13–1 Part of the Cabot, Arkansas topographic quadrangle, 1/24,000, C.I. 10 feet.

FIGURE 13–2 Part of the Chattanooga, Tennessee topographic quadrangle, 1/250,000, C.I. 100 feet.

FIGURE 13–3 Part of the Chattanooga, Tennessee topographic quadrangle, 1/250,000, C.I. 100 feet.

FIGURE 13–4 Part of the Fort Smith, Arkansas/Oklahoma topographic quadrangle, 1/250,000, C.I. 100 feet.

FIGURE 13–4a Part of the Fort Smith, Arkansas/Oklahoma topographic quadrangle, 1/250,000, C.I. 50 meters.

FIGURE 13–5 Part of the *Geologic Map of Oklahoma*, covering the area depicted in Figures 13–4 and 13–4a.

FIGURE 13–6 Part of the Alec Butte, Oregon topographic quadrangle, 1/24,000, C.I. 10 feet.

FIGURE 13–7 Part of the House Rock Spring NW, Arizona topographic quadrangle, 1/24,000, C.I. 40 feet.

FIGURE 13–8 Part of the Damascus, Arkansas topographic quadrangle, 1/24,000, C.I. 20 feet.

FIGURE 13–9 Part of the Hilton, Virginia topographic quadrangle, 1/24,000, C.I. 20 feet.

FIGURE 13–10 Part of the Mendota, Virginia topographic quadrangle, 1/24,000, C.I. 20 feet.

FIGURE 13–11 Part of the Brumley, Virginia topographic quadrangle, 1/24,000, C.I. 40 feet.

FIGURE 13–12 Part of the Hilton, Virginia topographic quadrangle, 1/24,000, C.I. 20 feet.

FIGURE 13–13 Part of the Bray, California topographic quadrangle, 1/62,500, C.I. 100 feet (solid contours) and 50 feet (dashed contours) (enlarged × 1.45).

FIGURE 13–14 Part of the Mt. Tom, California topographic quadrangle, 1/62,500, C.I. 80 feet.

FIGURE 13–15 Part of the Newark, New Jersey topographic quadrangle, 1/250,000, C.I. 100 feet.

FIGURE 13–15a Part of the *Geologic Map of New Jersey*, covering the area depicted in Figure 13–15, 1/250,000 (1910–12).

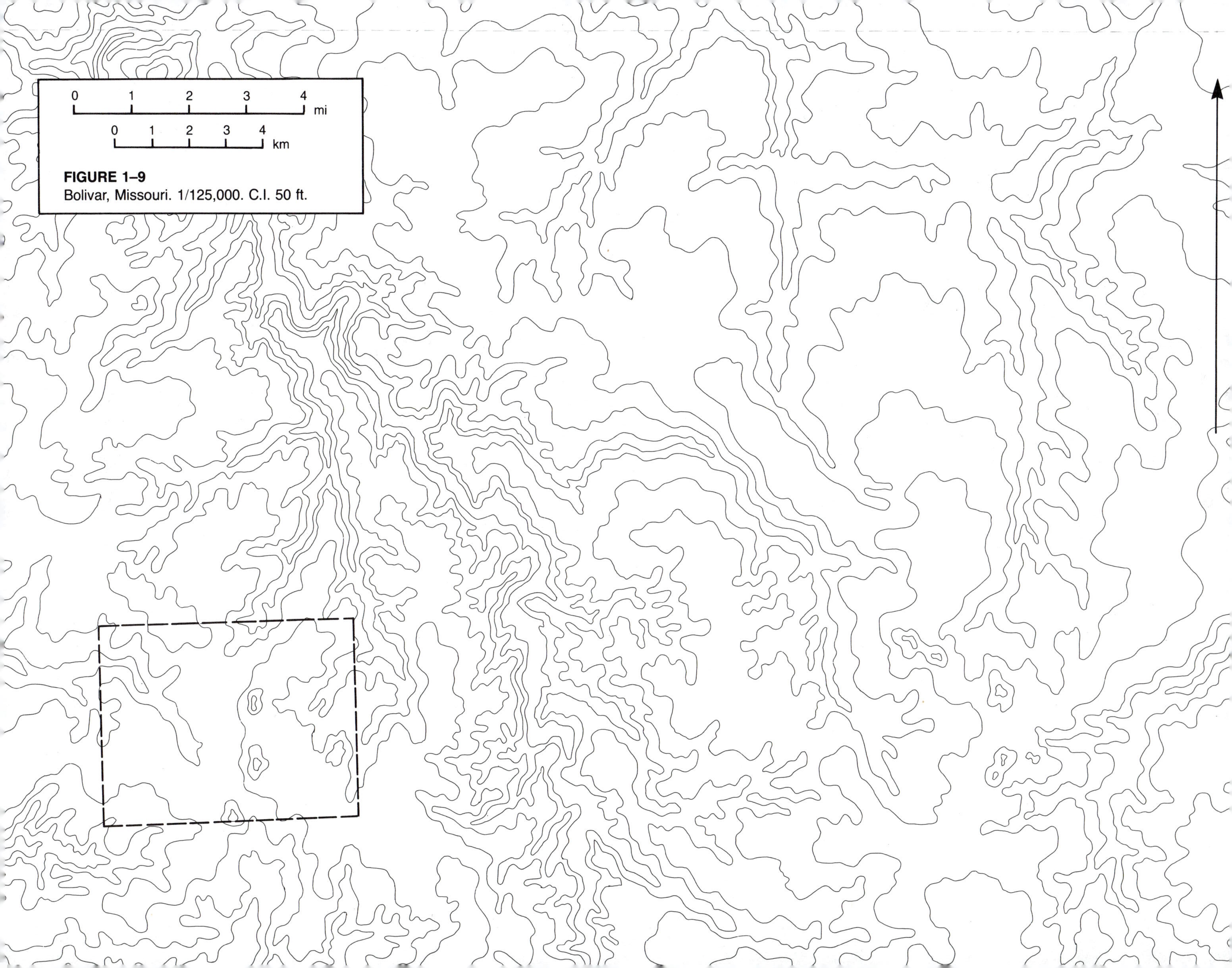

FIGURE 1–9
Bolivar, Missouri. 1/125,000. C.I. 50 ft.

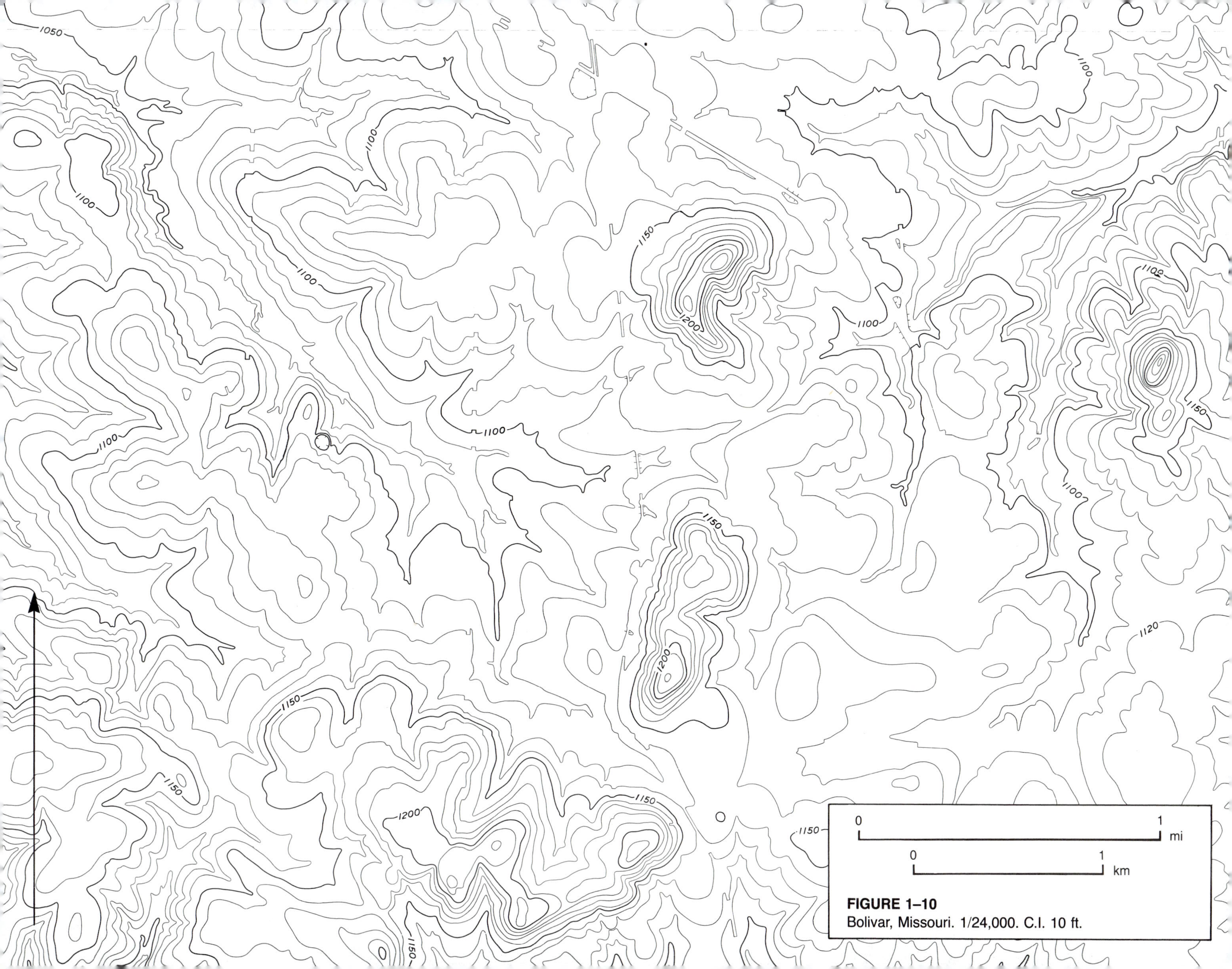

FIGURE 1–10
Bolivar, Missouri. 1/24,000. C.I. 10 ft.

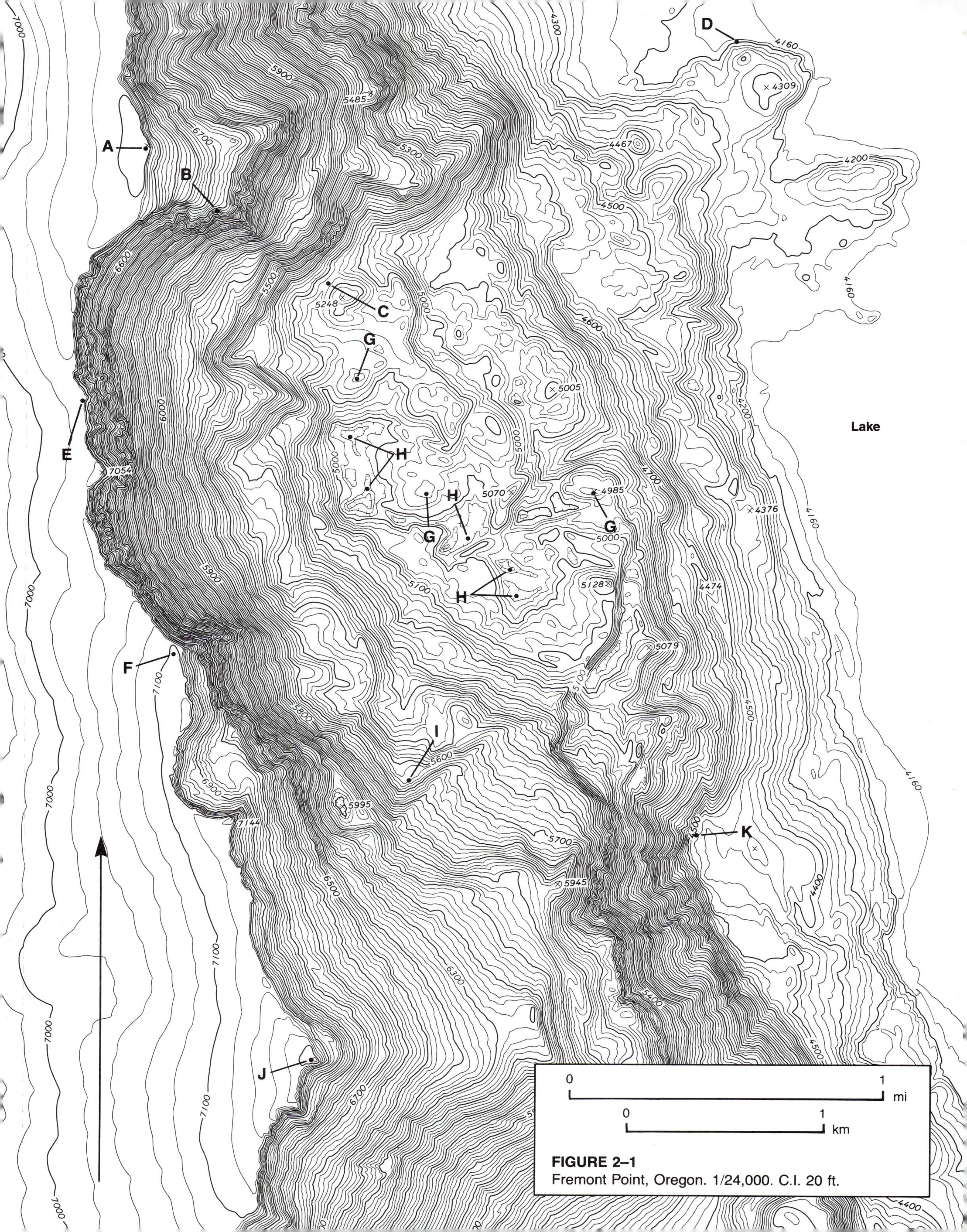

FIGURE 2–1
Fremont Point, Oregon. 1/24,000. C.I. 20 ft.

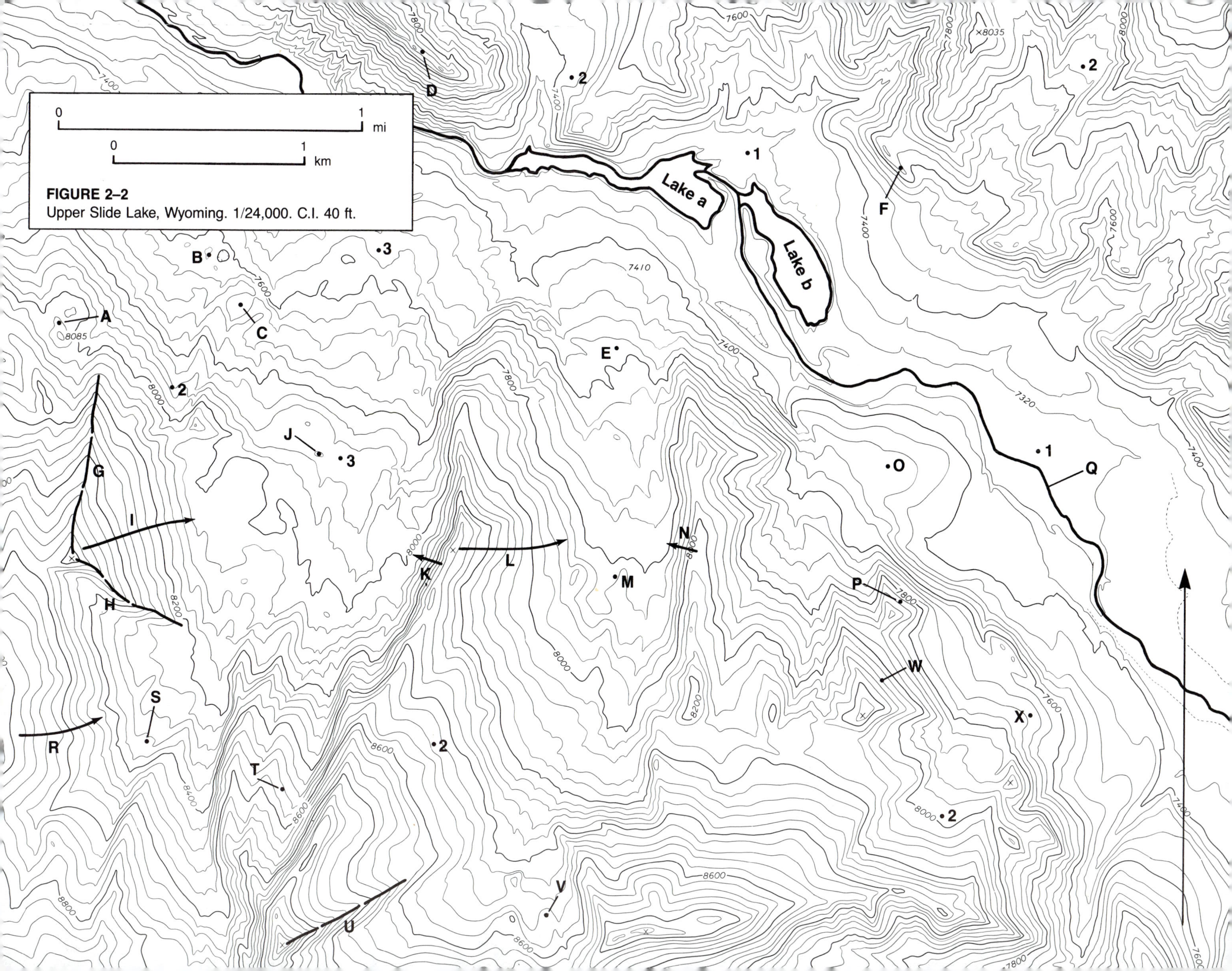

FIGURE 2–2
Upper Slide Lake, Wyoming. 1/24,000. C.I. 40 ft.

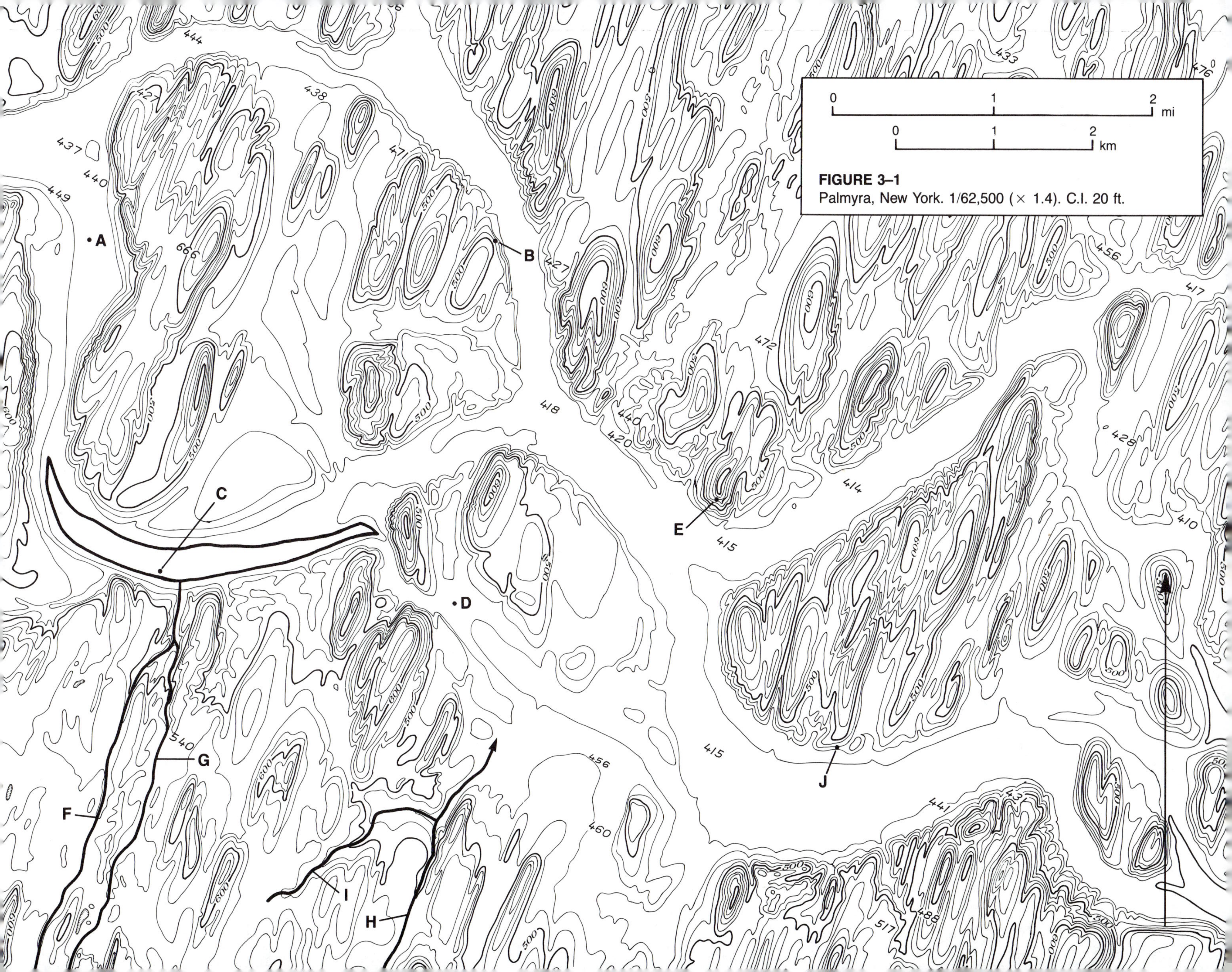

FIGURE 3–1
Palmyra, New York. 1/62,500 (× 1.4). C.I. 20 ft.

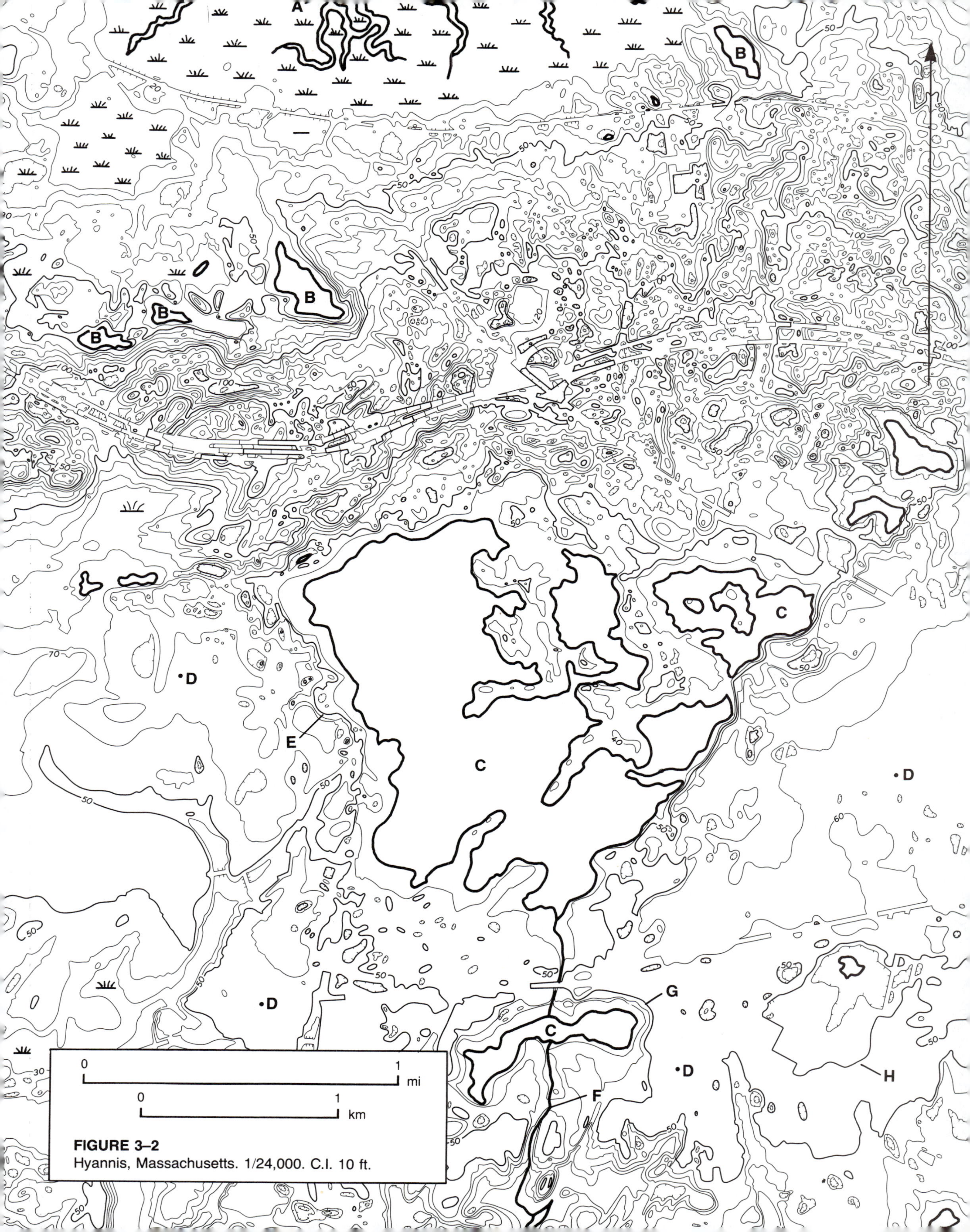

FIGURE 3–2
Hyannis, Massachusetts. 1/24,000. C.I. 10 ft.

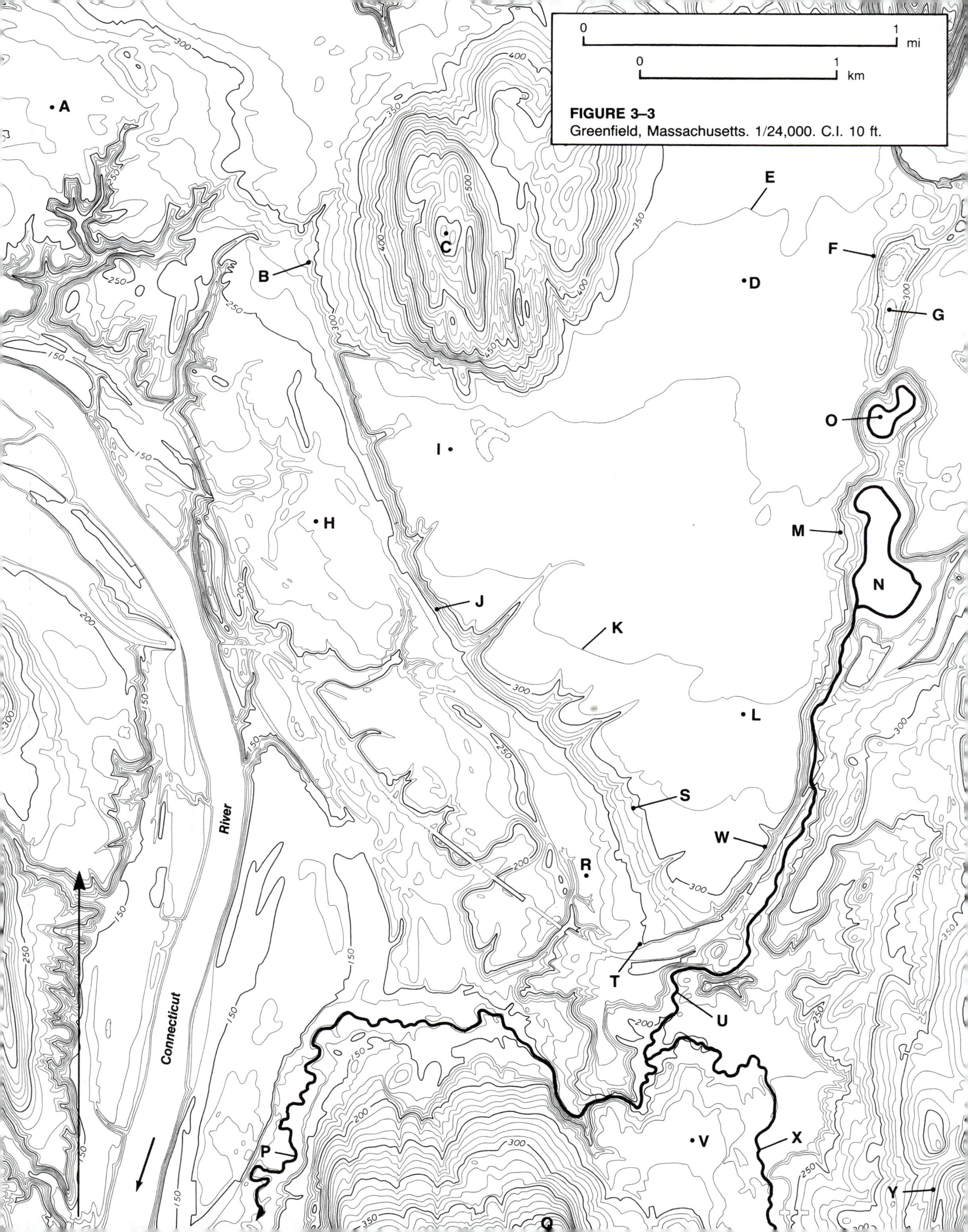

FIGURE 3–3
Greenfield, Massachusetts. 1/24,000. C.I. 10 ft.

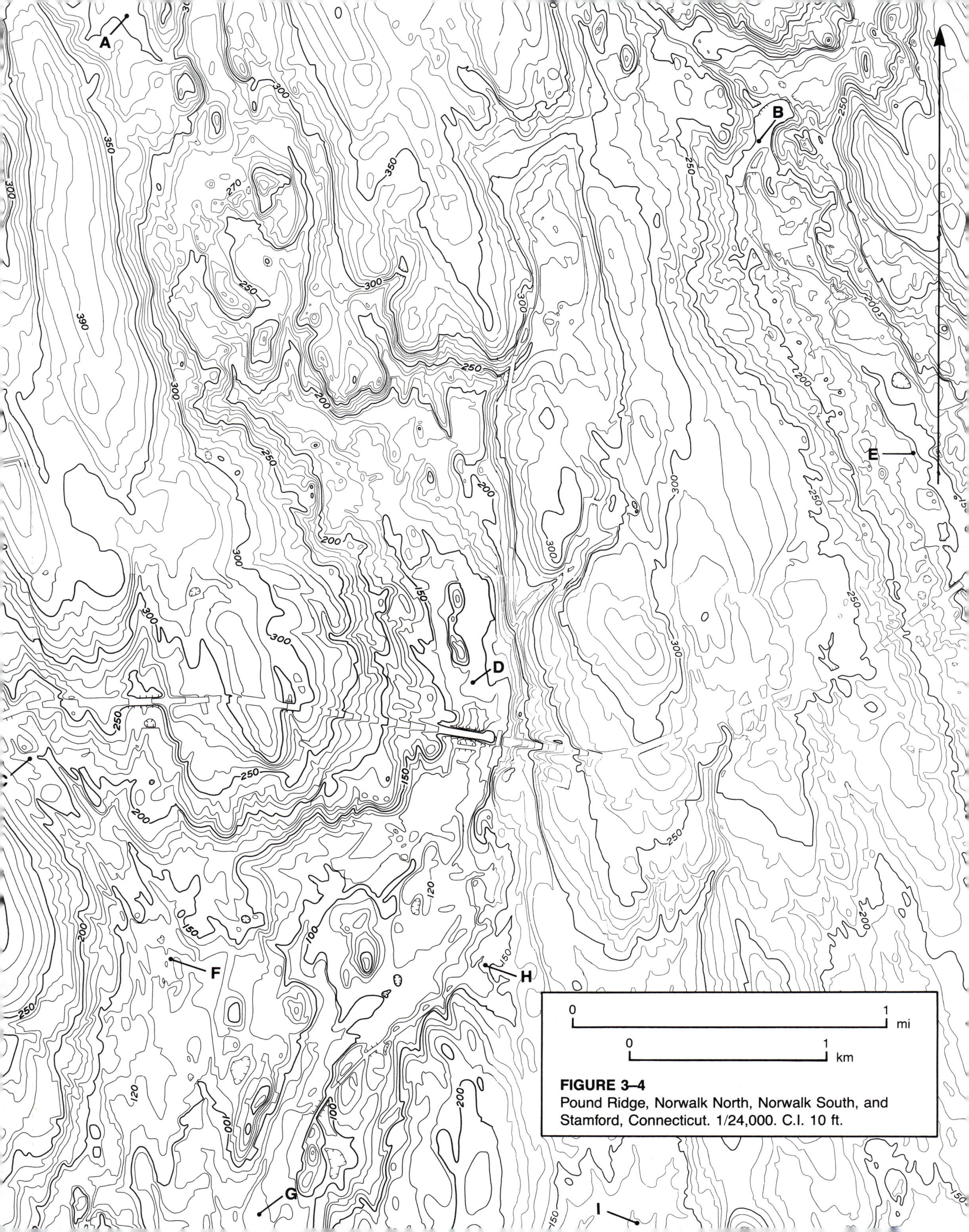

FIGURE 3–4
Pound Ridge, Norwalk North, Norwalk South, and Stamford, Connecticut. 1/24,000. C.I. 10 ft.

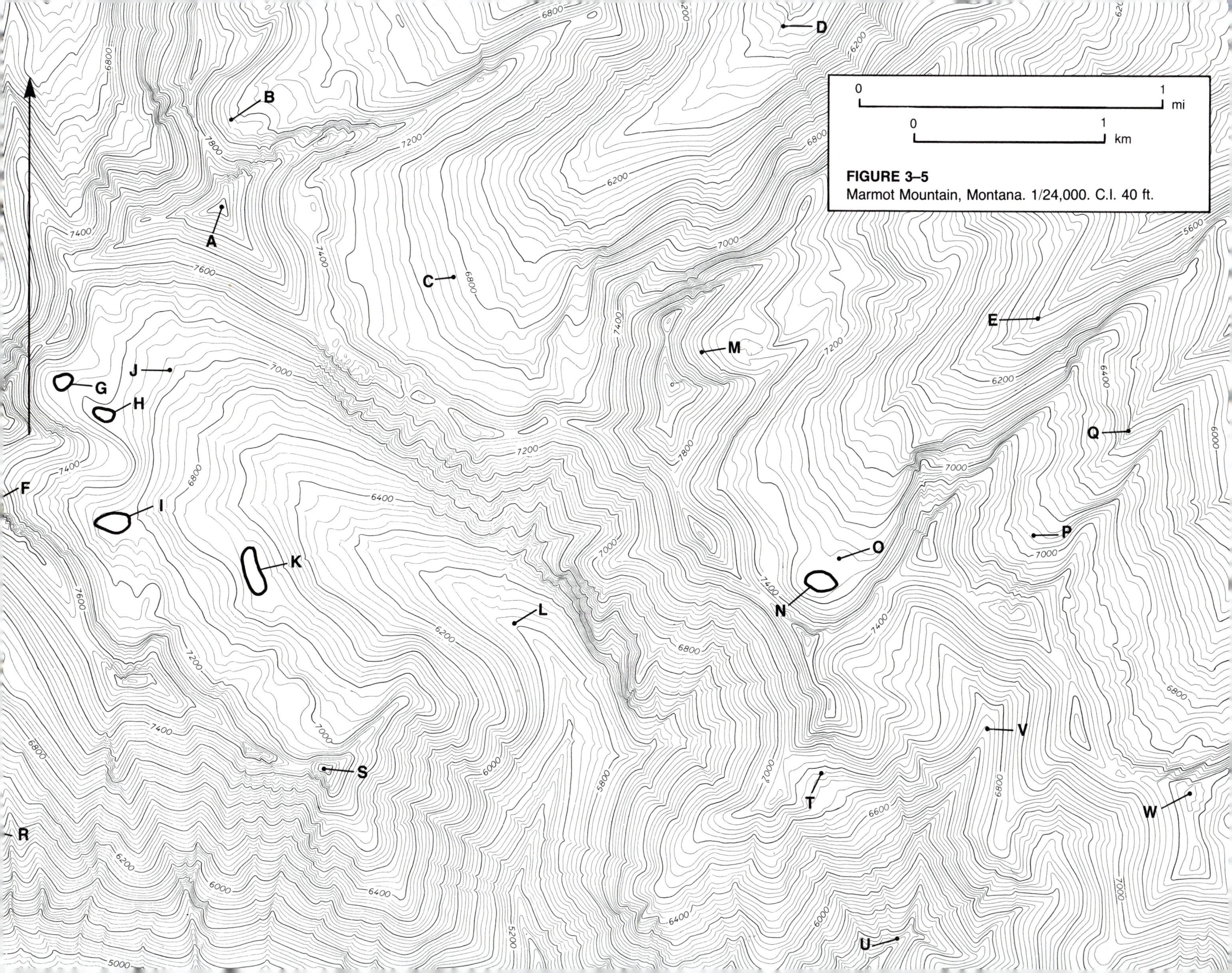

FIGURE 3–5
Marmot Mountain, Montana. 1/24,000. C.I. 40 ft.

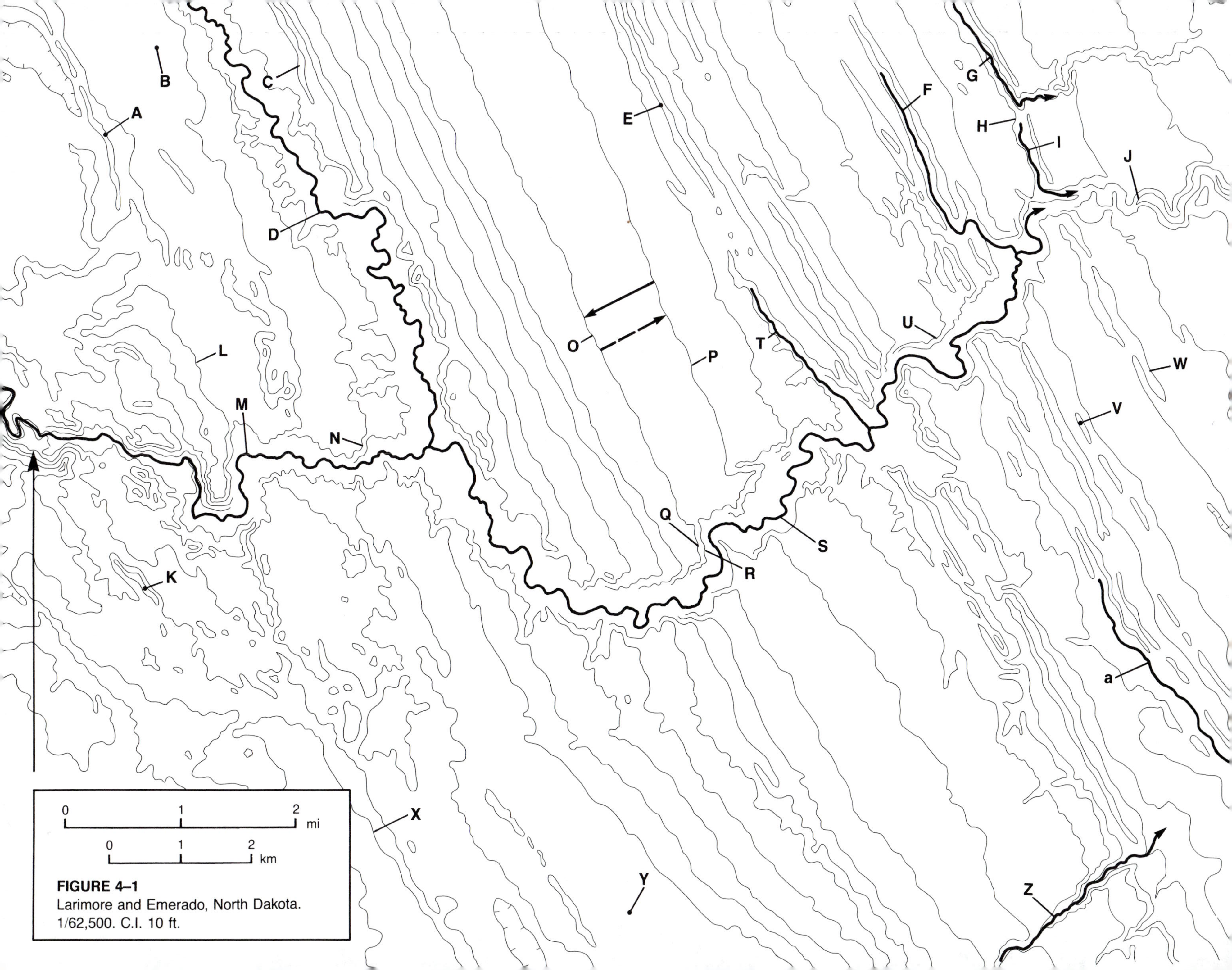

FIGURE 4–1
Larimore and Emerado, North Dakota.
1/62,500. C.I. 10 ft.

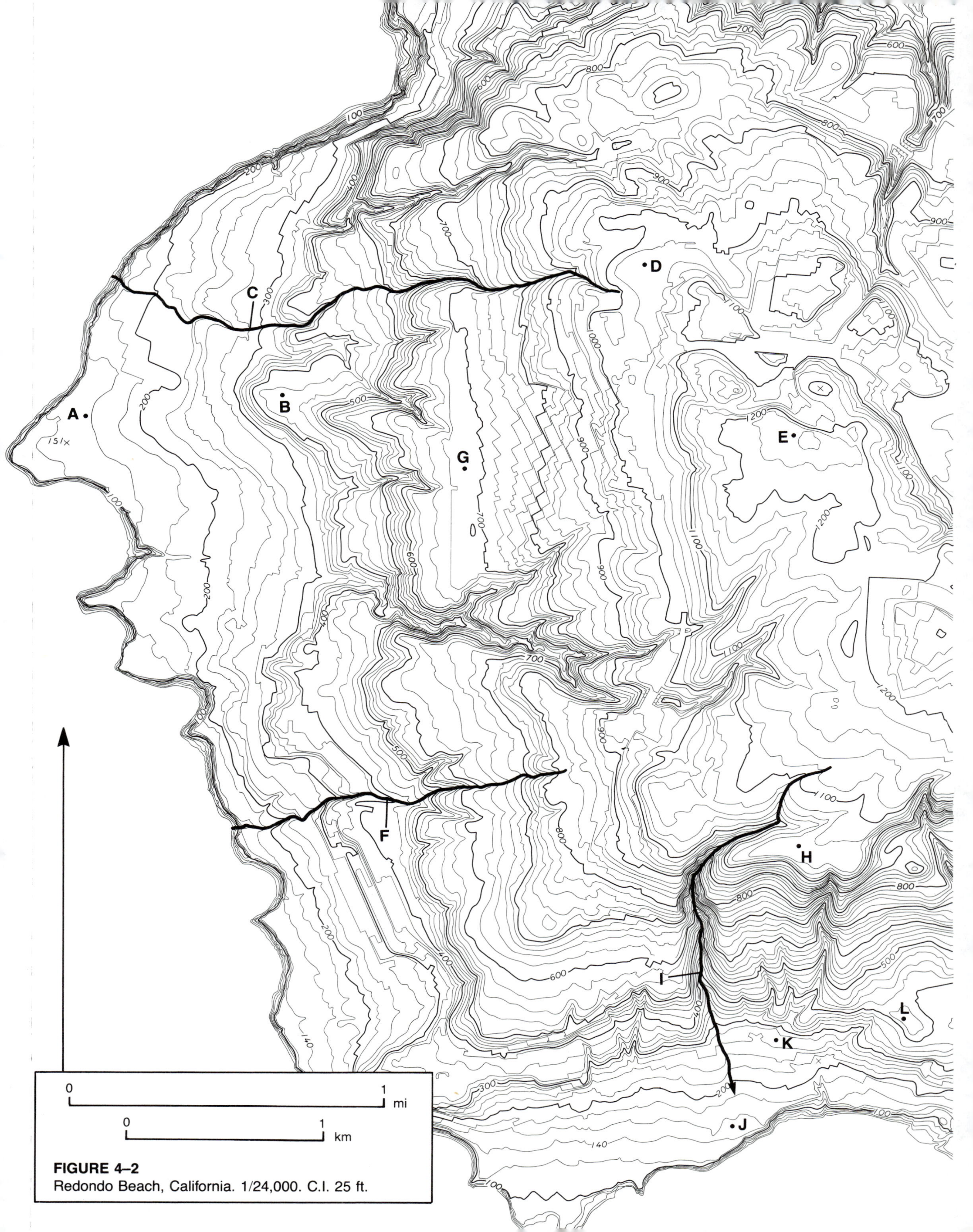

FIGURE 4–2
Redondo Beach, California. 1/24,000. C.I. 25 ft.

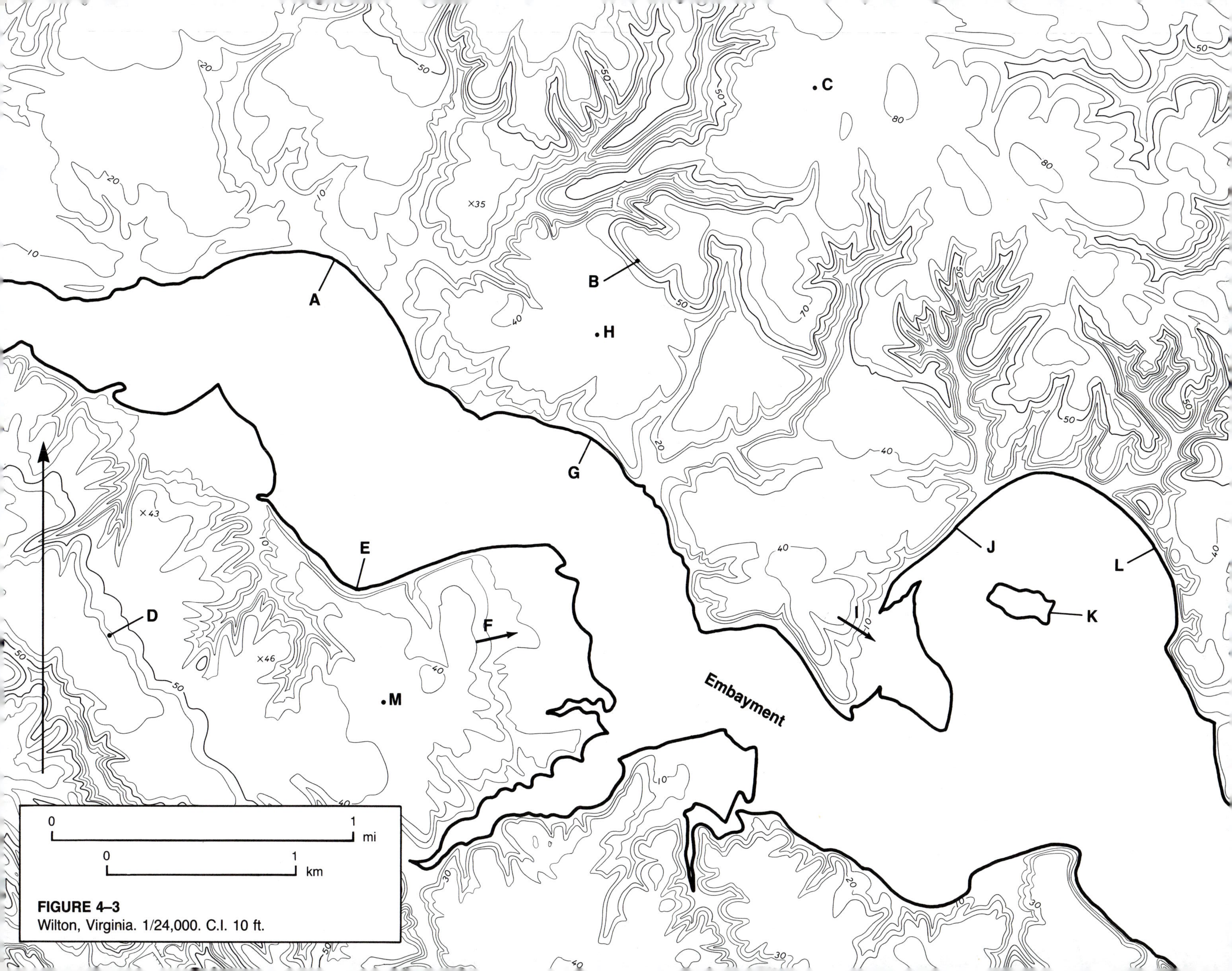

FIGURE 4–3
Wilton, Virginia. 1/24,000. C.I. 10 ft.

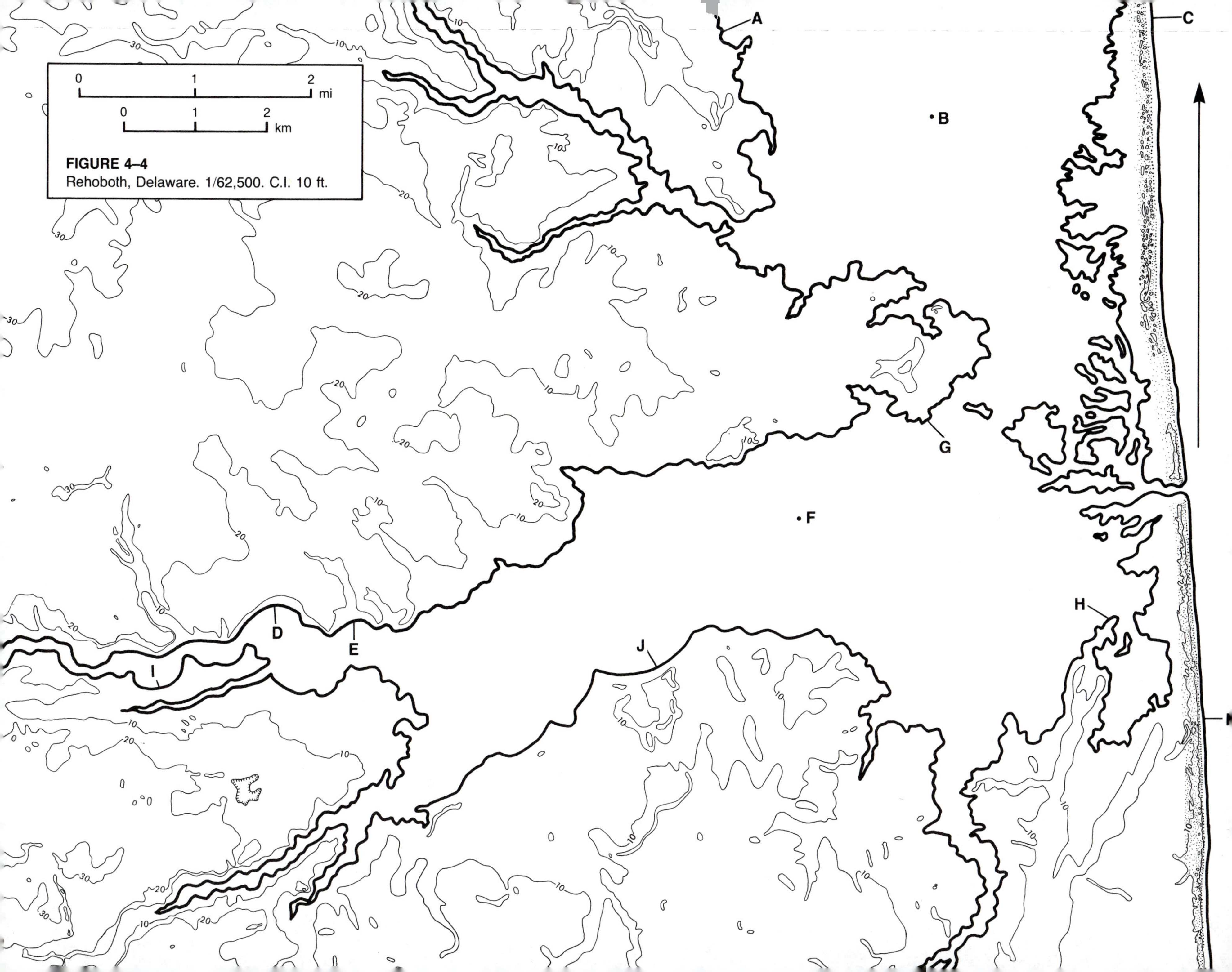

FIGURE 4–4
Rehoboth, Delaware. 1/62,500. C.I. 10 ft.

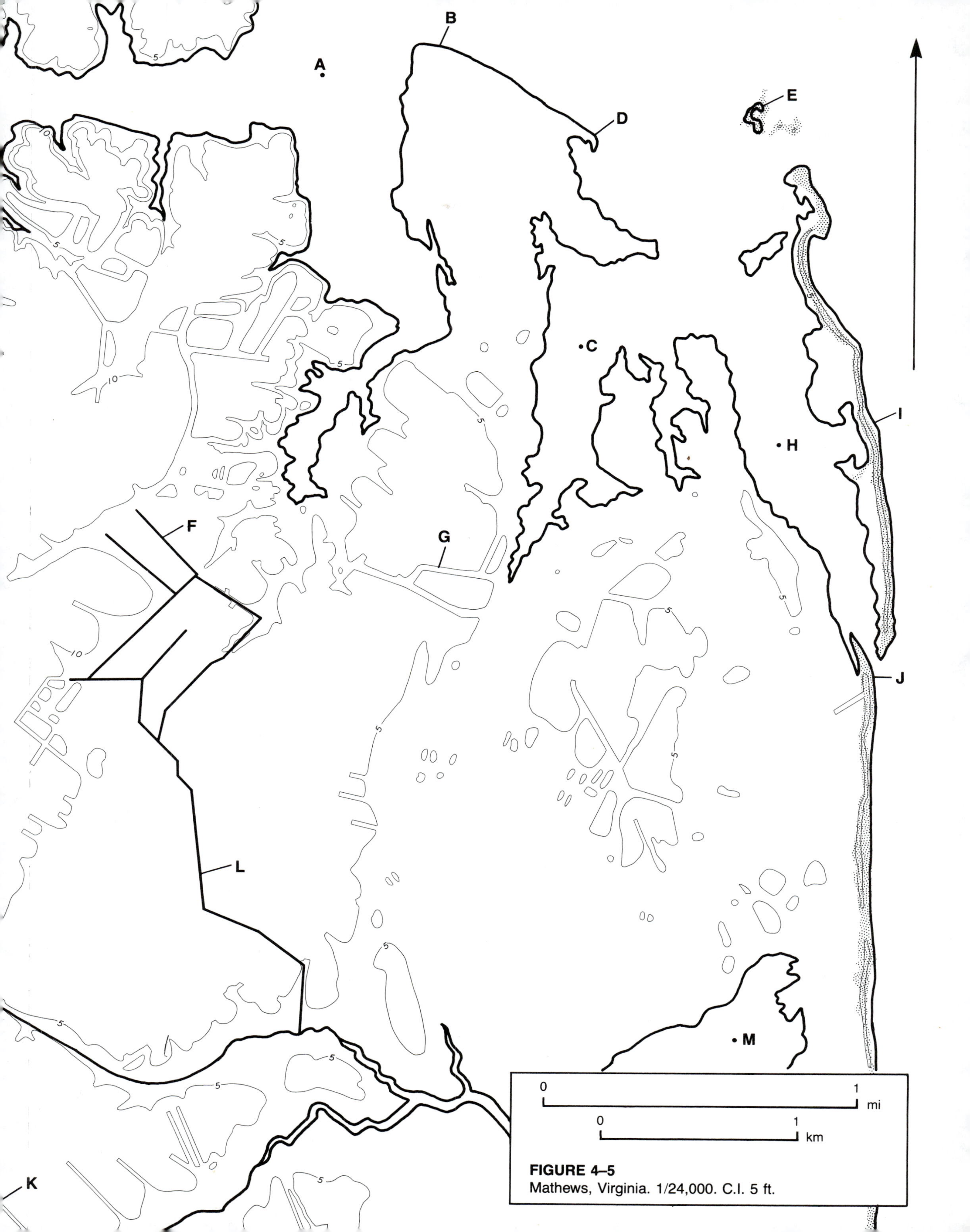

FIGURE 4–5
Mathews, Virginia. 1/24,000. C.I. 5 ft.

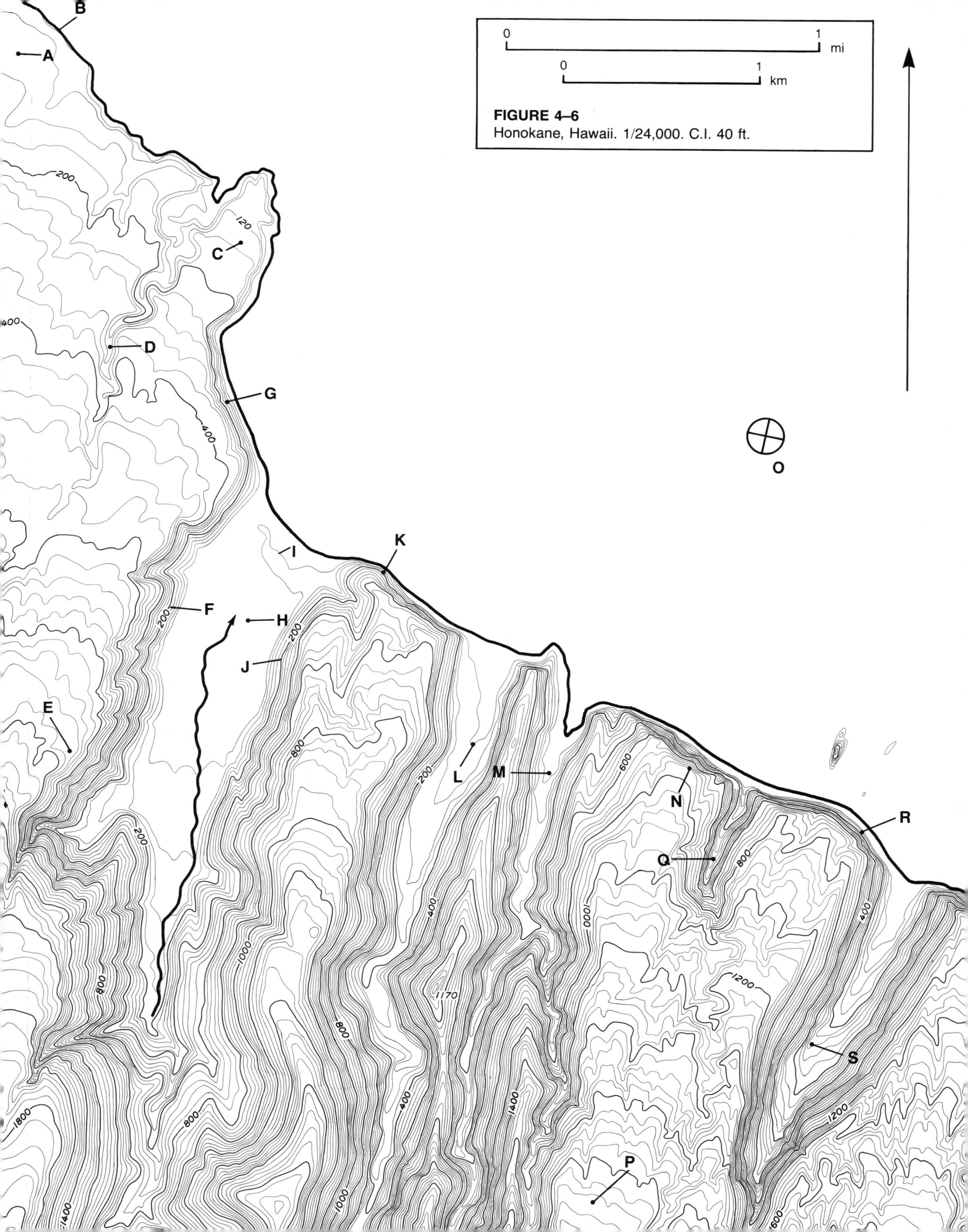

FIGURE 4–6
Honokane, Hawaii. 1/24,000. C.I. 40 ft.

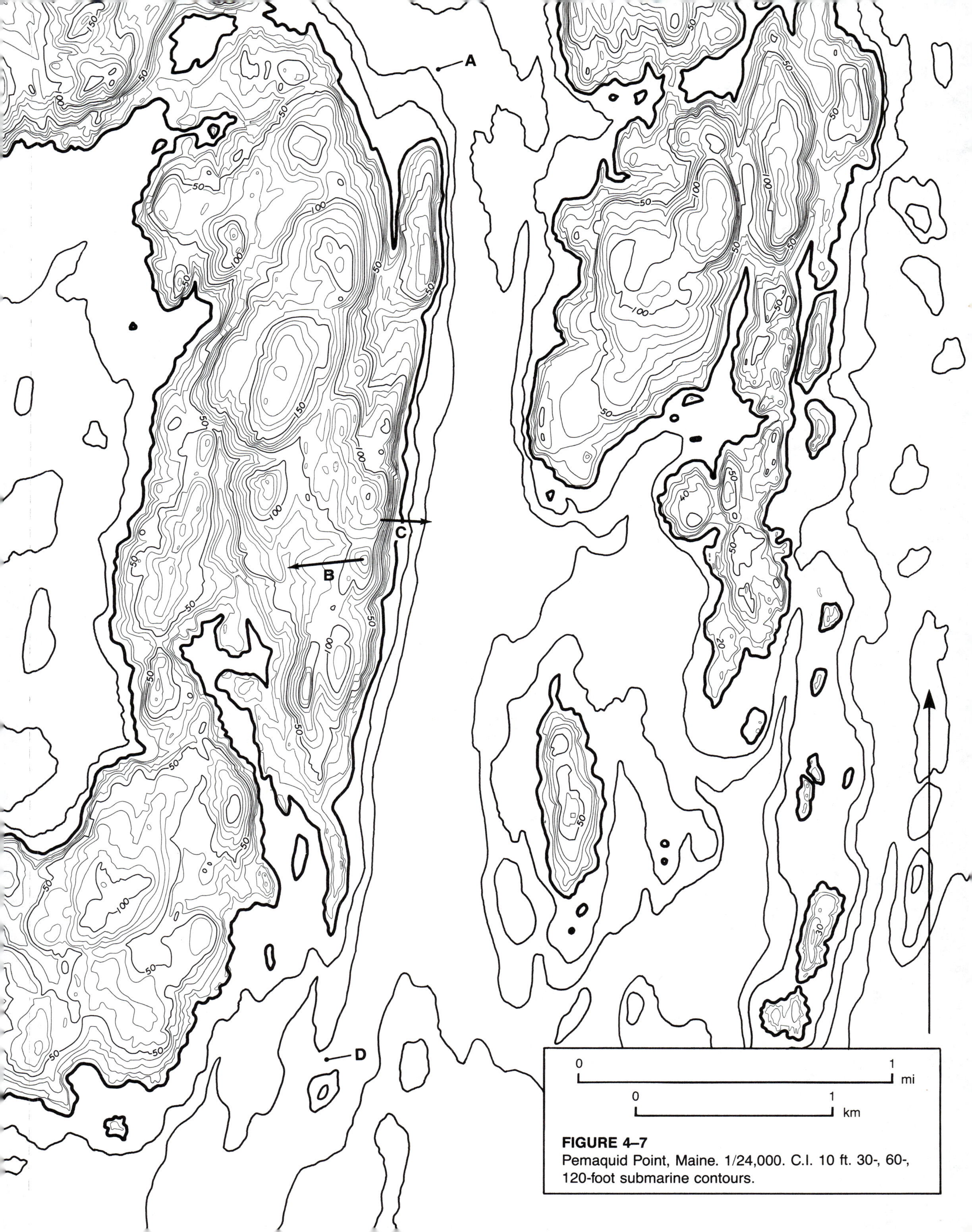

FIGURE 4–7
Pemaquid Point, Maine. 1/24,000. C.I. 10 ft. 30-, 60-, 120-foot submarine contours.

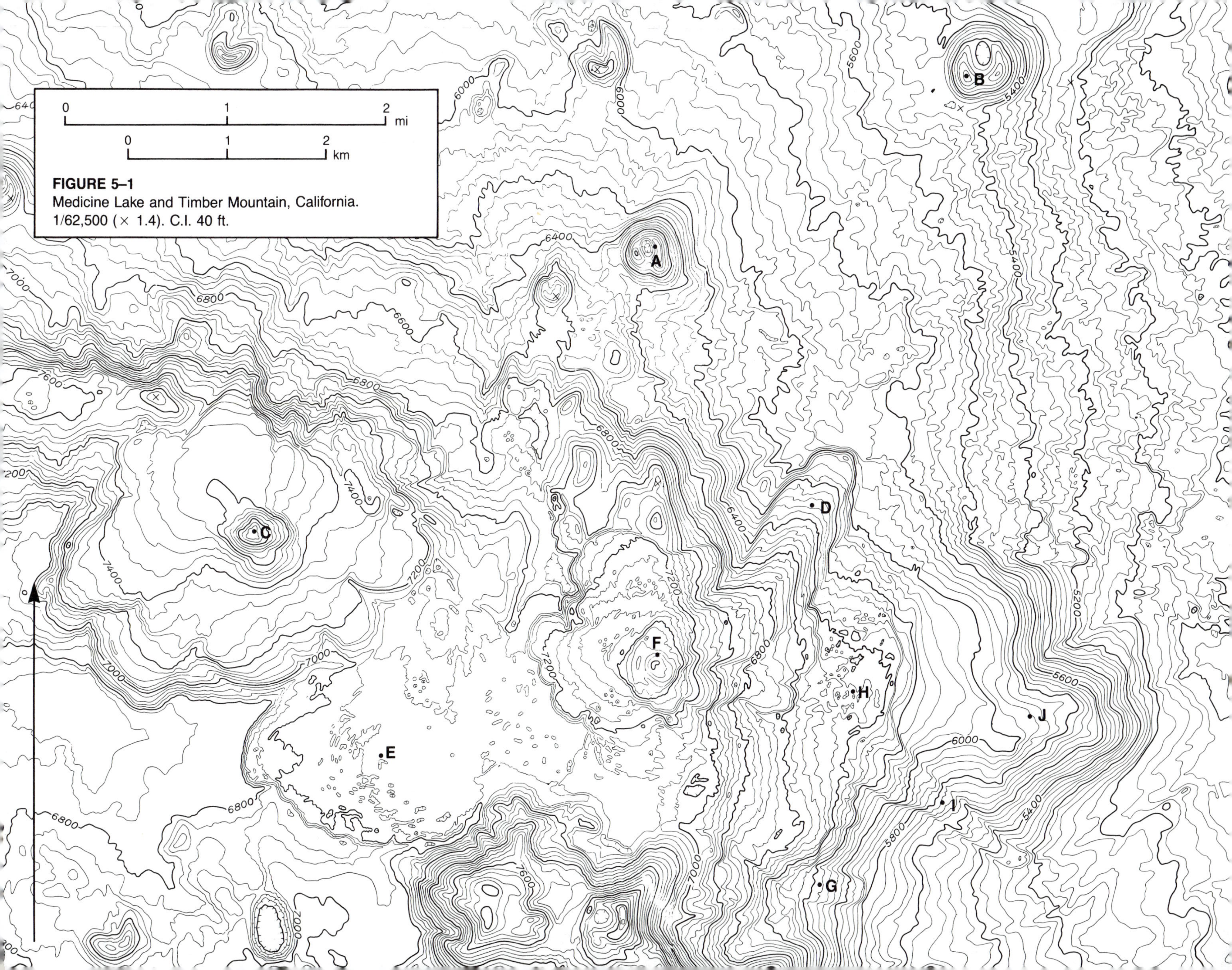

FIGURE 5–1
Medicine Lake and Timber Mountain, California.
1/62,500 (× 1.4). C.I. 40 ft.

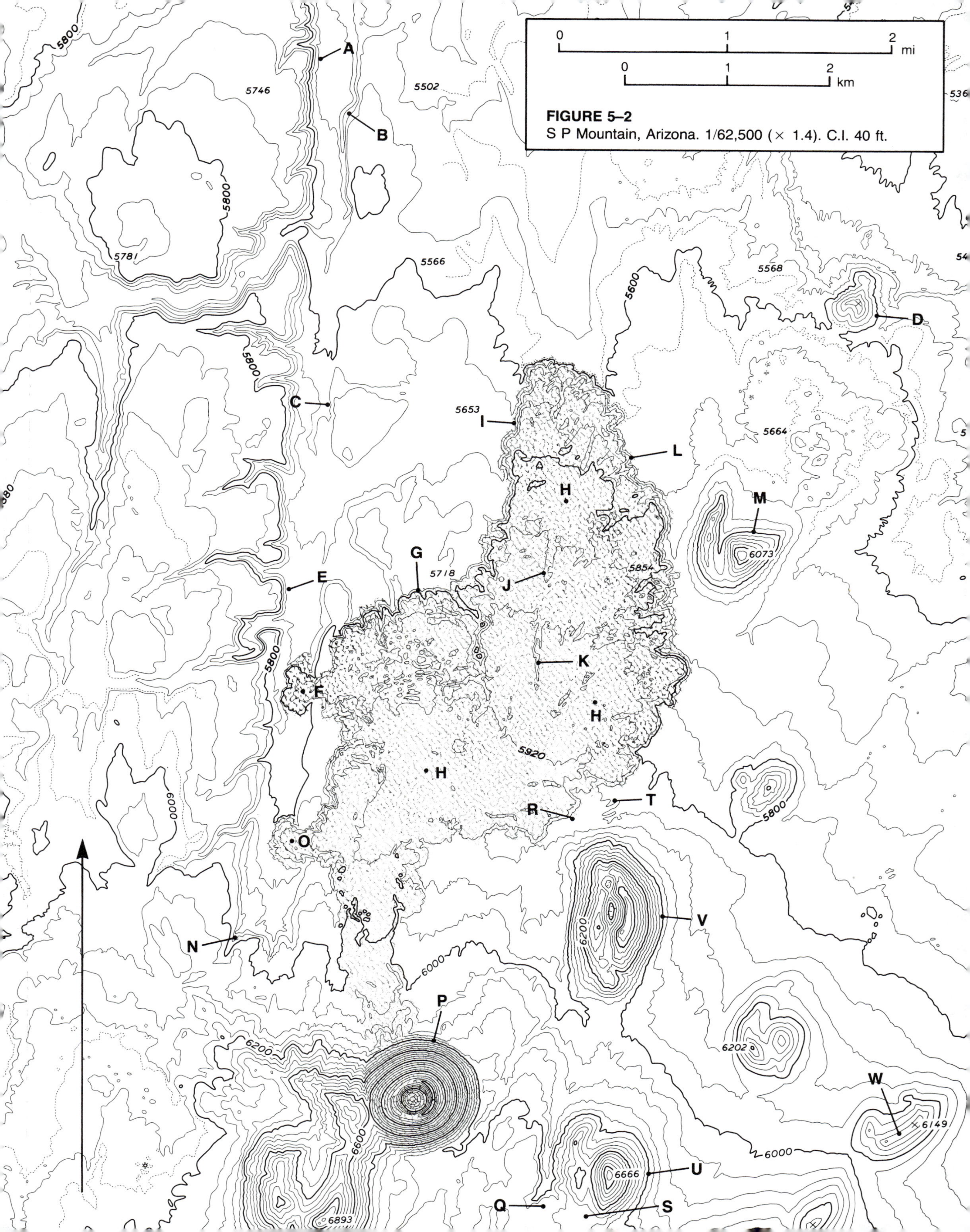

FIGURE 5–2
S P Mountain, Arizona. 1/62,500 (× 1.4). C.I. 40 ft.

FIGURE 5–3
Grand Canyon, Arizona. 1/250,000 (× 1.4).
C.I. 200 ft.

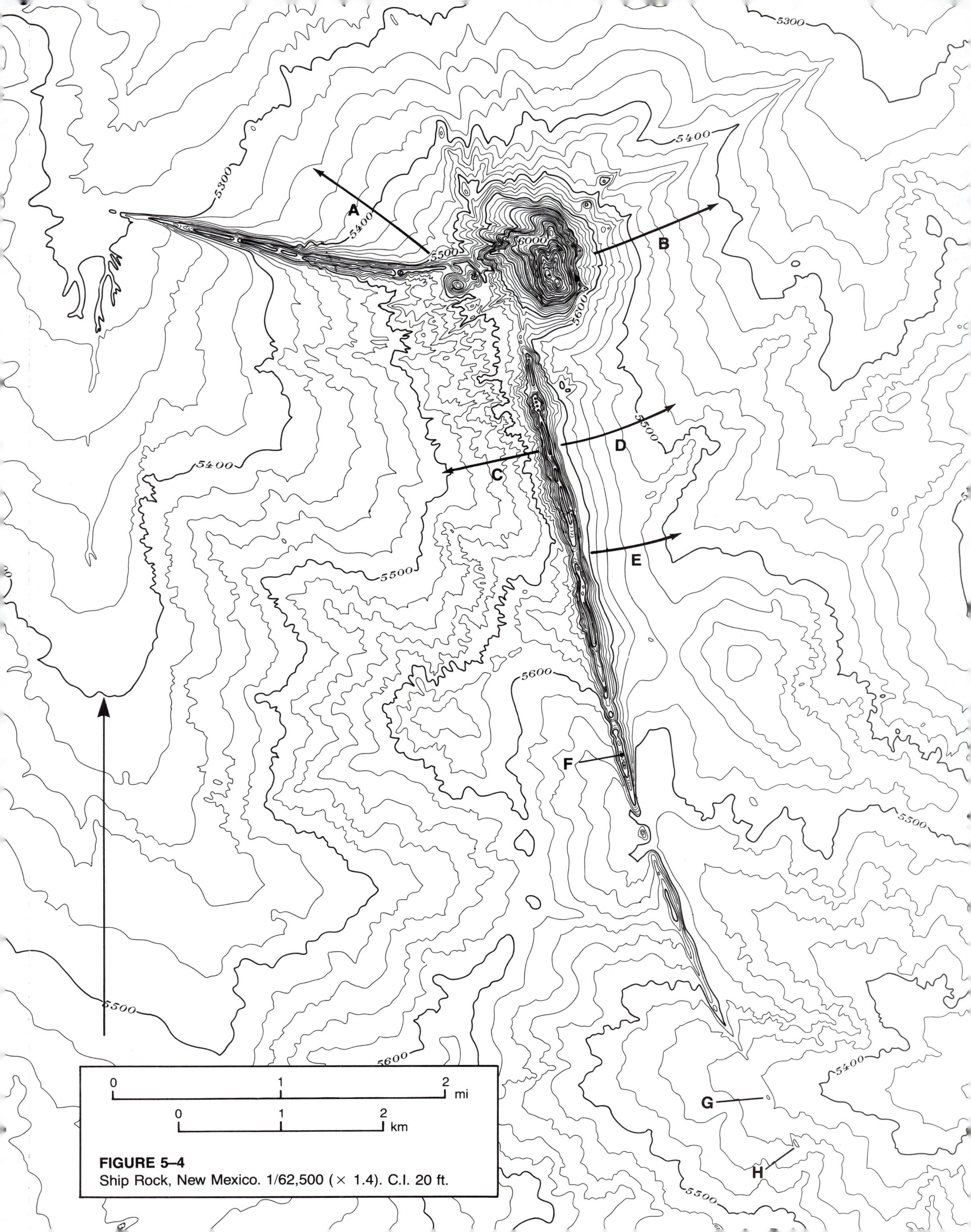

FIGURE 5–4
Ship Rock, New Mexico. 1/62,500 (× 1.4). C.I. 20 ft.

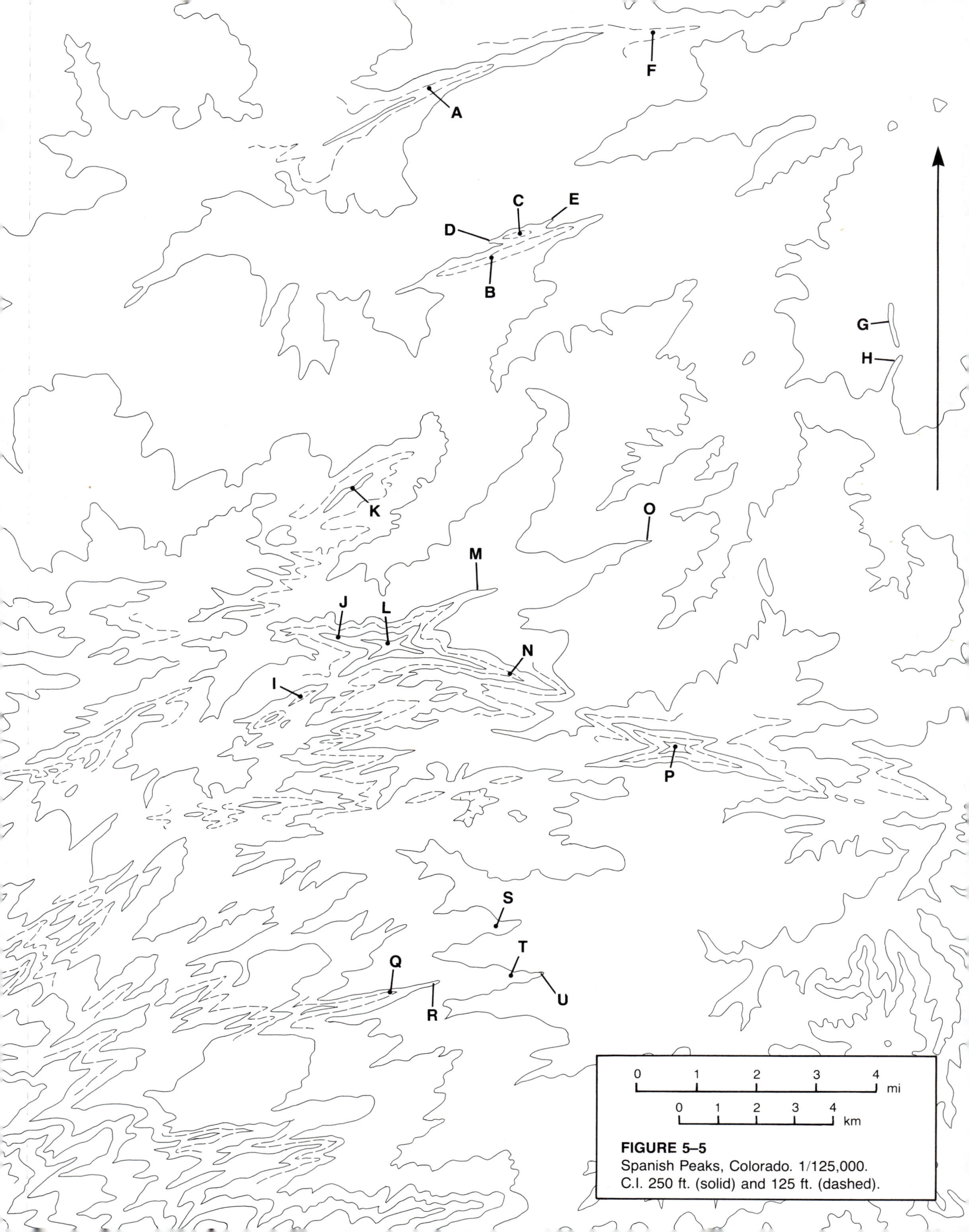

FIGURE 5–5
Spanish Peaks, Colorado. 1/125,000.
C.I. 250 ft. (solid) and 125 ft. (dashed).

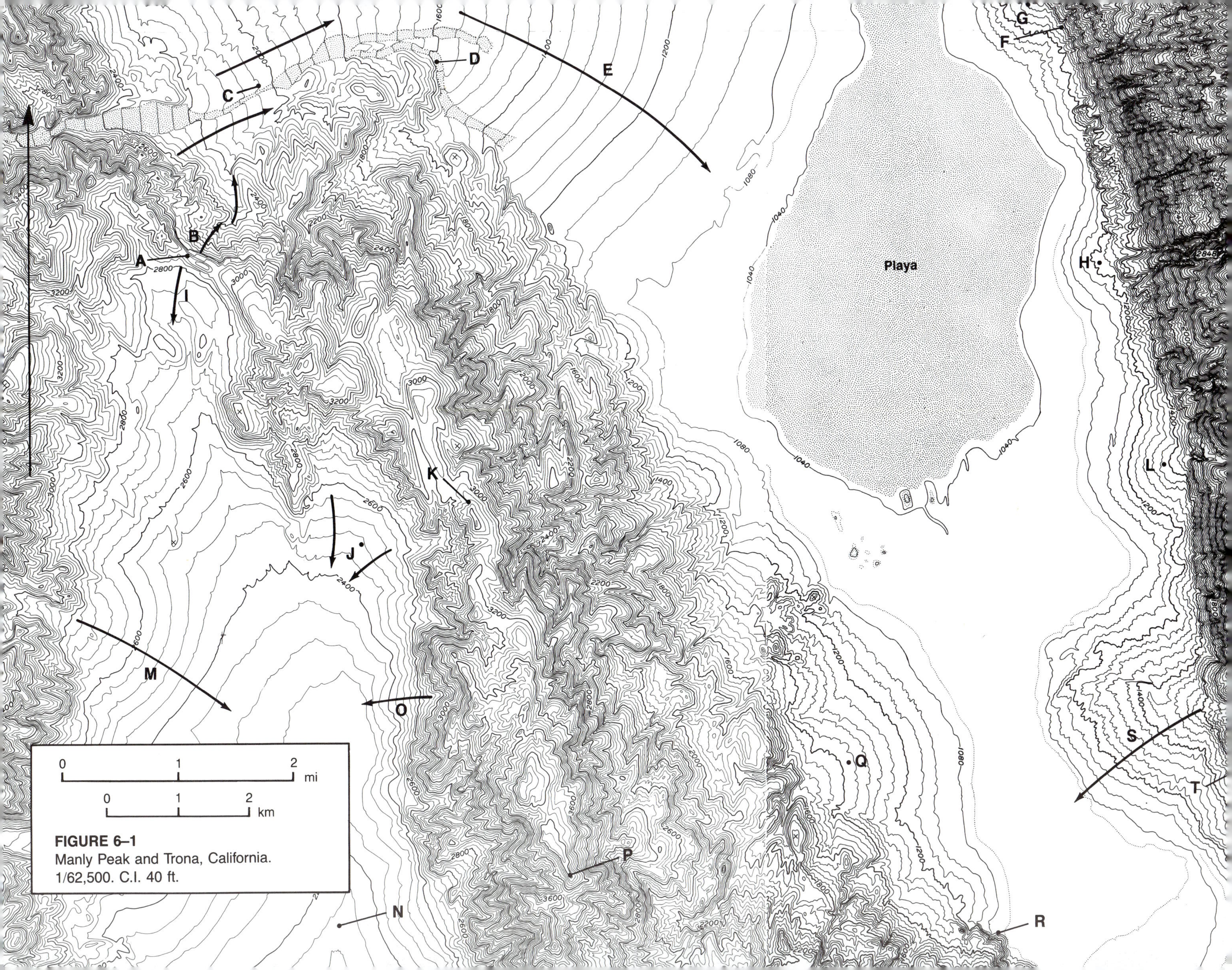

FIGURE 6–1
Manly Peak and Trona, California.
1/62,500. C.I. 40 ft.

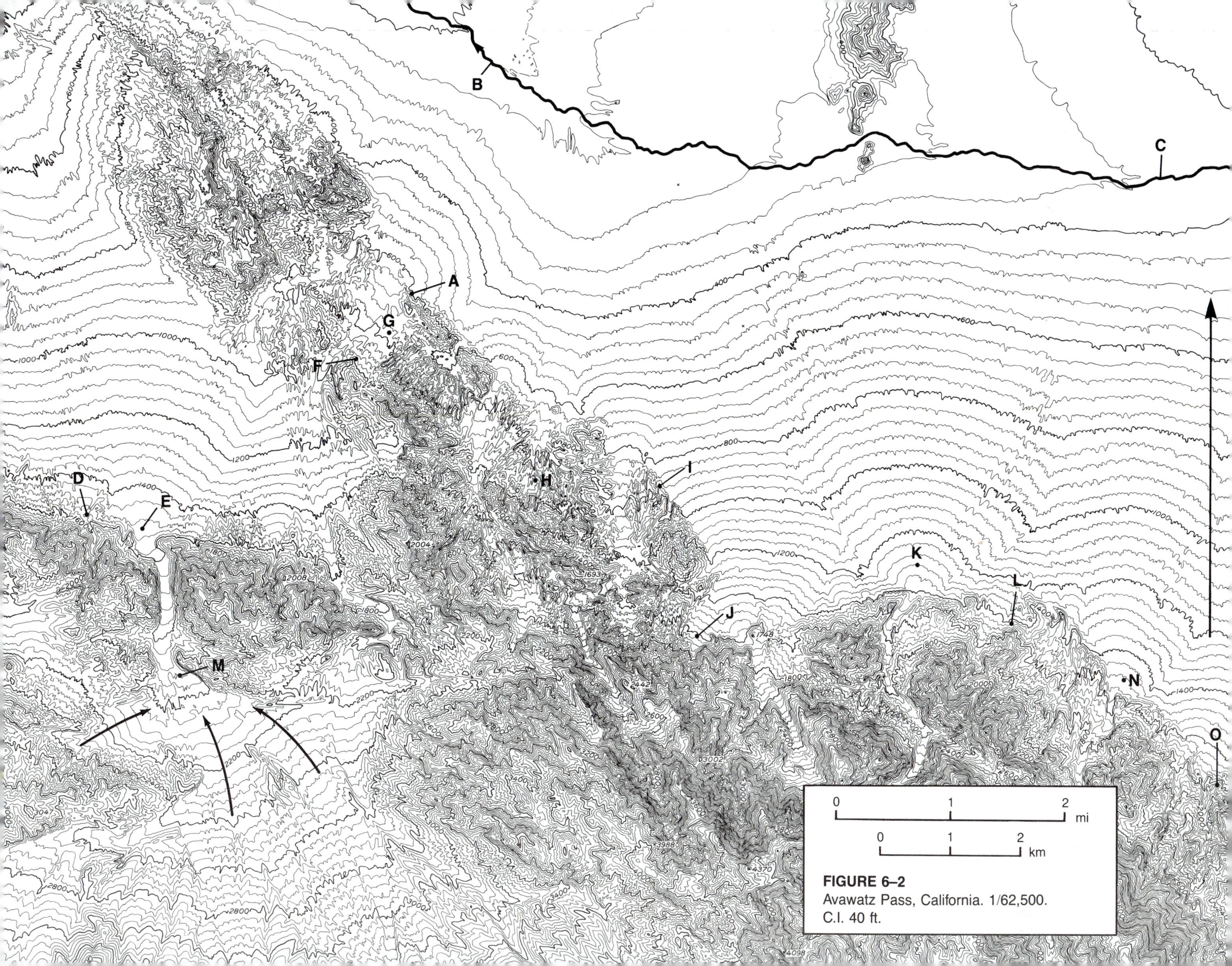

FIGURE 6–2
Avawatz Pass, California. 1/62,500.
C.I. 40 ft.

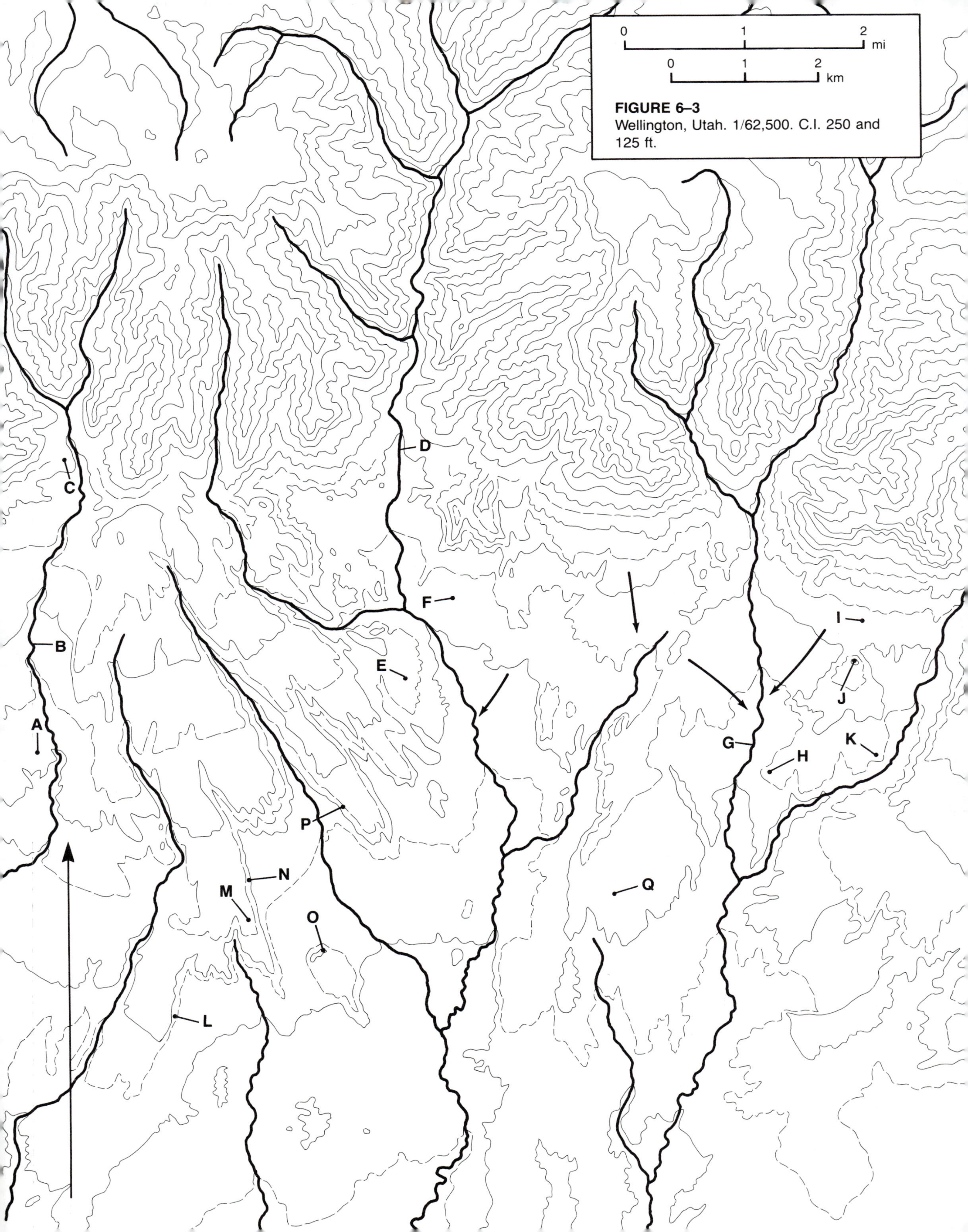

FIGURE 6–3
Wellington, Utah. 1/62,500. C.I. 250 and 125 ft.

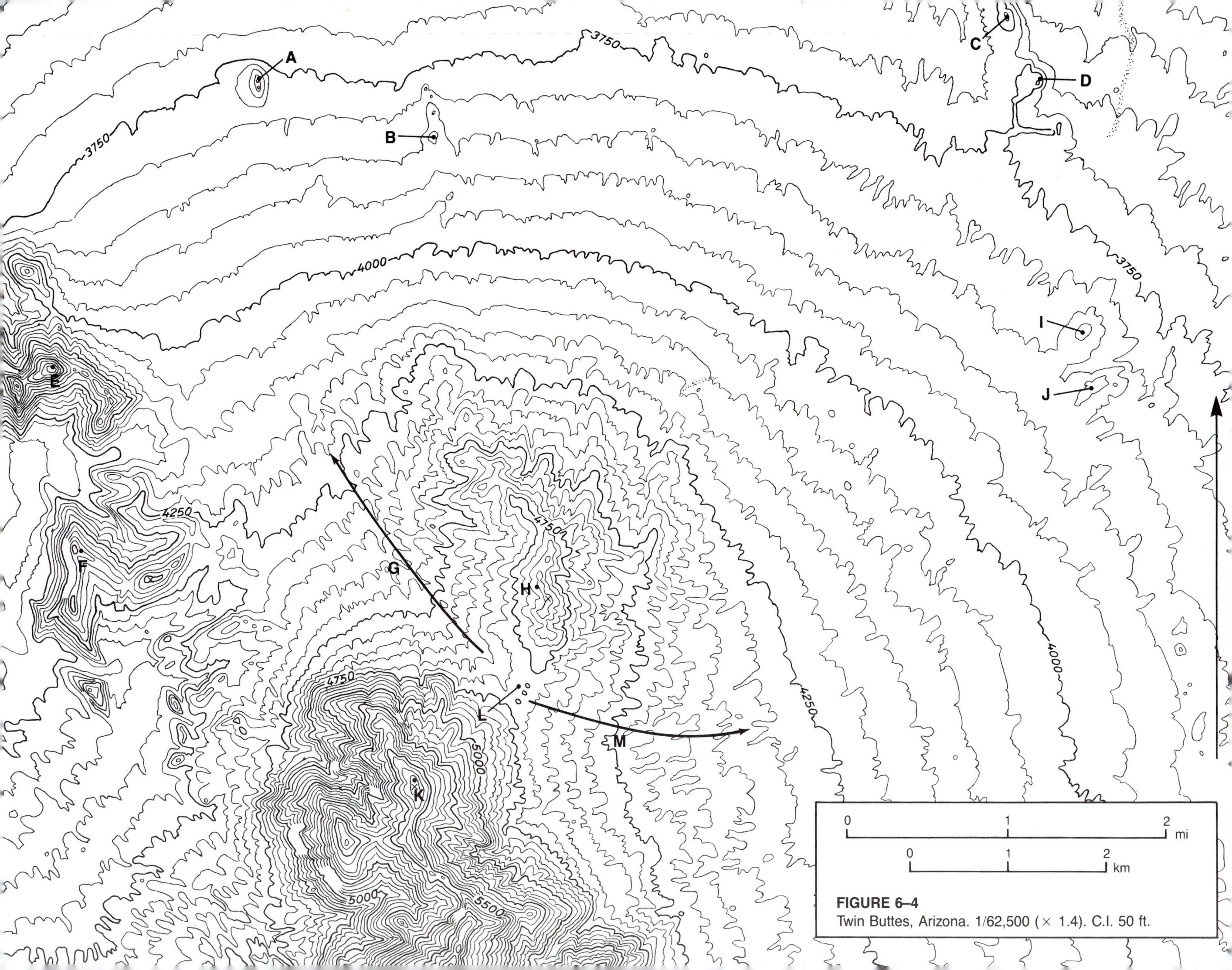

FIGURE 6–4
Twin Buttes, Arizona. 1/62,500 (× 1.4). C.I. 50 ft.

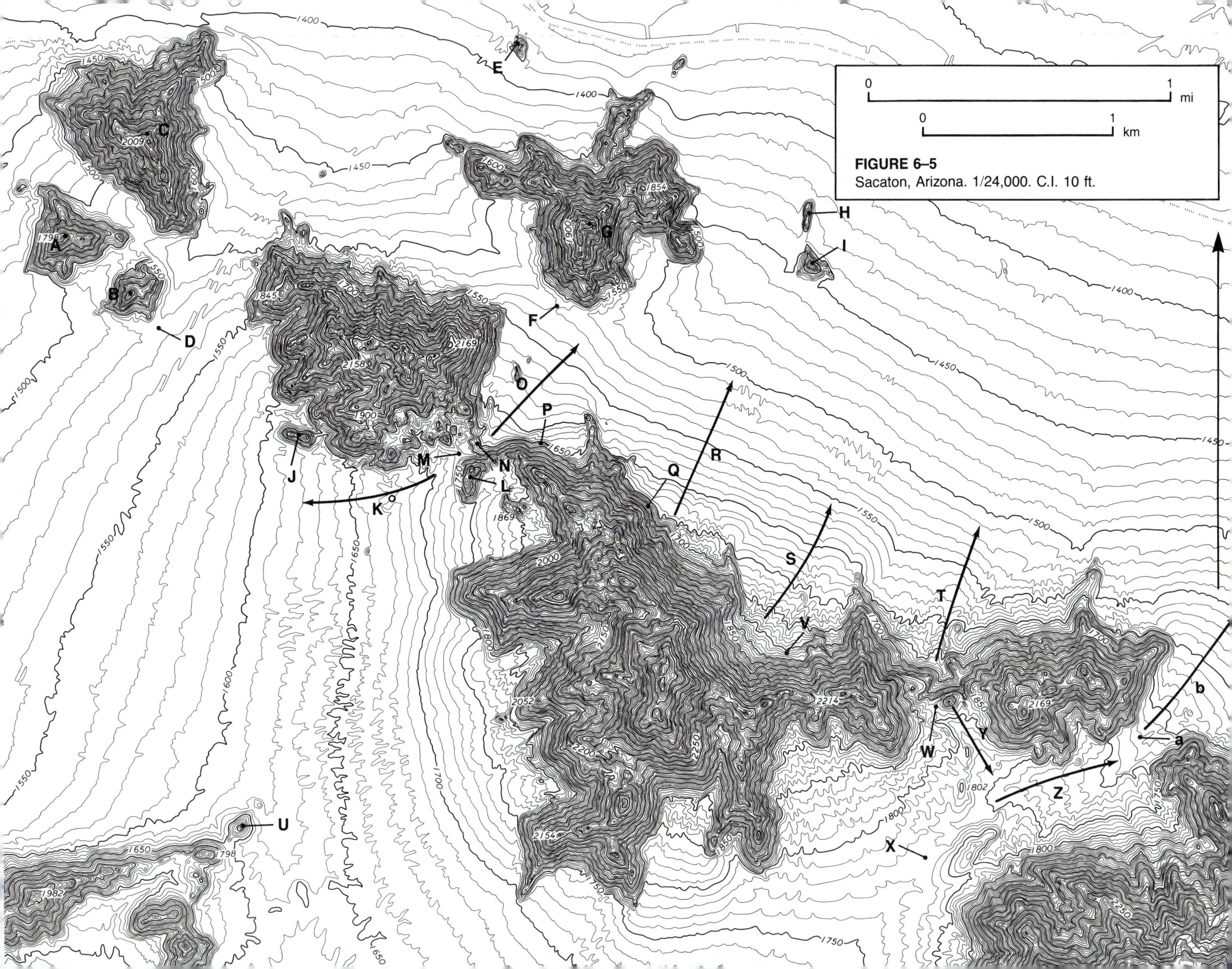

FIGURE 6–5
Sacaton, Arizona. 1/24,000. C.I. 10 ft.

FIGURE 6–6
Sacaton, Arizona. 1/62,500 (× 1.7). C.I. 50 ft.

FIGURE 7–1
Anton, Texas. 1/62,500. C.I. 10 ft.

FIGURE 7–2
Mammoth Cave, Kentucky. 1/62,500 (× 1.7).
C.I. 20 ft.

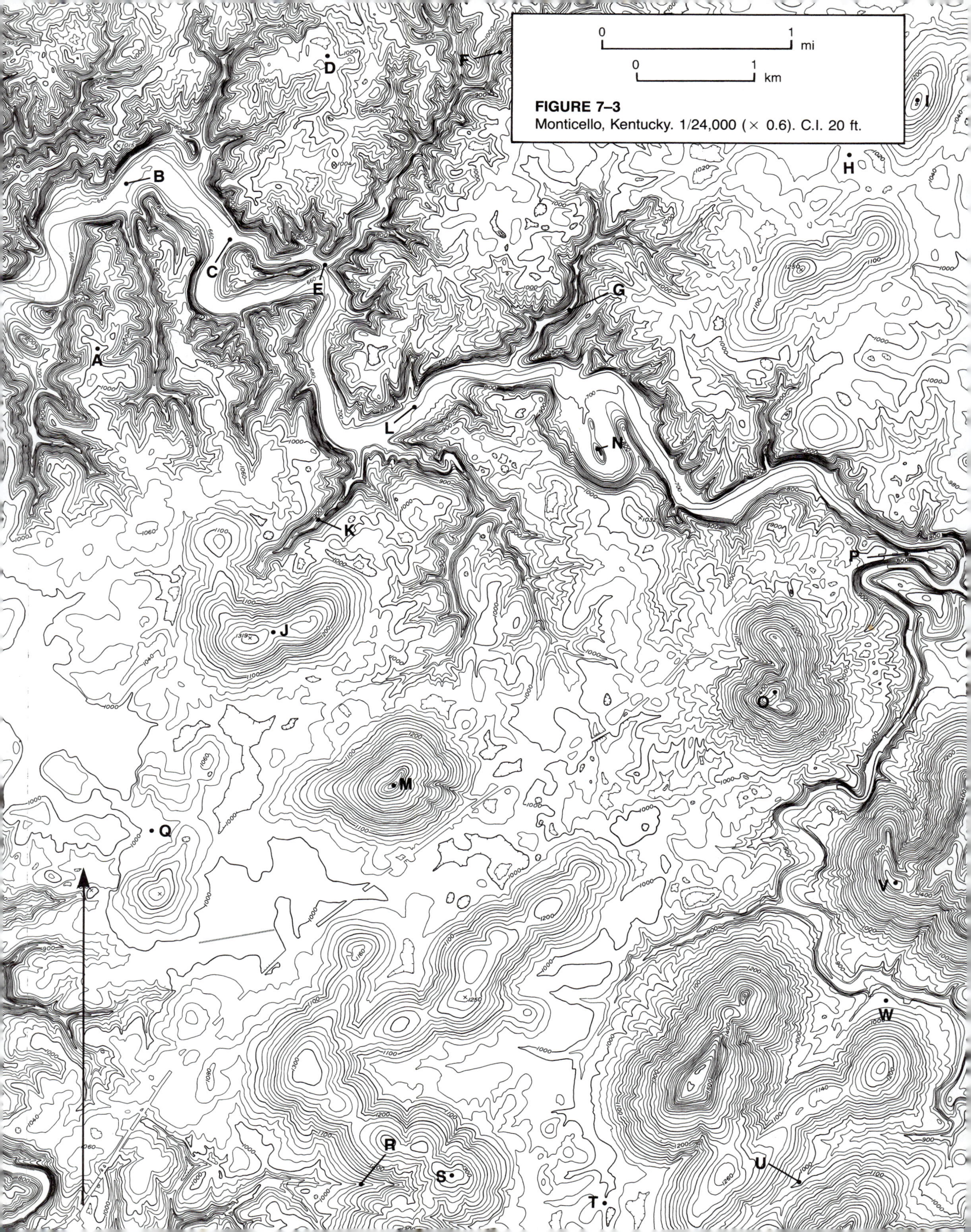

FIGURE 7–3
Monticello, Kentucky. 1/24,000 (× 0.6). C.I. 20 ft.

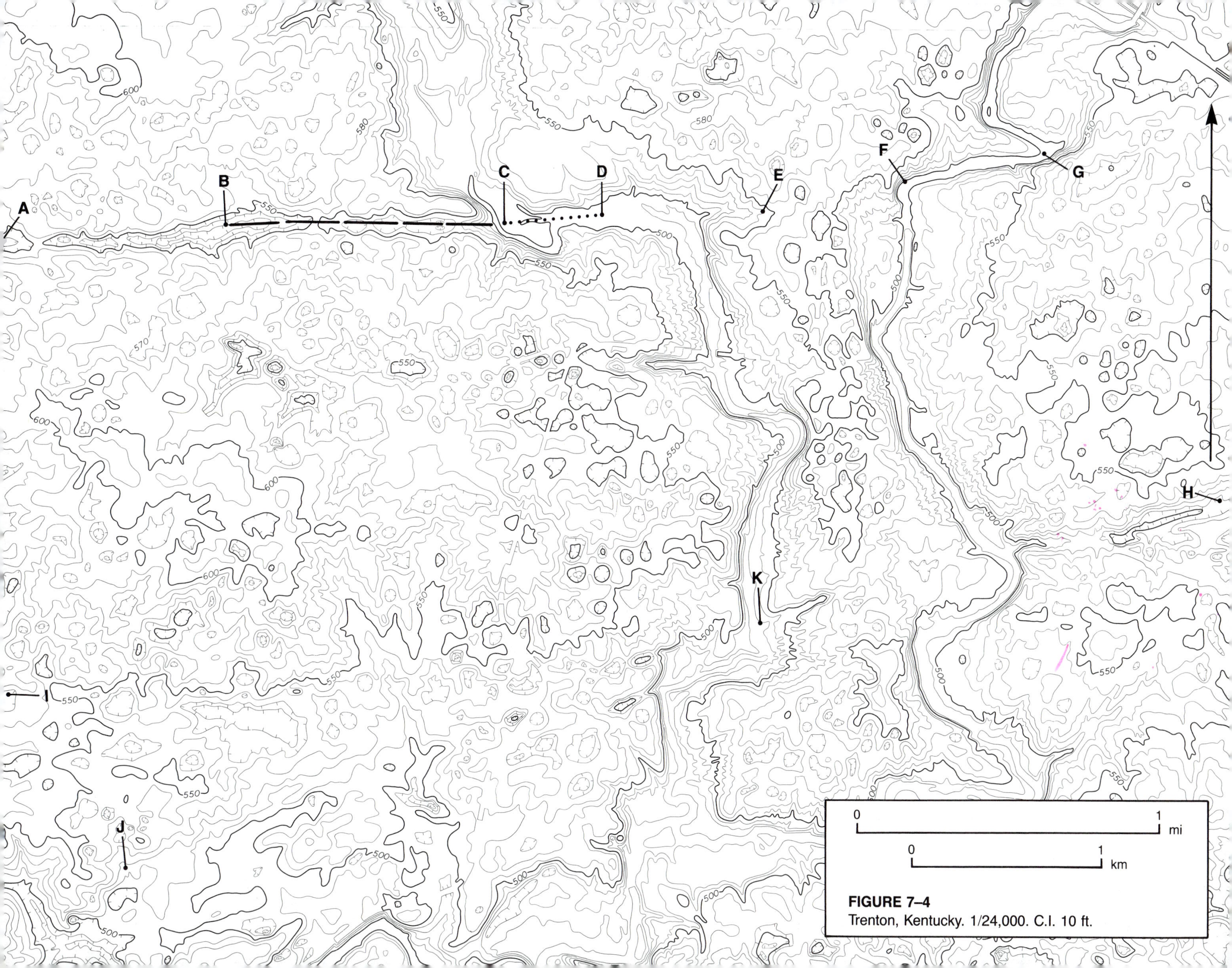

FIGURE 7–4
Trenton, Kentucky. 1/24,000. C.I. 10 ft.

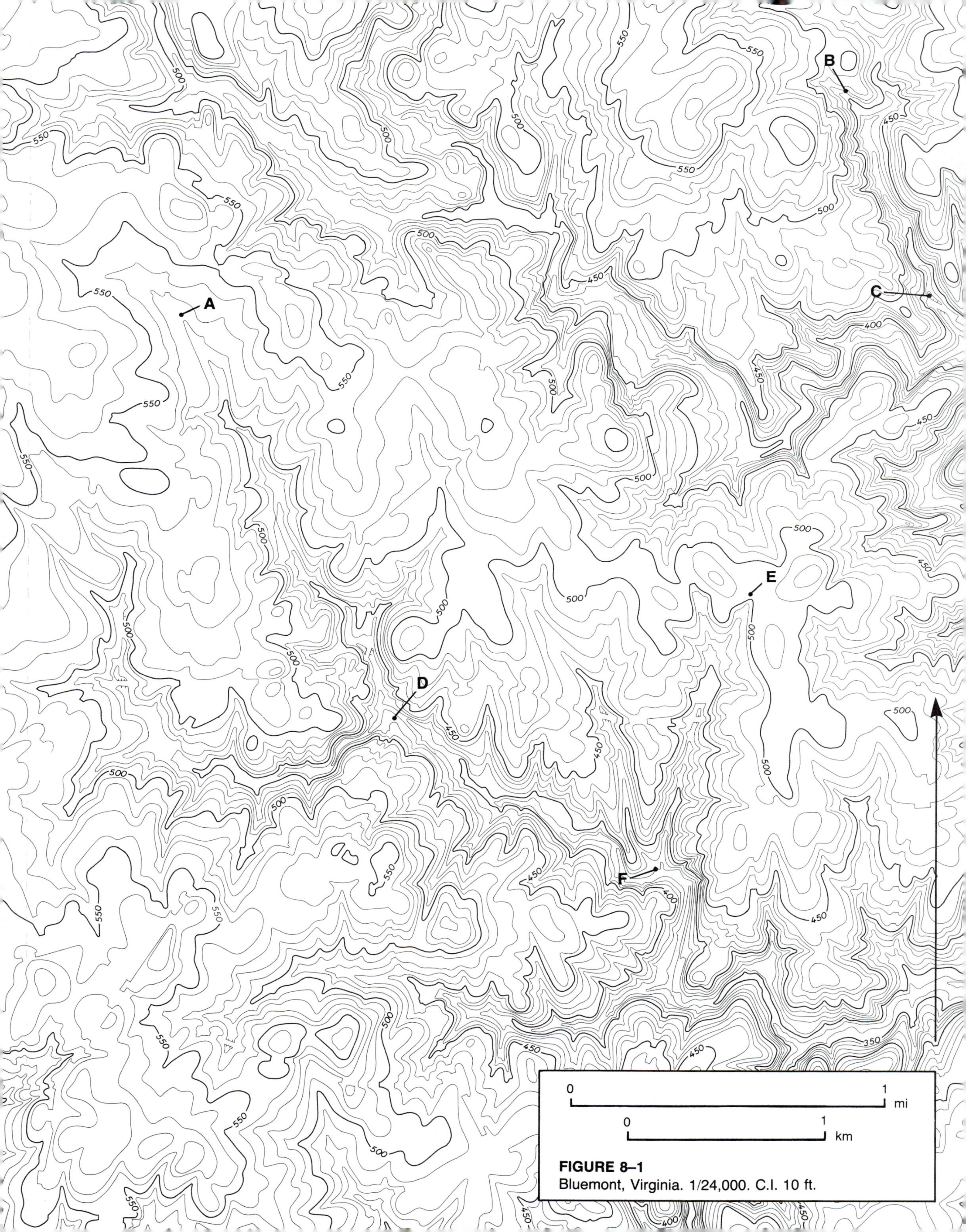

FIGURE 8–1
Bluemont, Virginia. 1/24,000. C.I. 10 ft.

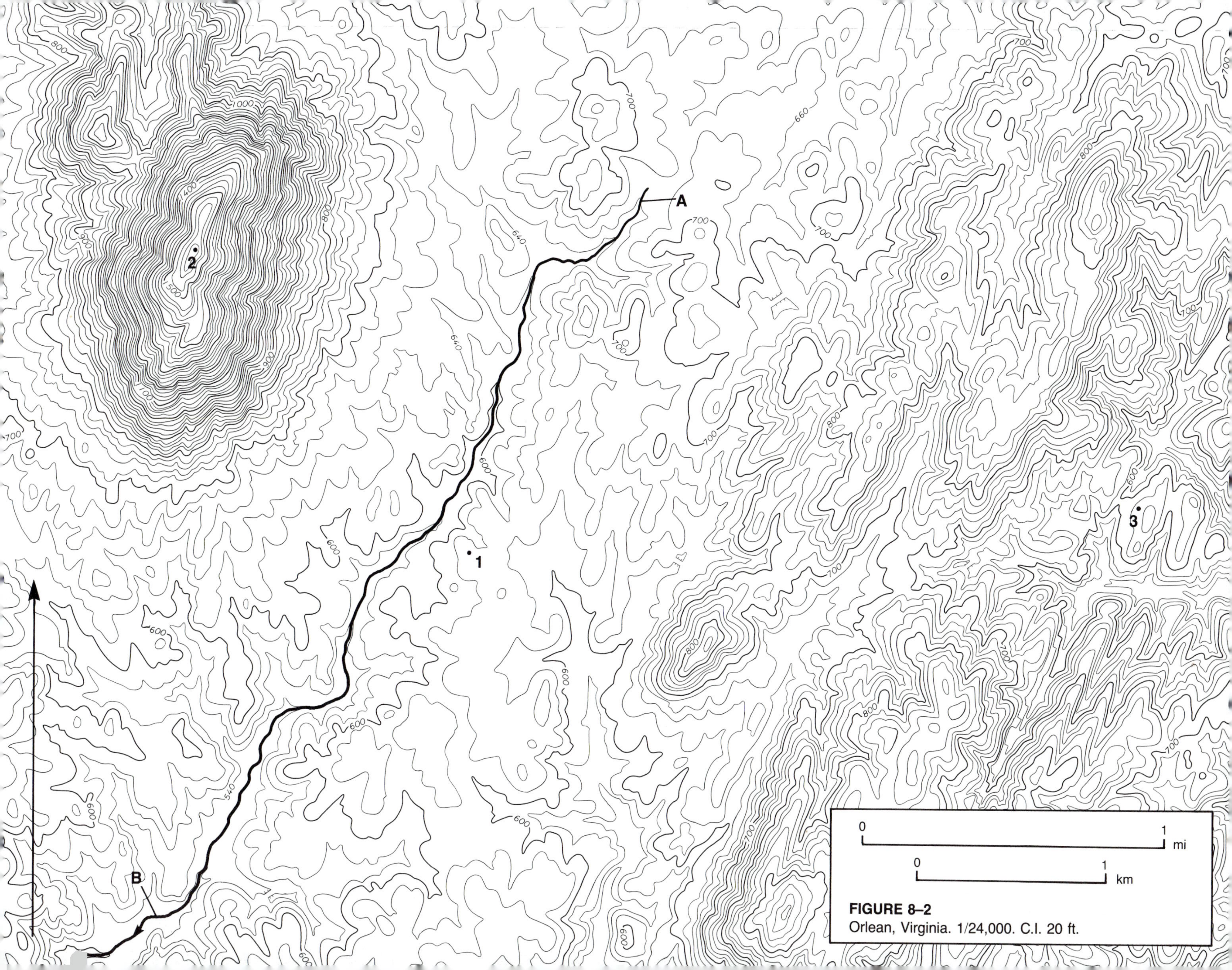

FIGURE 8–2
Orlean, Virginia. 1/24,000. C.I. 20 ft.

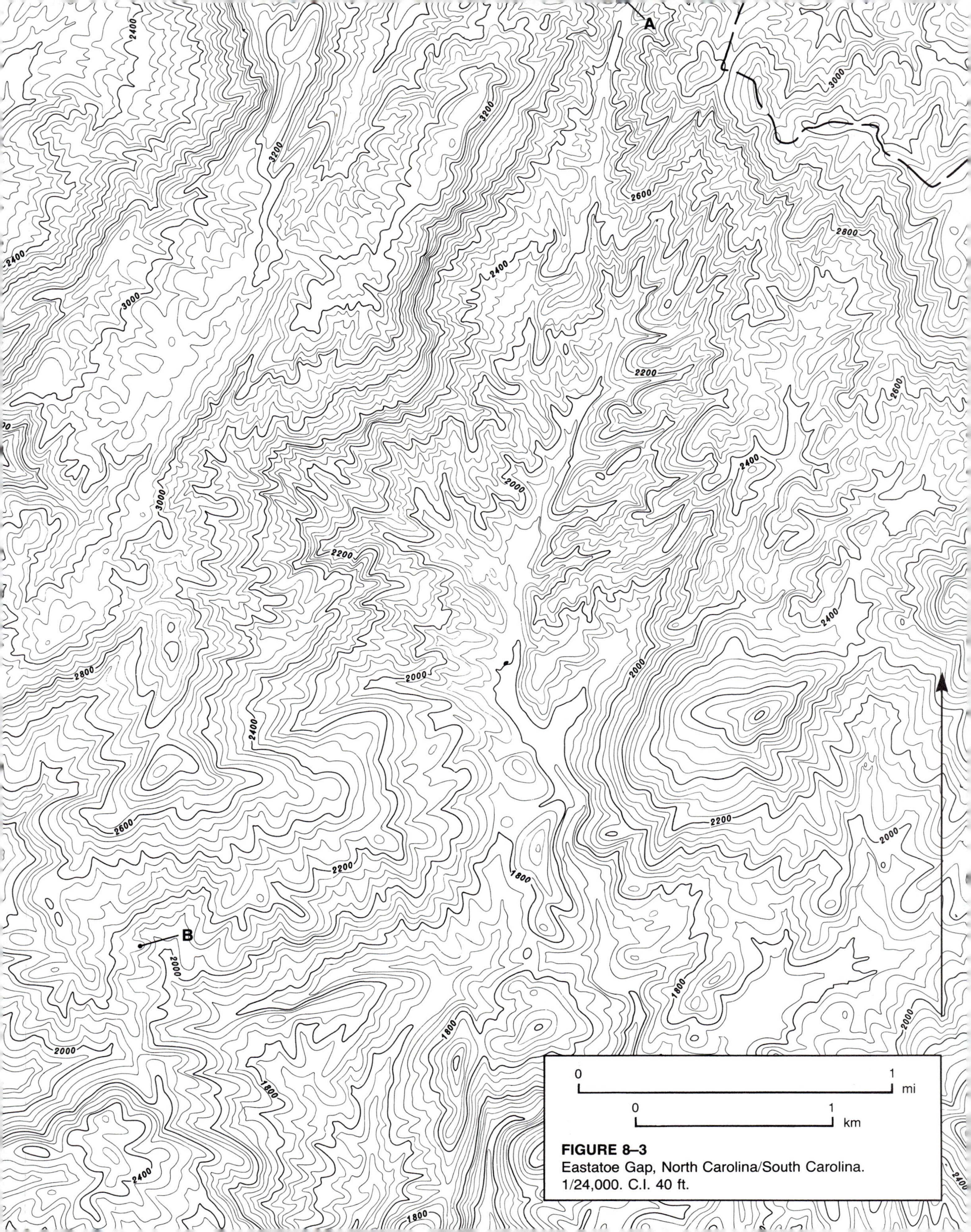

FIGURE 8–3
Eastatoe Gap, North Carolina/South Carolina. 1/24,000. C.I. 40 ft.

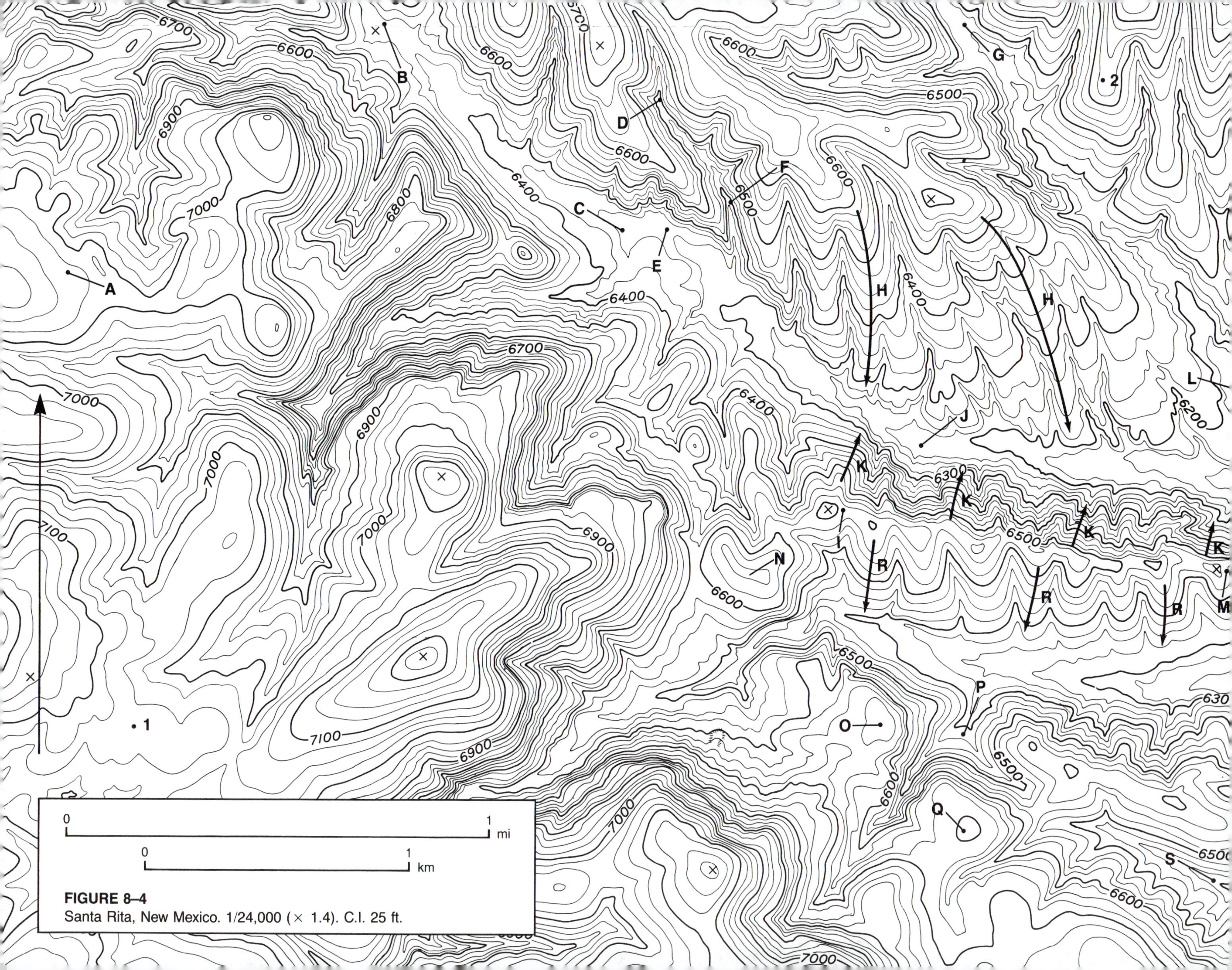

FIGURE 8–4
Santa Rita, New Mexico. 1/24,000 (× 1.4). C.I. 25 ft.

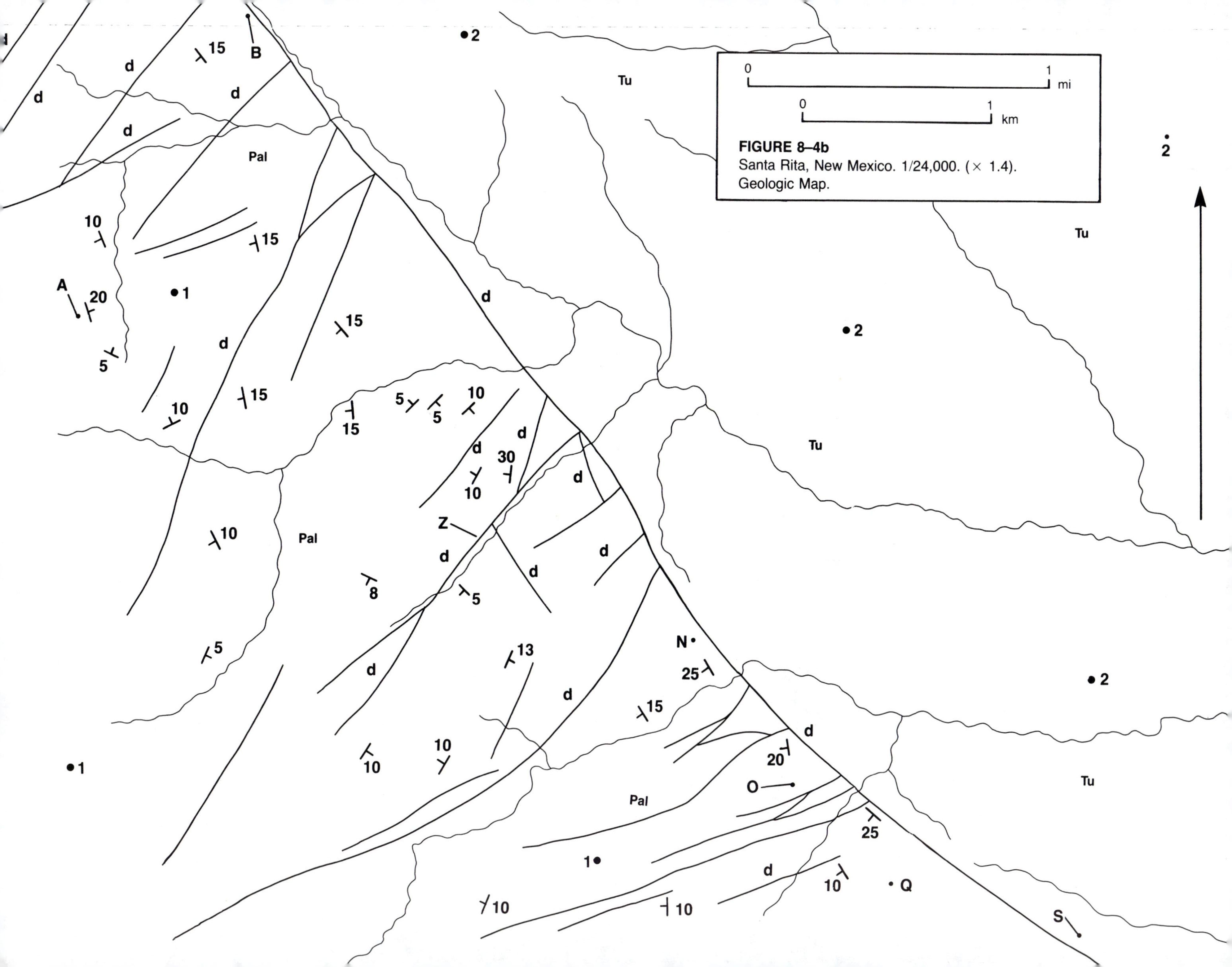

FIGURE 8–4b
Santa Rita, New Mexico. 1/24,000. (× 1.4).
Geologic Map.

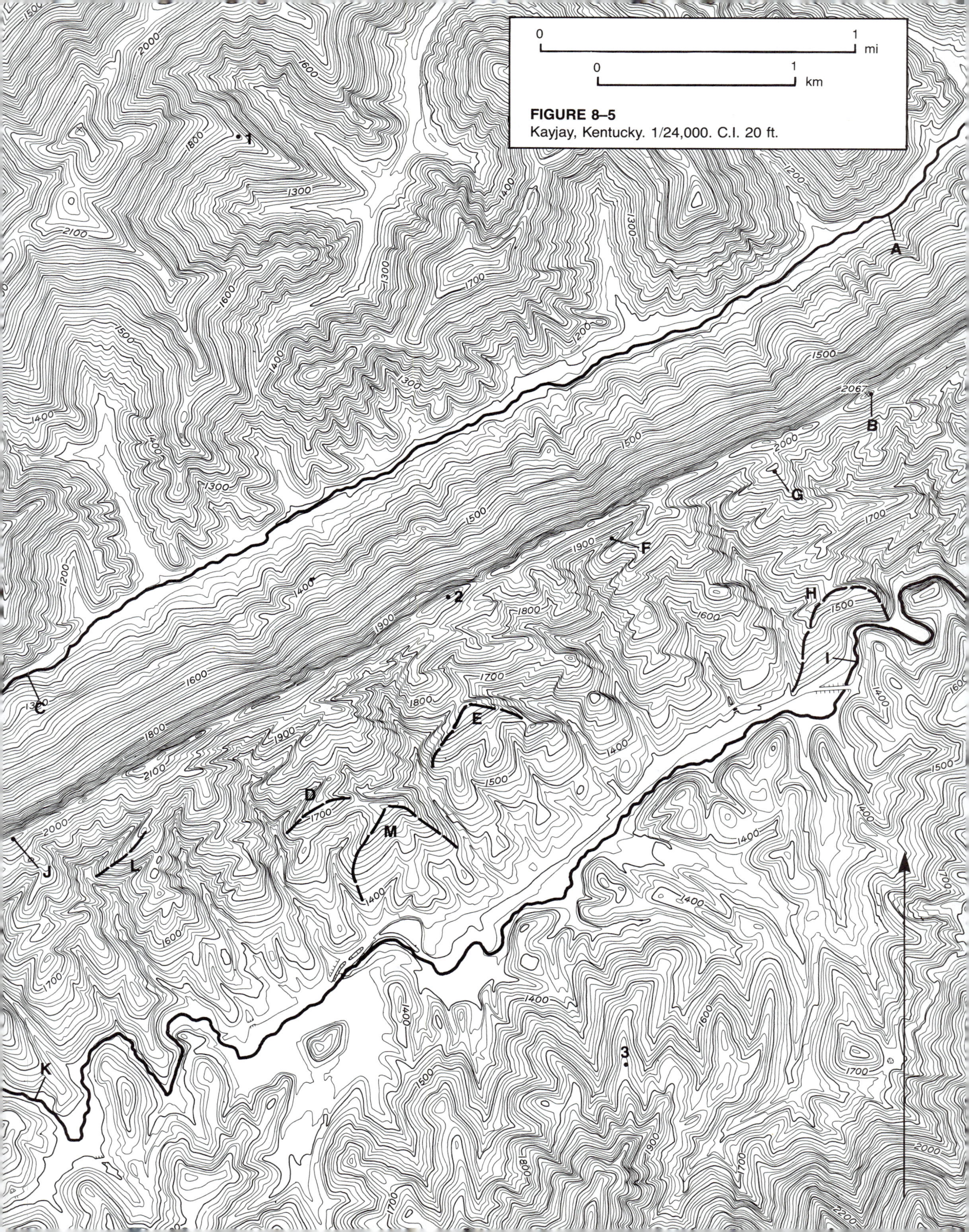

FIGURE 8–5
Kayjay, Kentucky. 1/24,000. C.I. 20 ft.

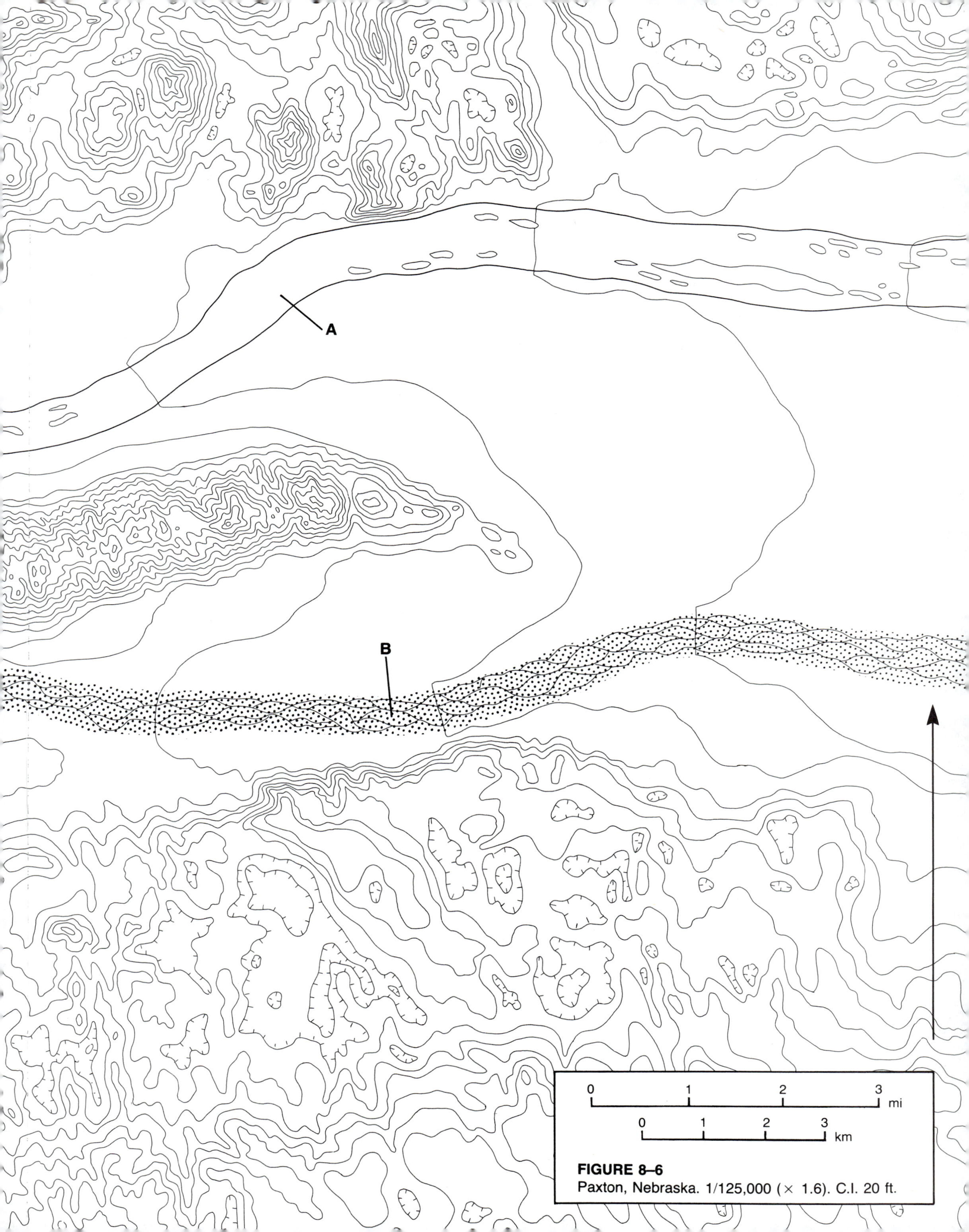

FIGURE 8–6
Paxton, Nebraska. 1/125,000 (× 1.6). C.I. 20 ft.

FIGURE 8–7
Frankfort East and Frankfort West, Kentucky. 1/24,000. C.I. 10 ft.

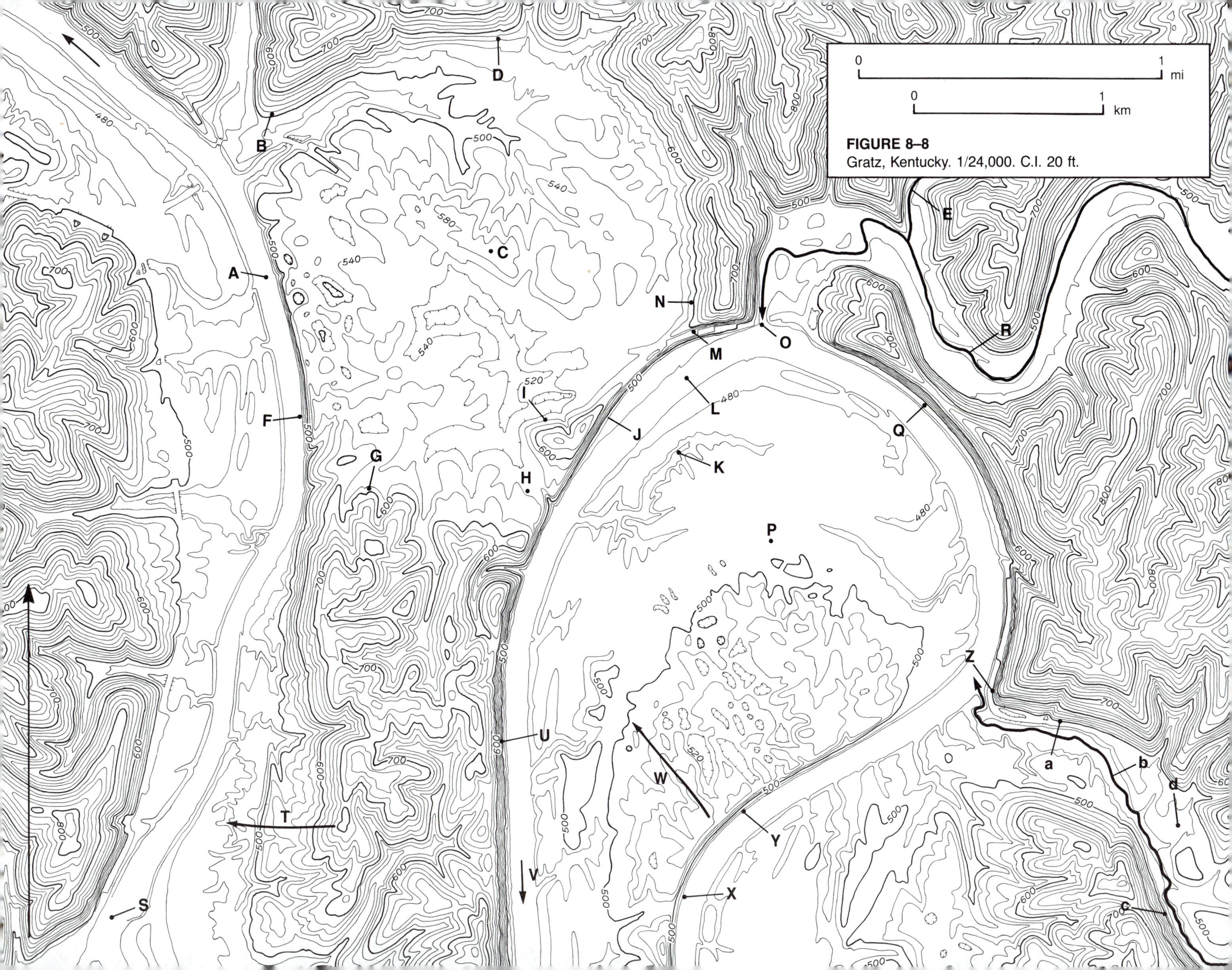

FIGURE 8–8
Gratz, Kentucky. 1/24,000. C.I. 20 ft.

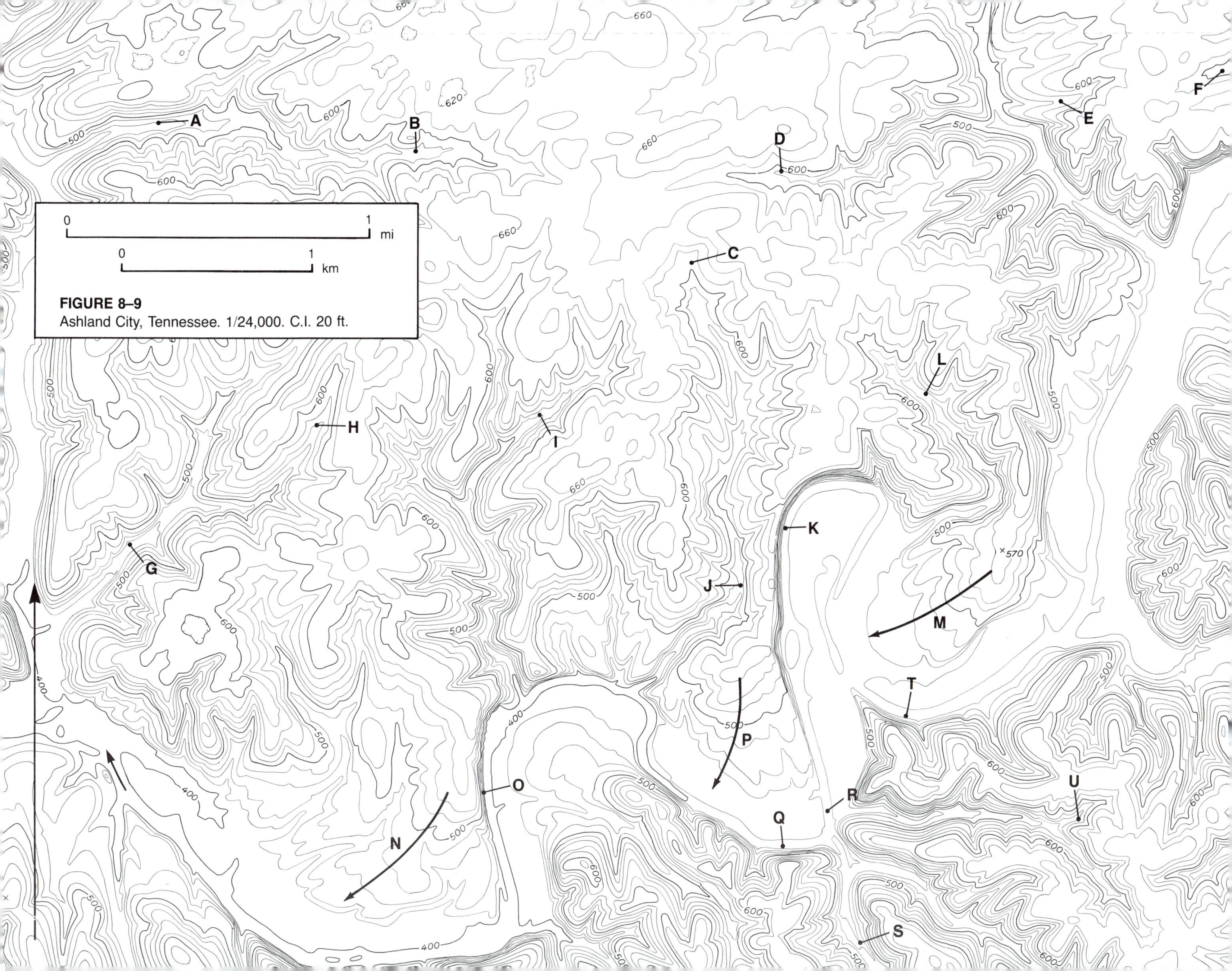

FIGURE 8–9
Ashland City, Tennessee. 1/24,000. C.I. 20 ft.

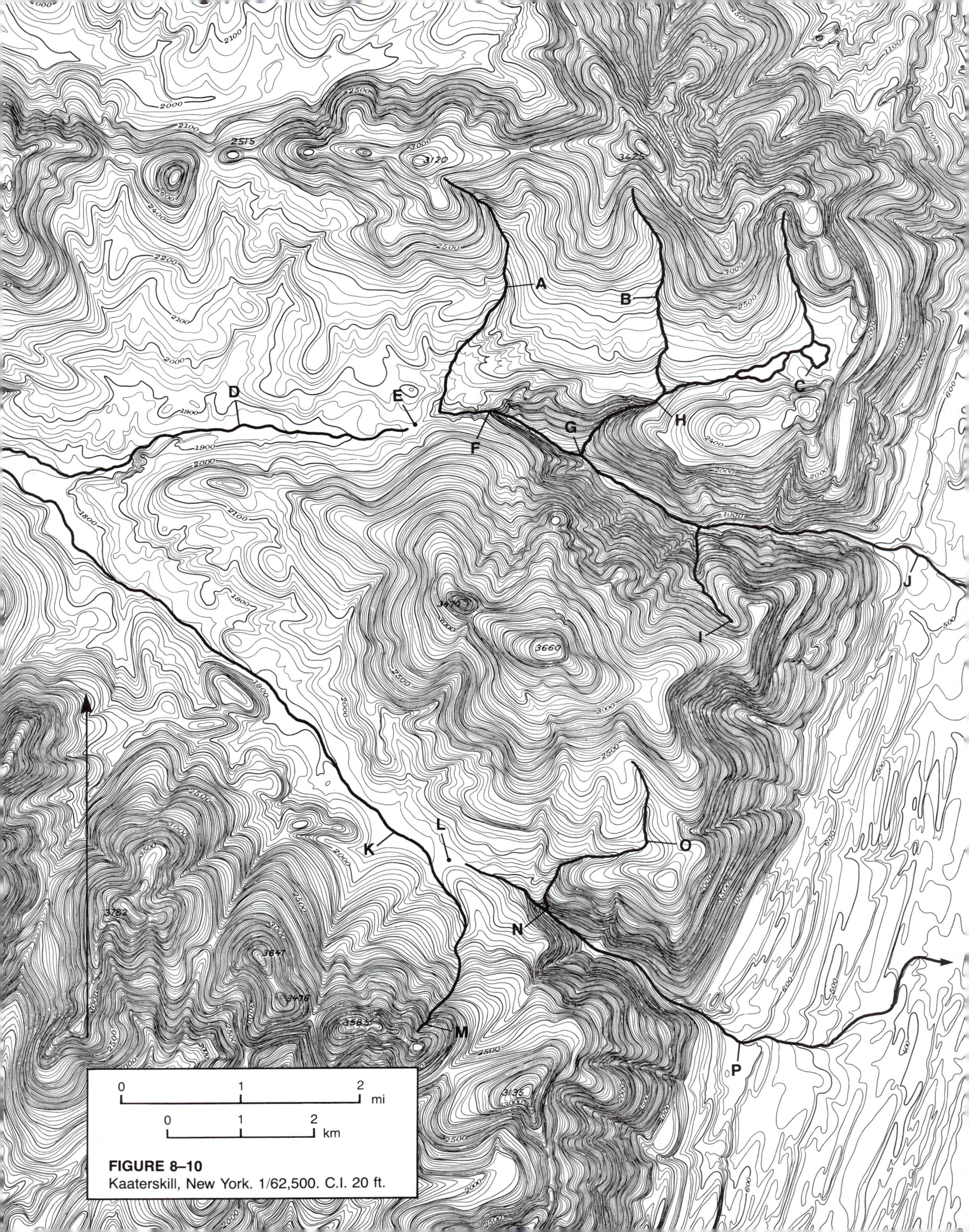

FIGURE 8–10
Kaaterskill, New York. 1/62,500. C.I. 20 ft.

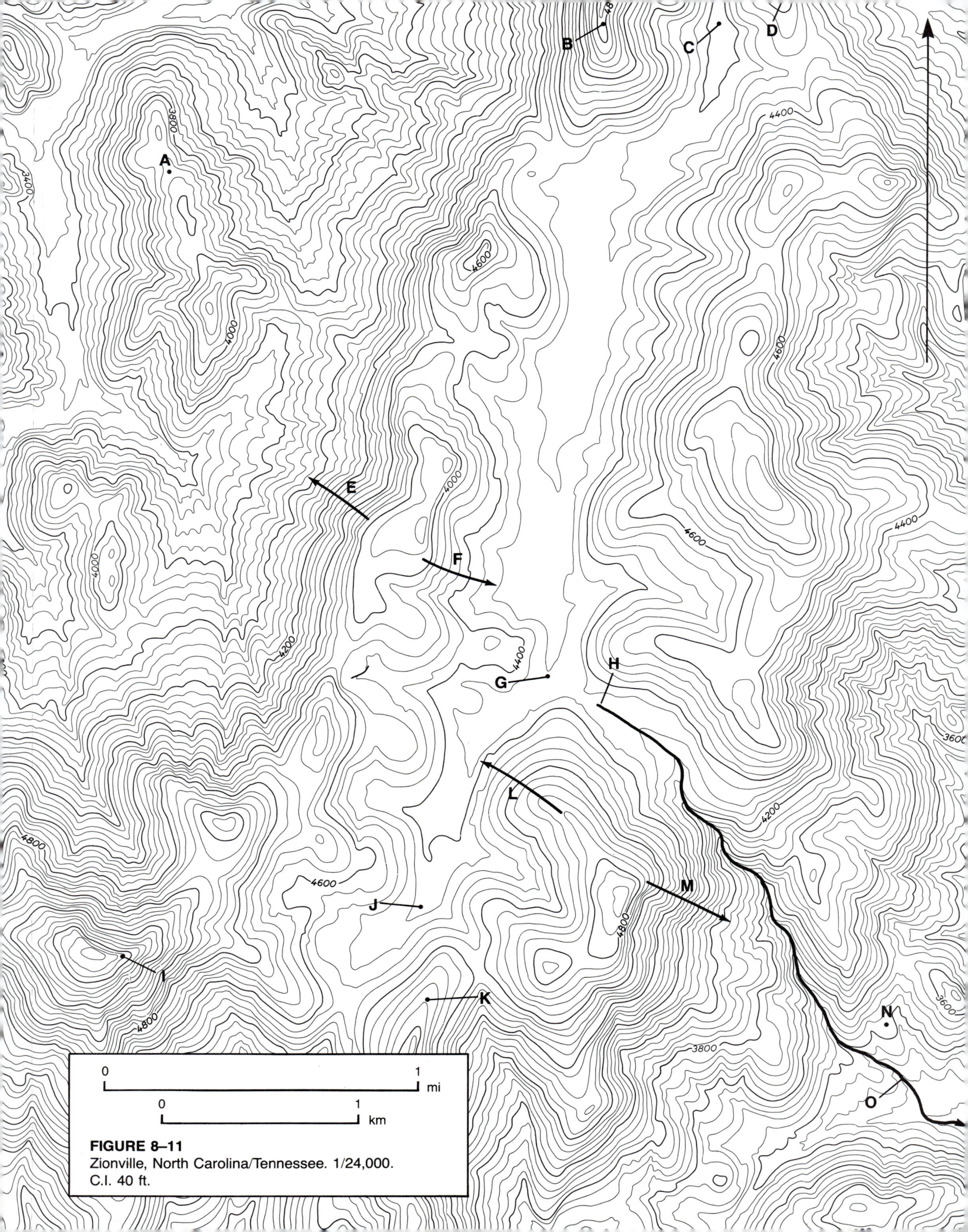

FIGURE 8–11
Zionville, North Carolina/Tennessee. 1/24,000.
C.I. 40 ft.

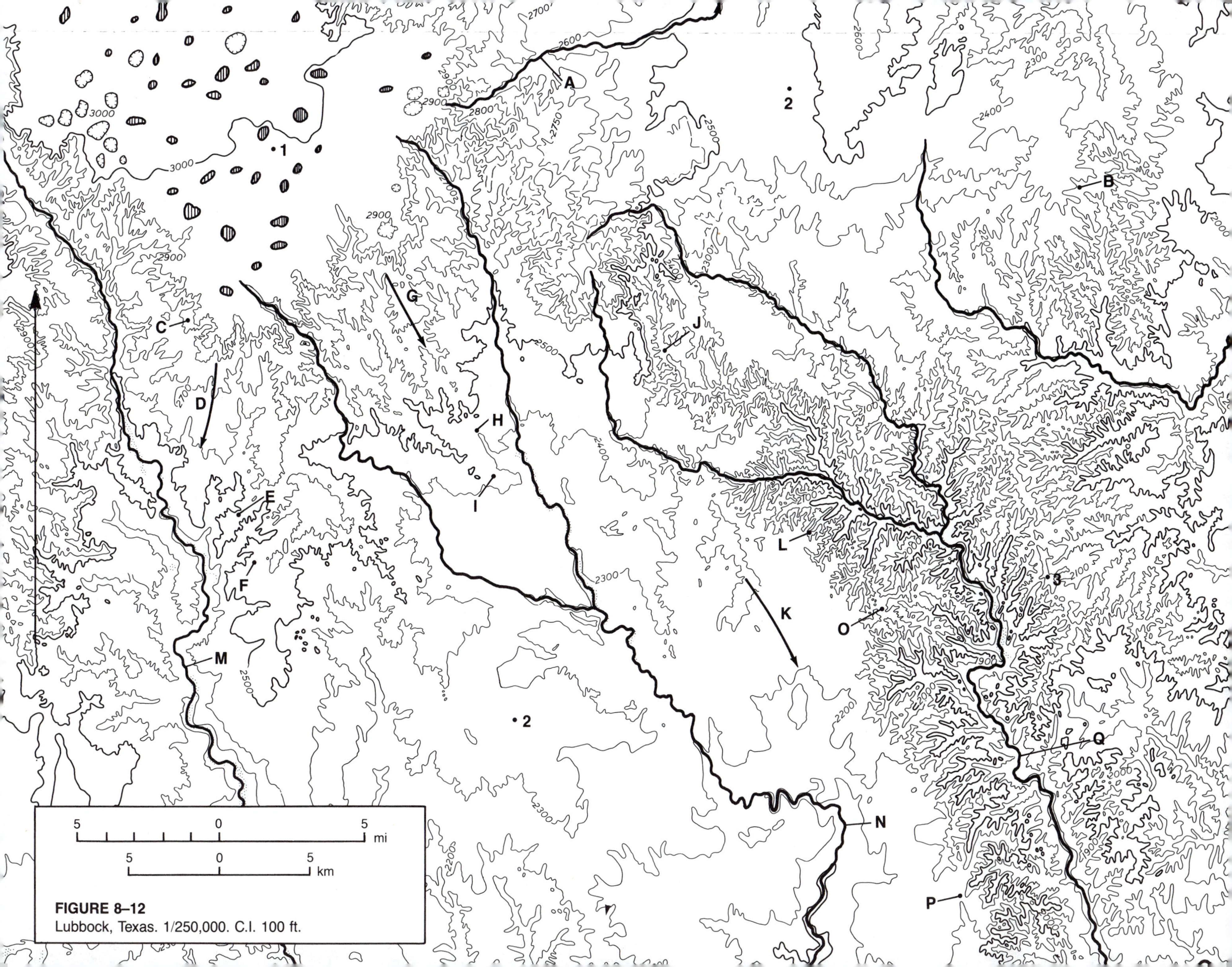

FIGURE 8–12
Lubbock, Texas. 1/250,000. C.I. 100 ft.

FIGURE 8–13
Lookout Ridge, Alaska. 1/250,000. C.I. 100 ft.

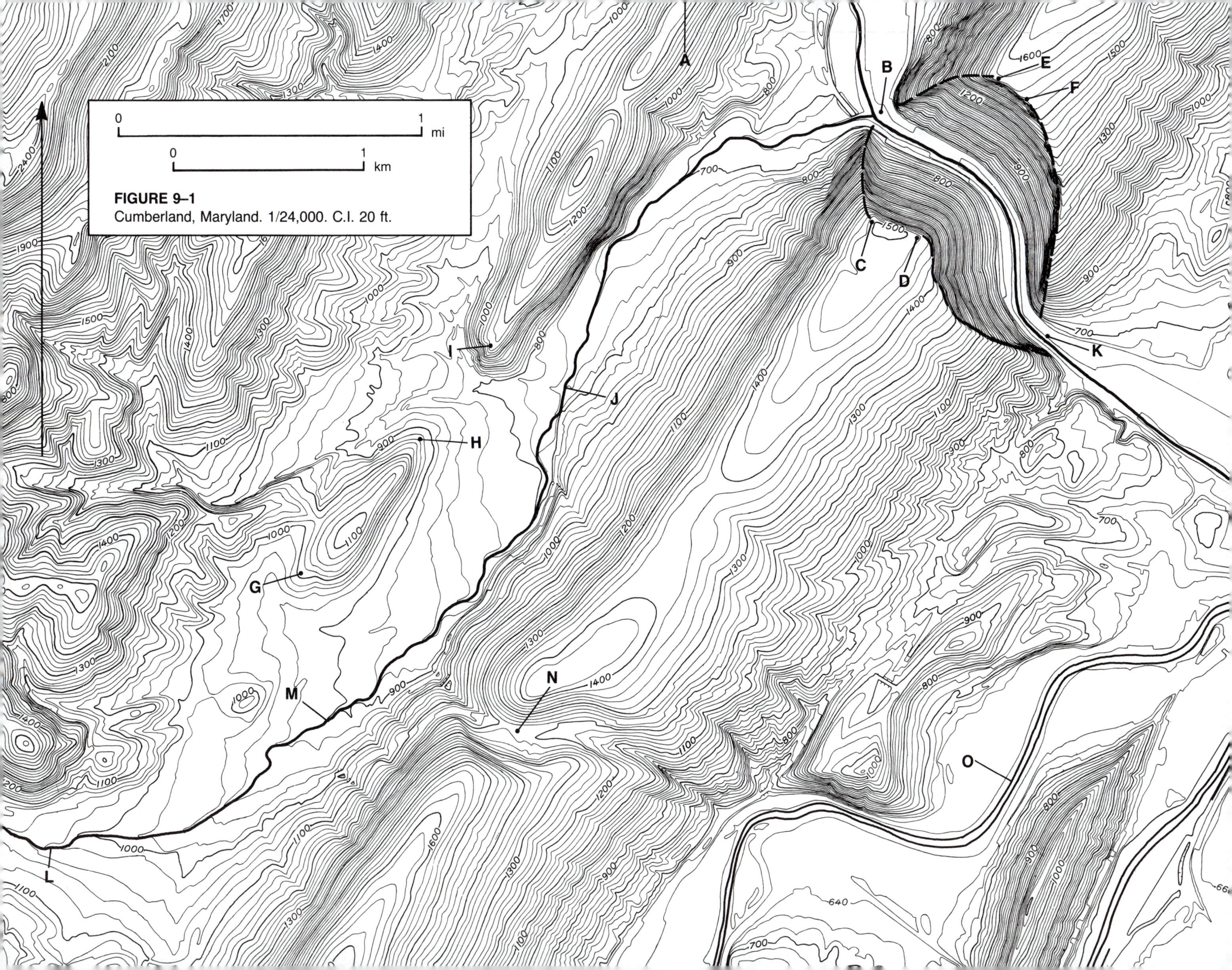

FIGURE 9–1
Cumberland, Maryland. 1/24,000. C.I. 20 ft.

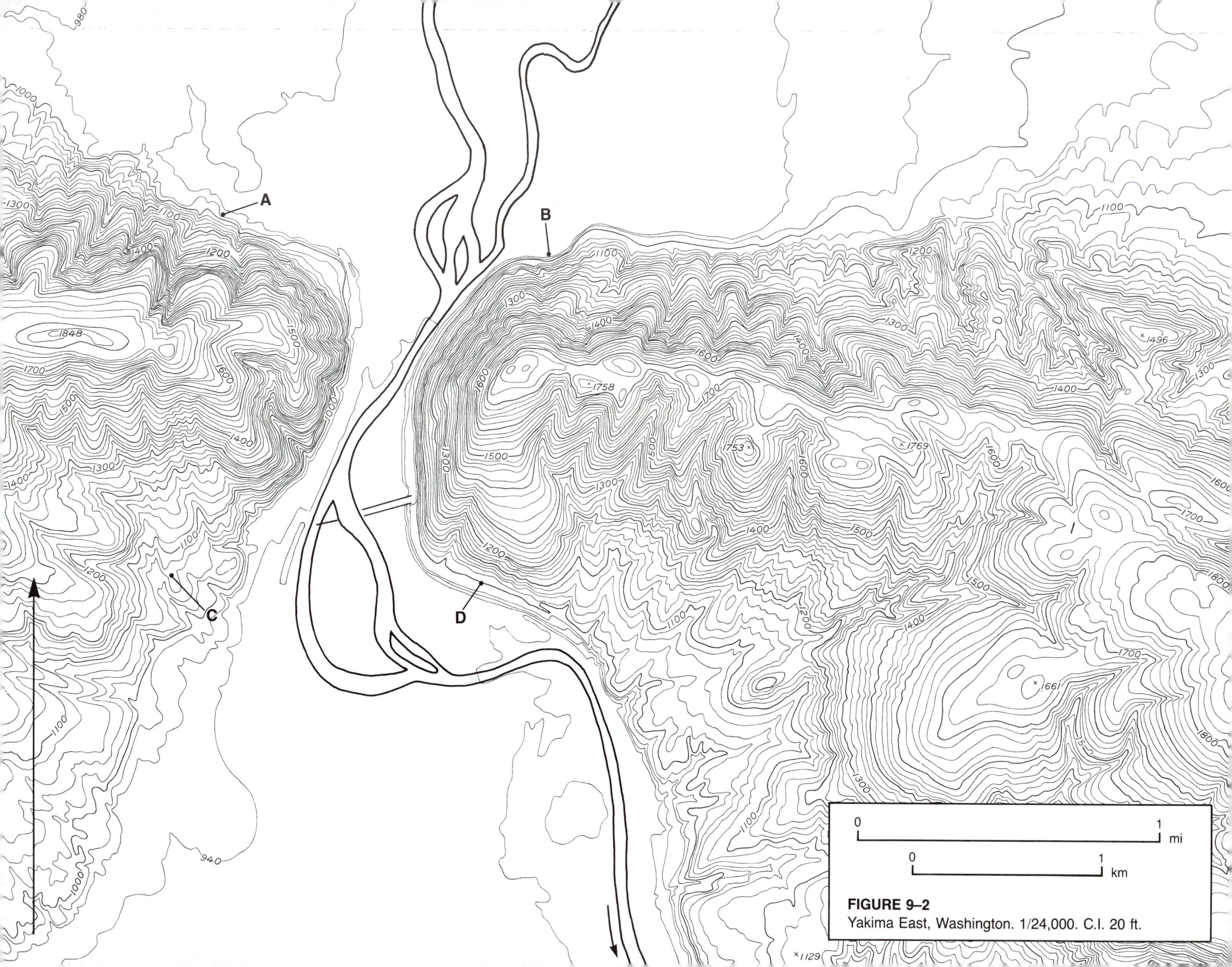

FIGURE 9–2
Yakima East, Washington. 1/24,000. C.I. 20 ft.

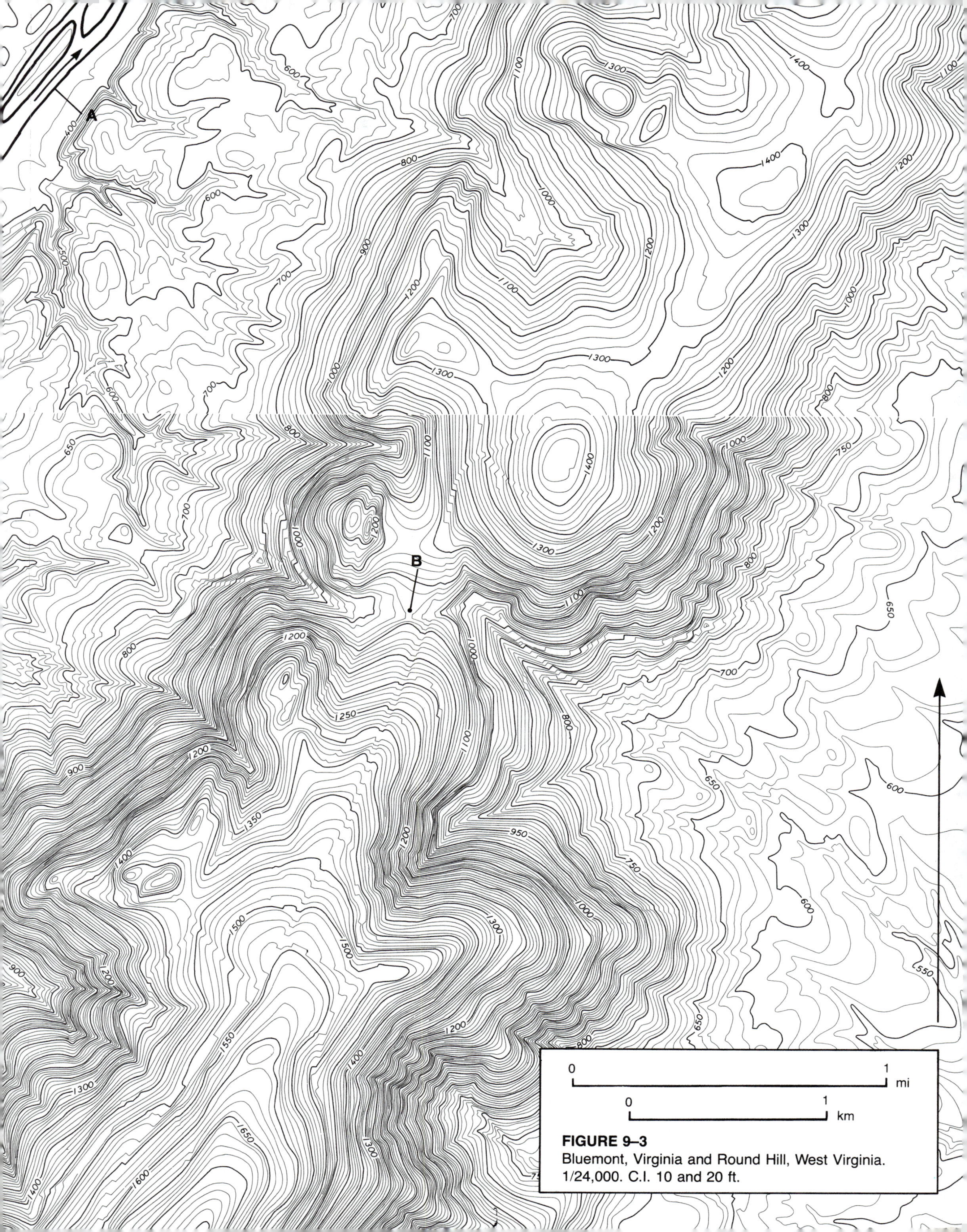

FIGURE 9–3
Bluemont, Virginia and Round Hill, West Virginia.
1/24,000. C.I. 10 and 20 ft.

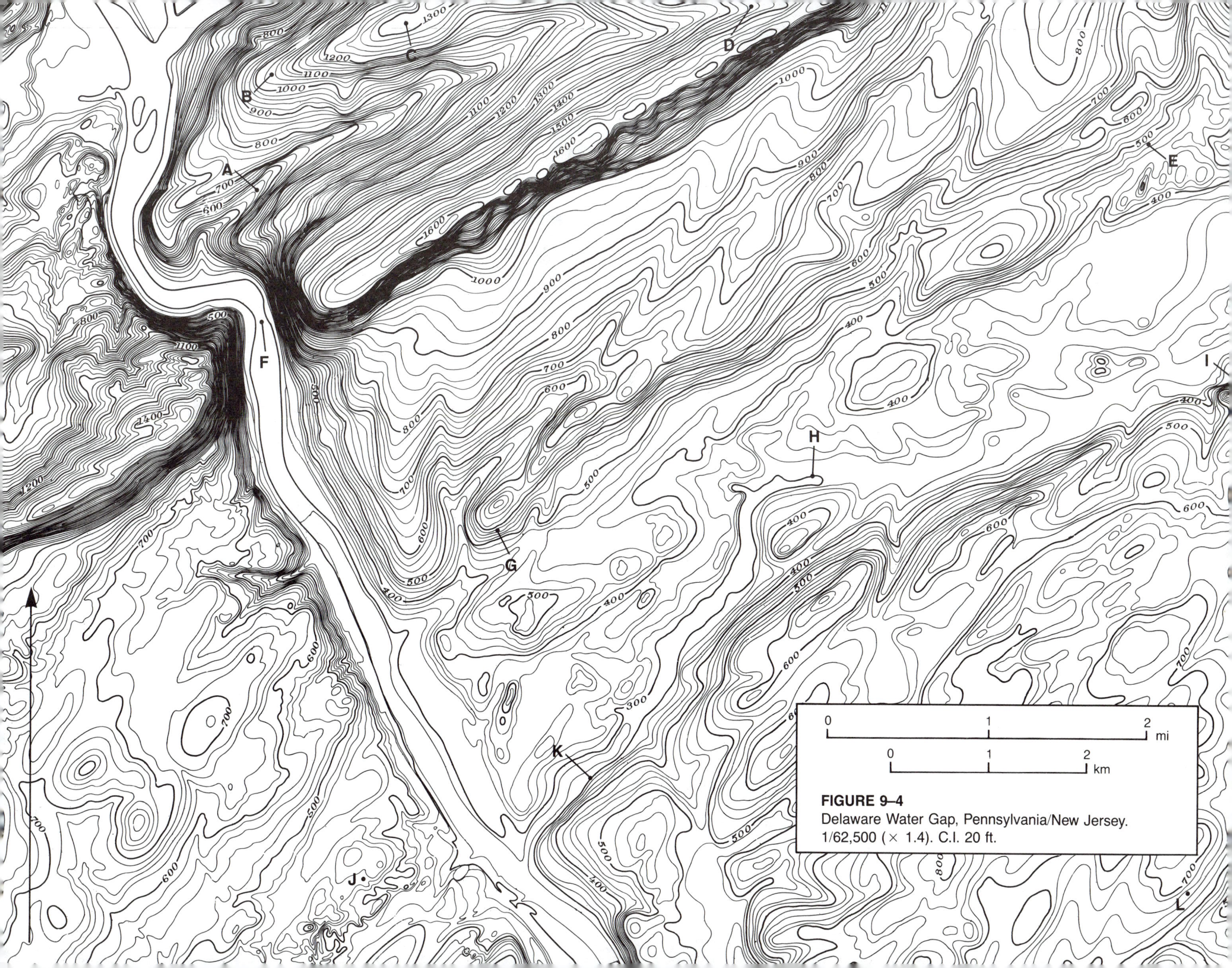

FIGURE 9–4
Delaware Water Gap, Pennsylvania/New Jersey. 1/62,500 (× 1.4). C.I. 20 ft.

FIGURE 9–5
Independence Rock, Wyoming. 1/24,000. C.I. 20 ft.

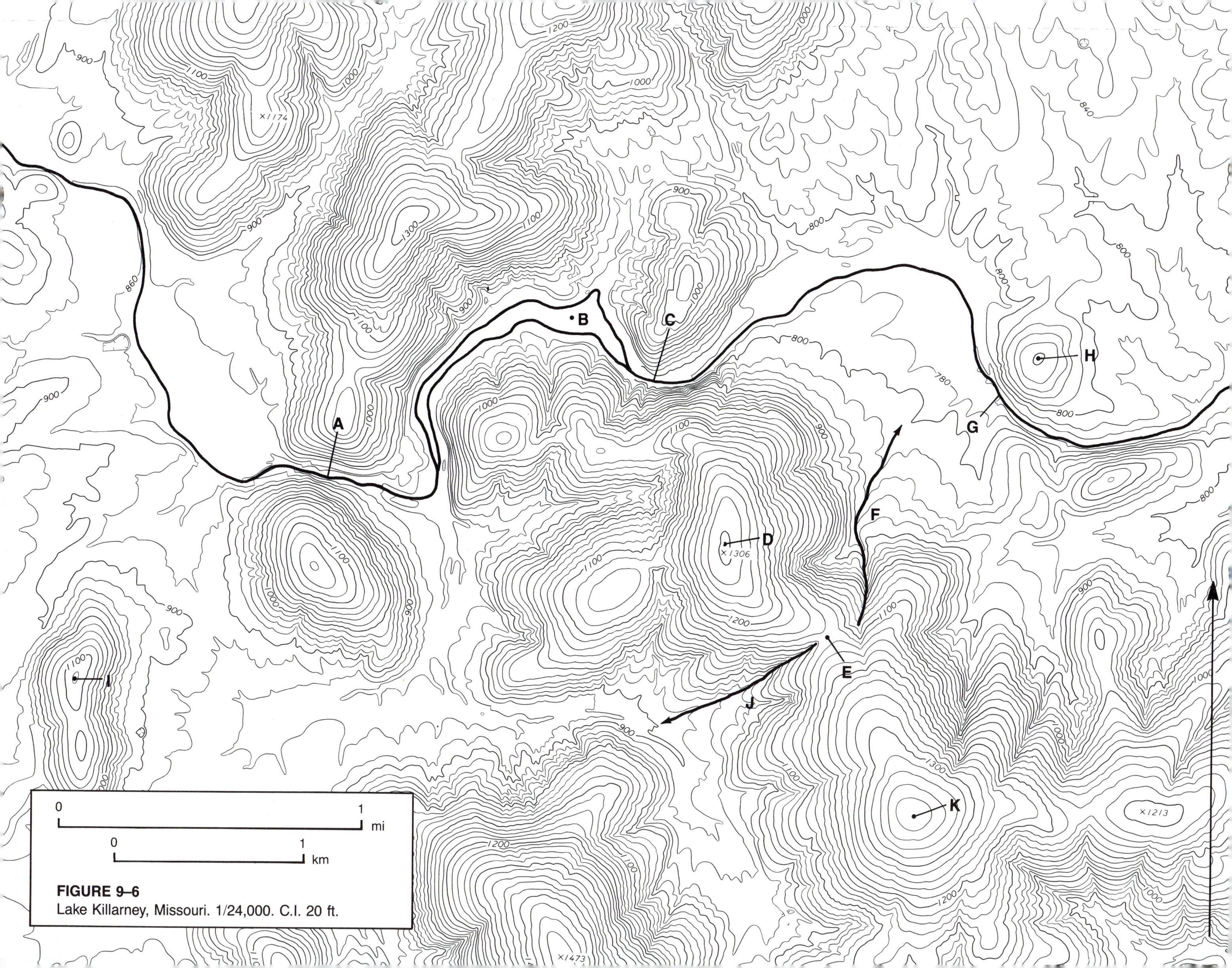

FIGURE 9–6
Lake Killarney, Missouri. 1/24,000. C.I. 20 ft.

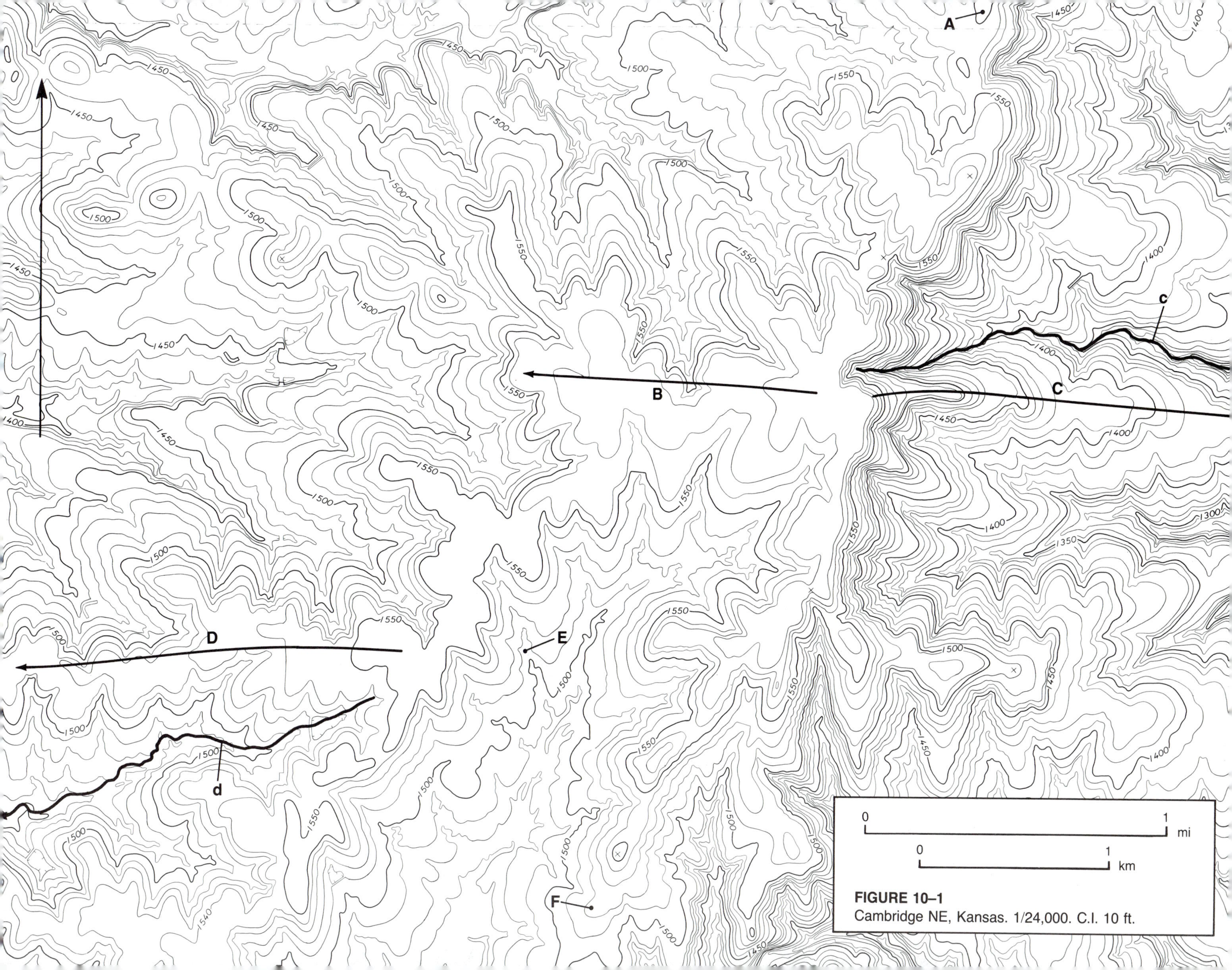

FIGURE 10–1
Cambridge NE, Kansas. 1/24,000. C.I. 10 ft.

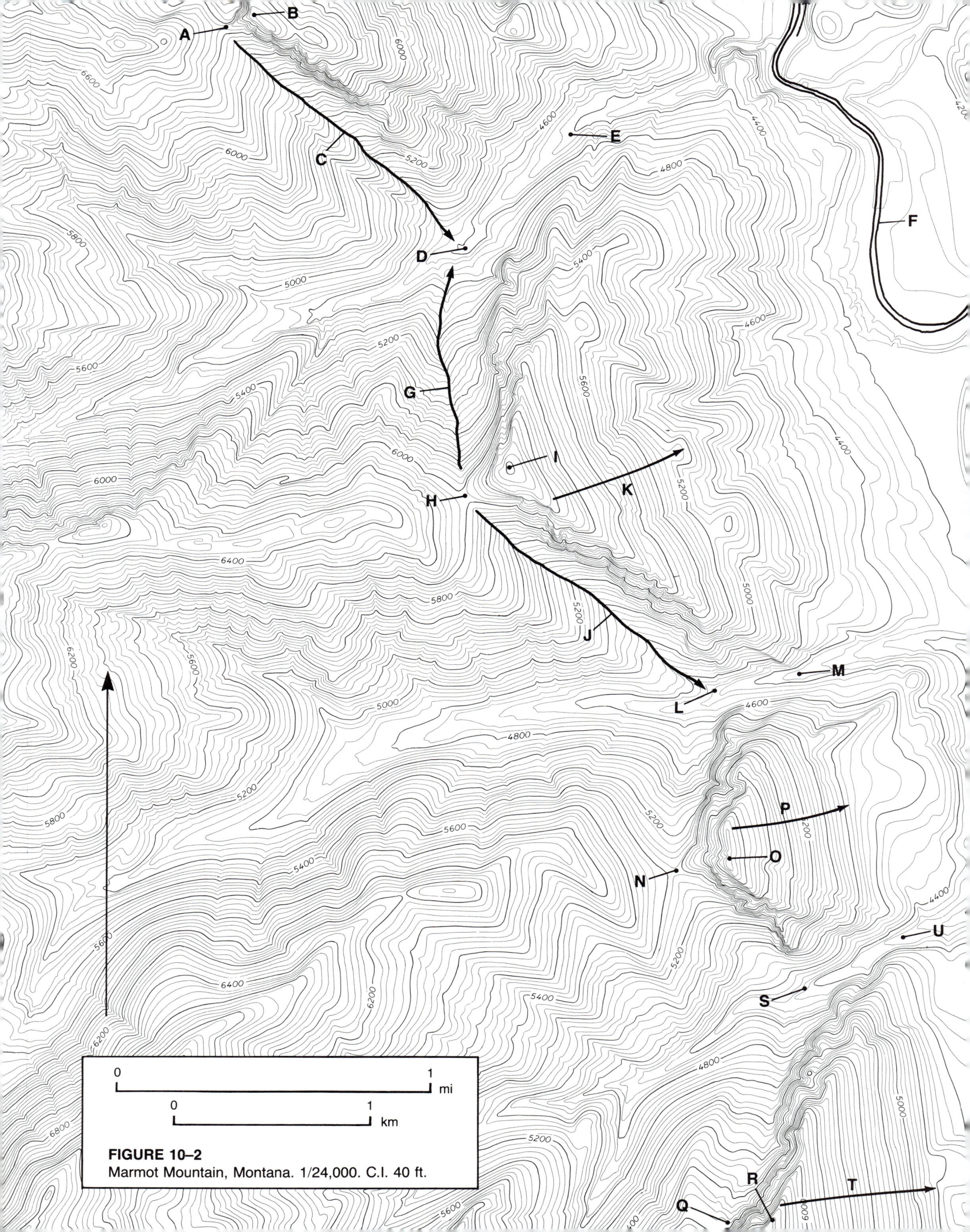

FIGURE 10–2
Marmot Mountain, Montana. 1/24,000. C.I. 40 ft.

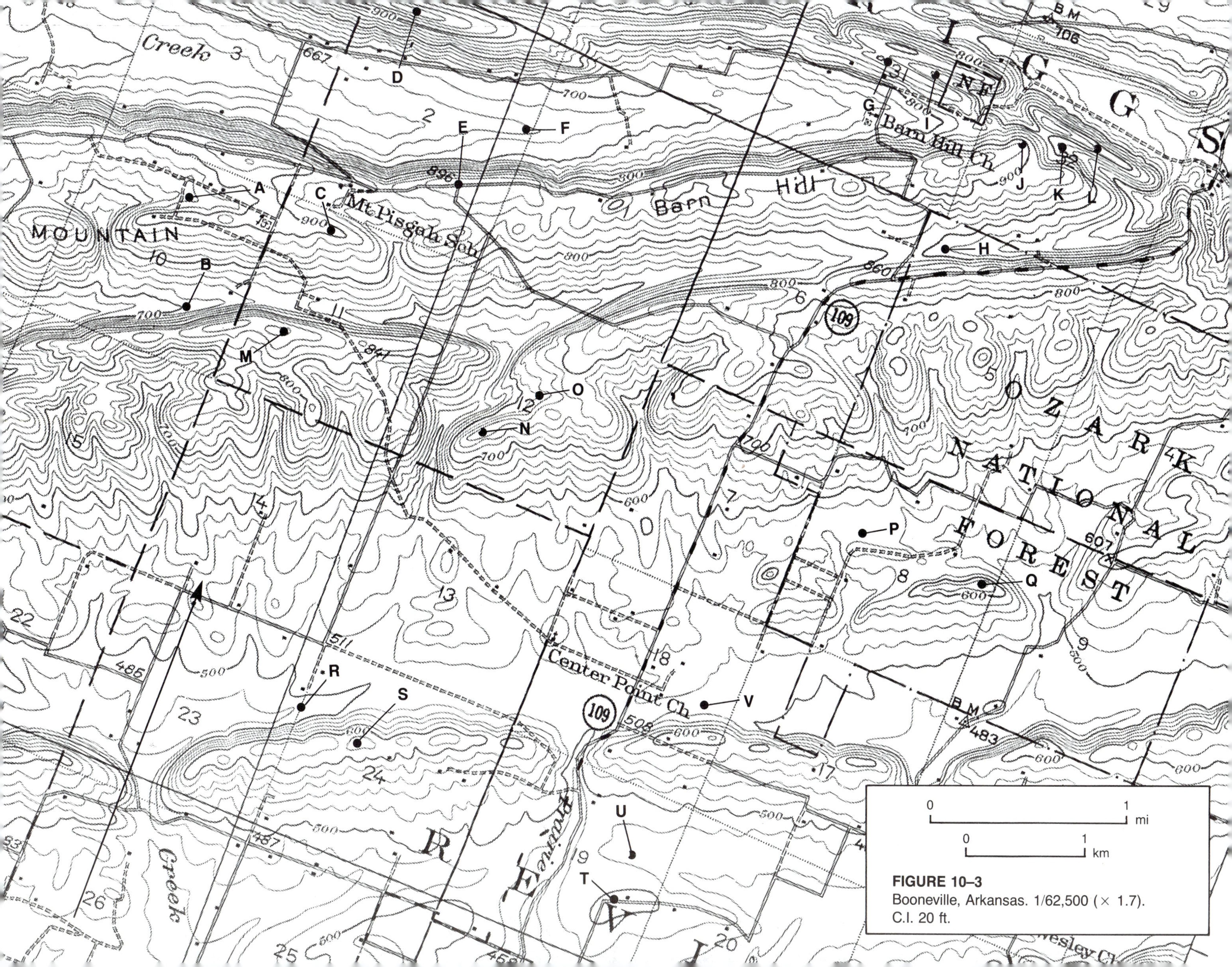

FIGURE 10–3
Booneville, Arkansas. 1/62,500 (× 1.7).
C.I. 20 ft.

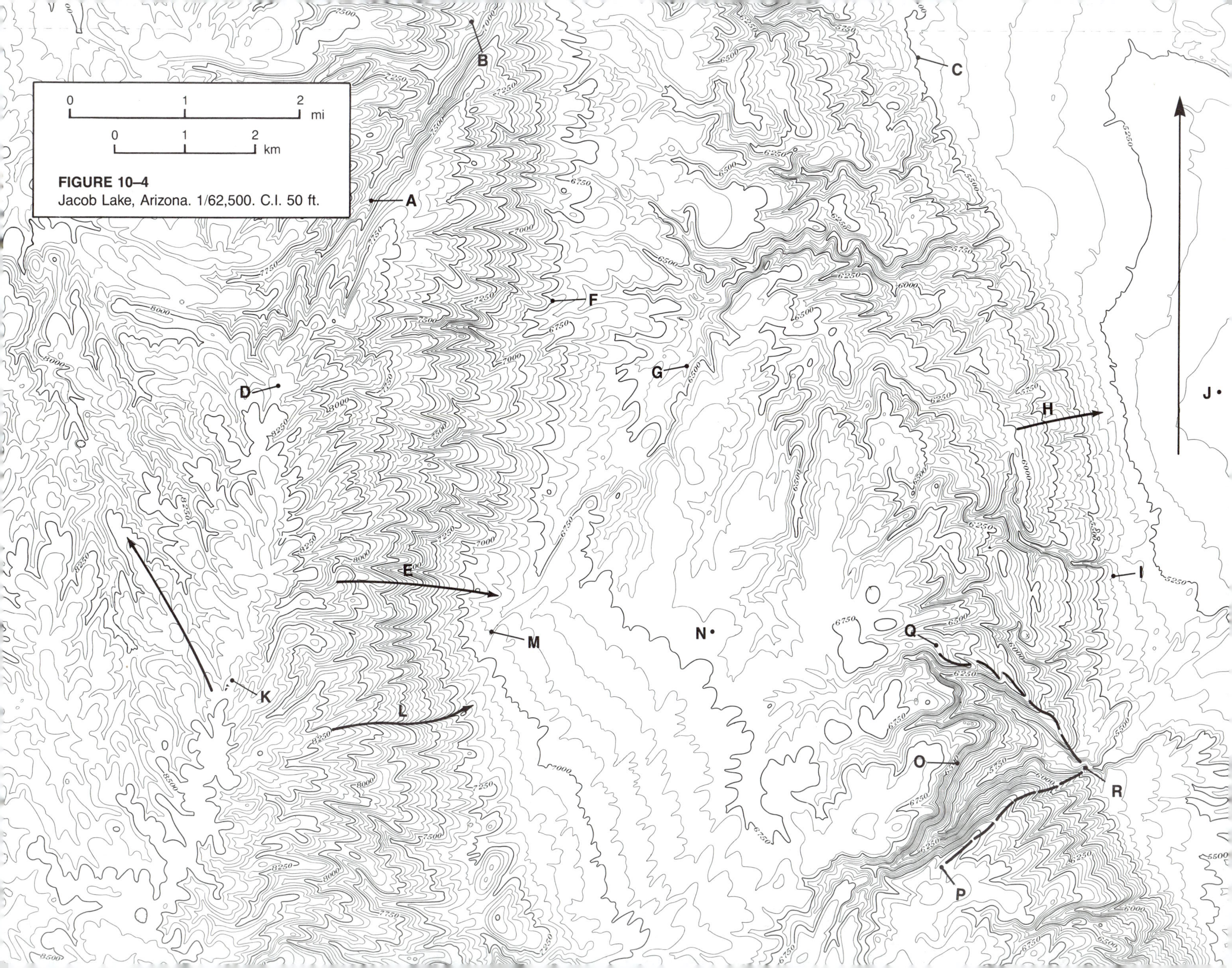

FIGURE 10–4
Jacob Lake, Arizona. 1/62,500. C.I. 50 ft.

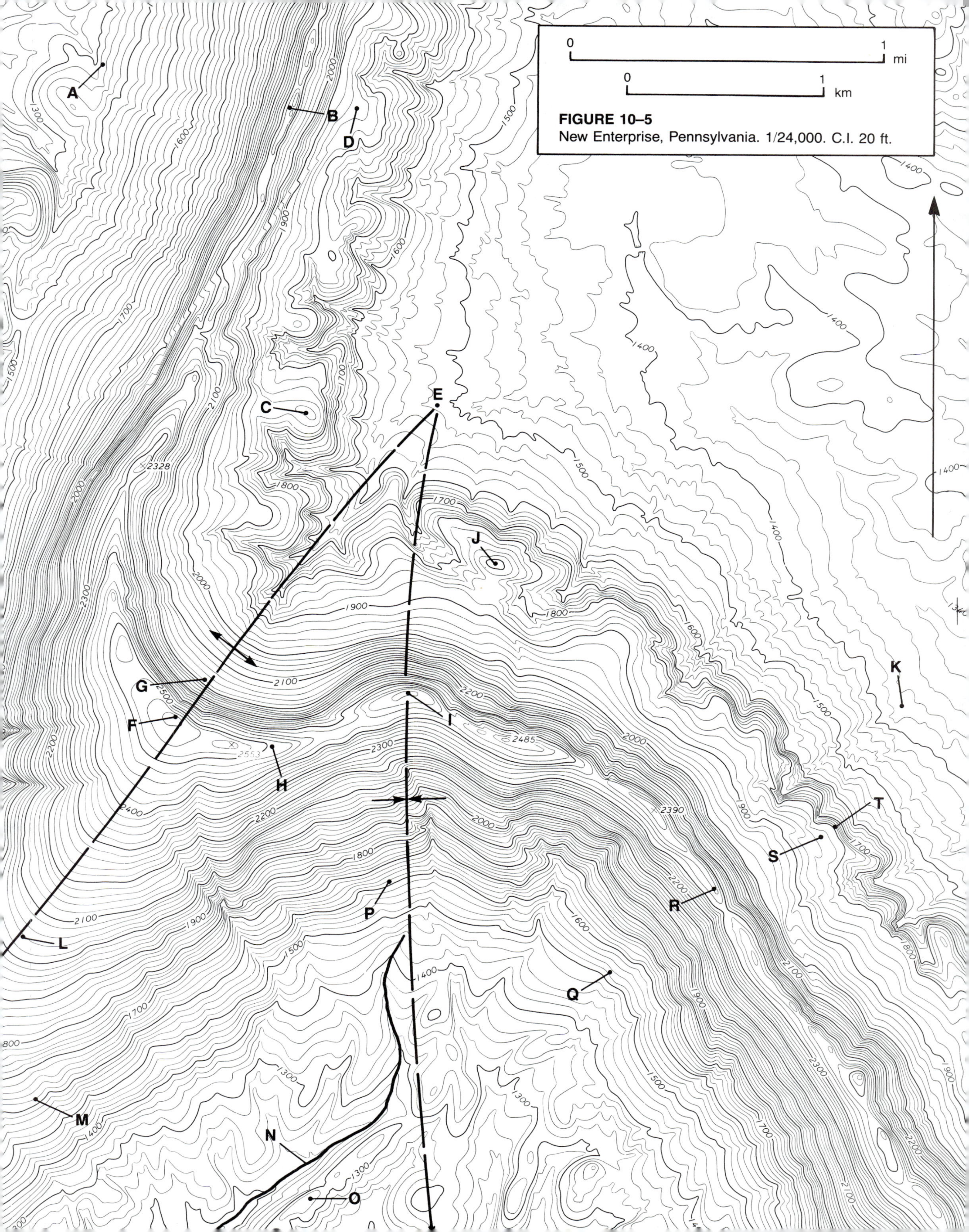

FIGURE 10–5
New Enterprise, Pennsylvania. 1/24,000. C.I. 20 ft.

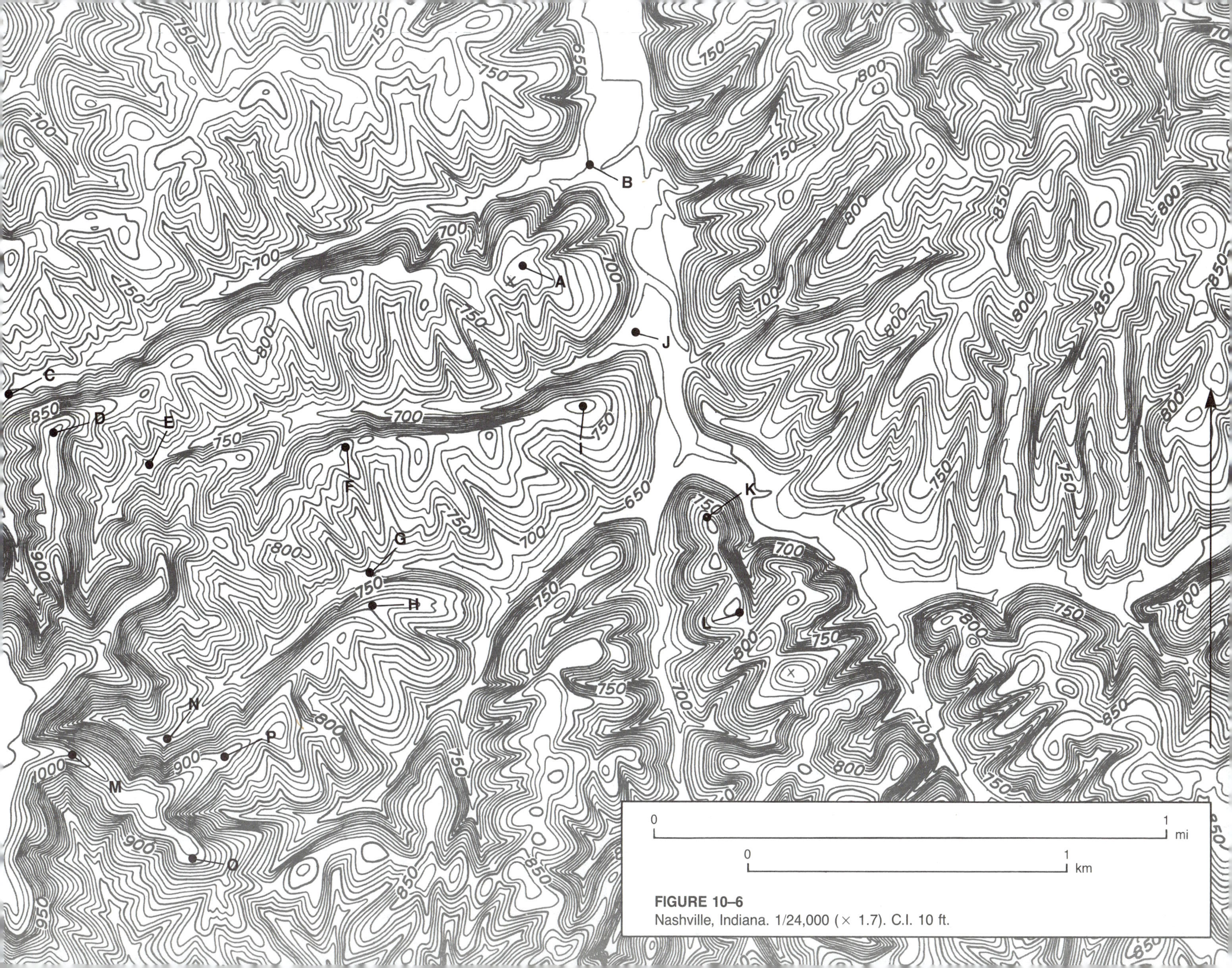

FIGURE 10–6
Nashville, Indiana. 1/24,000 (× 1.7). C.I. 10 ft.

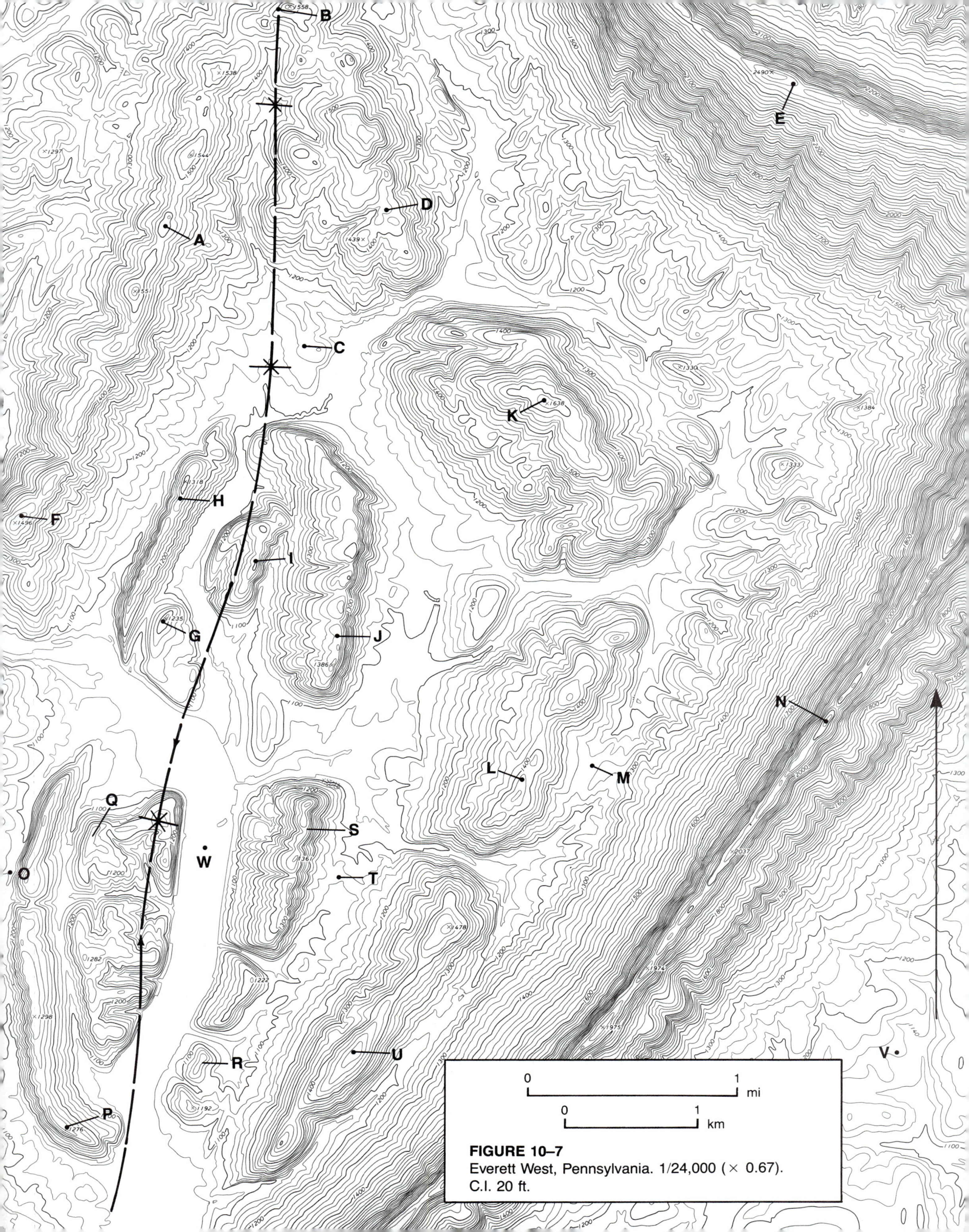

FIGURE 10–7
Everett West, Pennsylvania. 1/24,000 (× 0.67).
C.I. 20 ft.

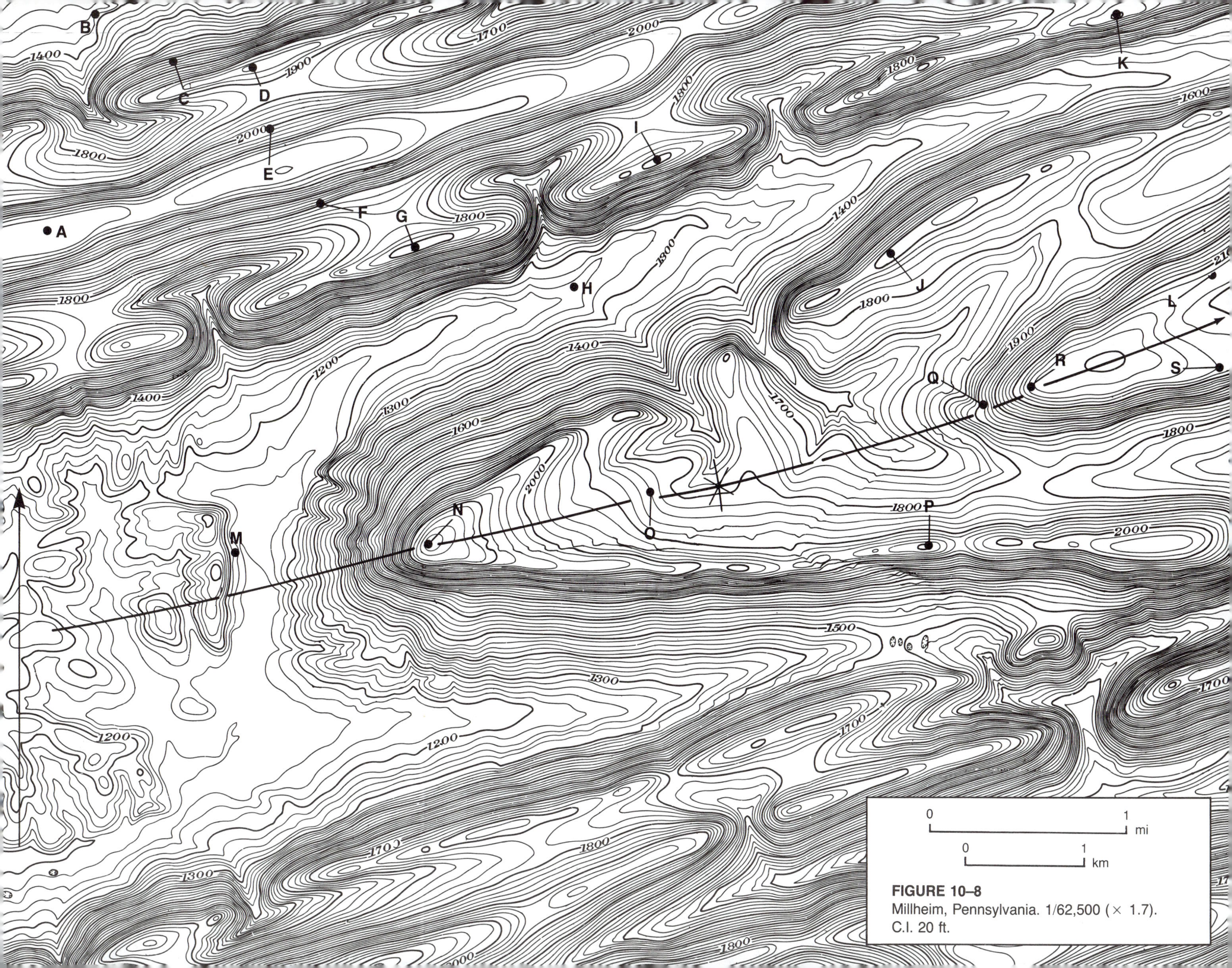

FIGURE 10–8
Millheim, Pennsylvania. 1/62,500 (× 1.7). C.I. 20 ft.

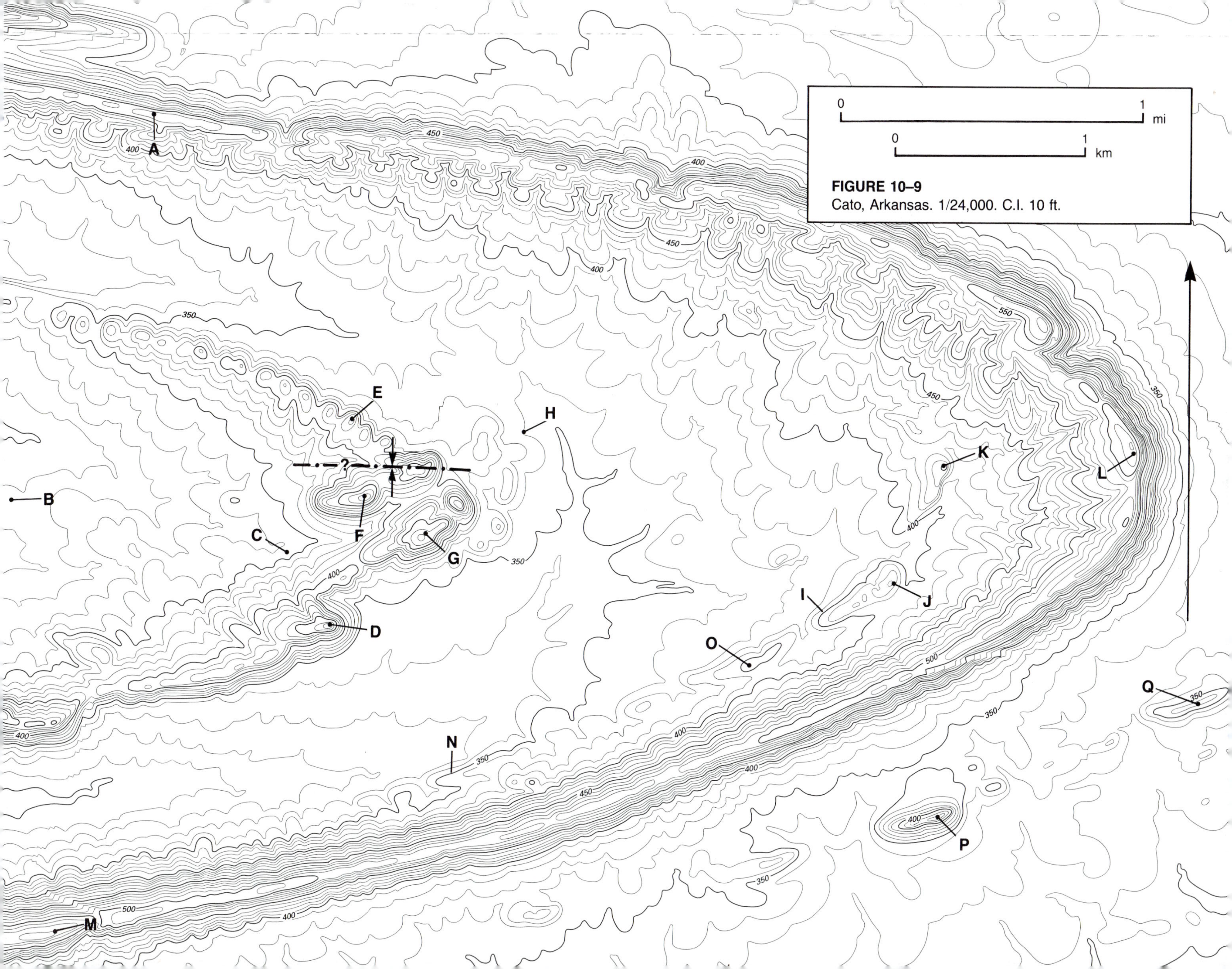

FIGURE 10–9
Cato, Arkansas. 1/24,000. C.I. 10 ft.

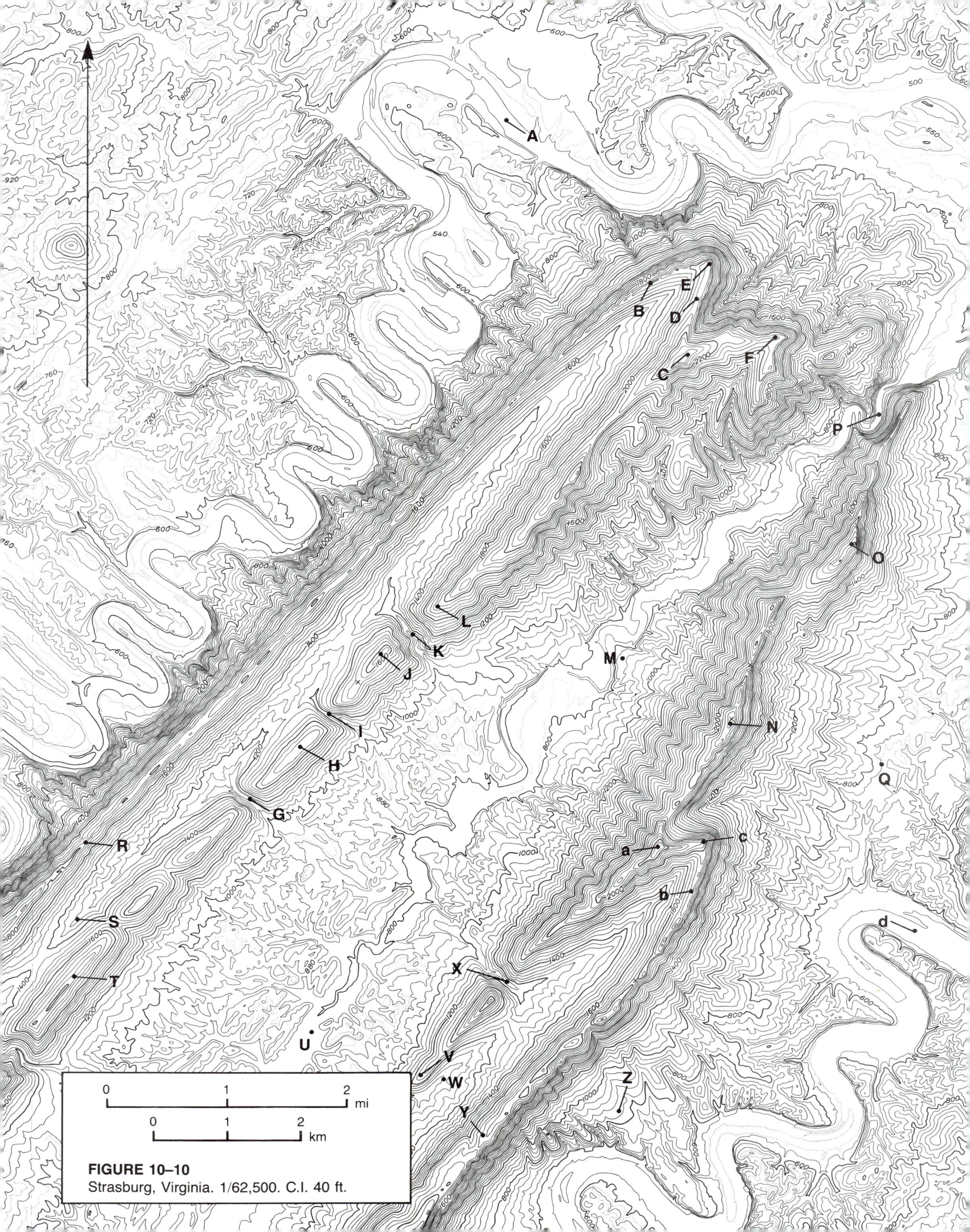

FIGURE 10–10
Strasburg, Virginia. 1/62,500. C.I. 40 ft.

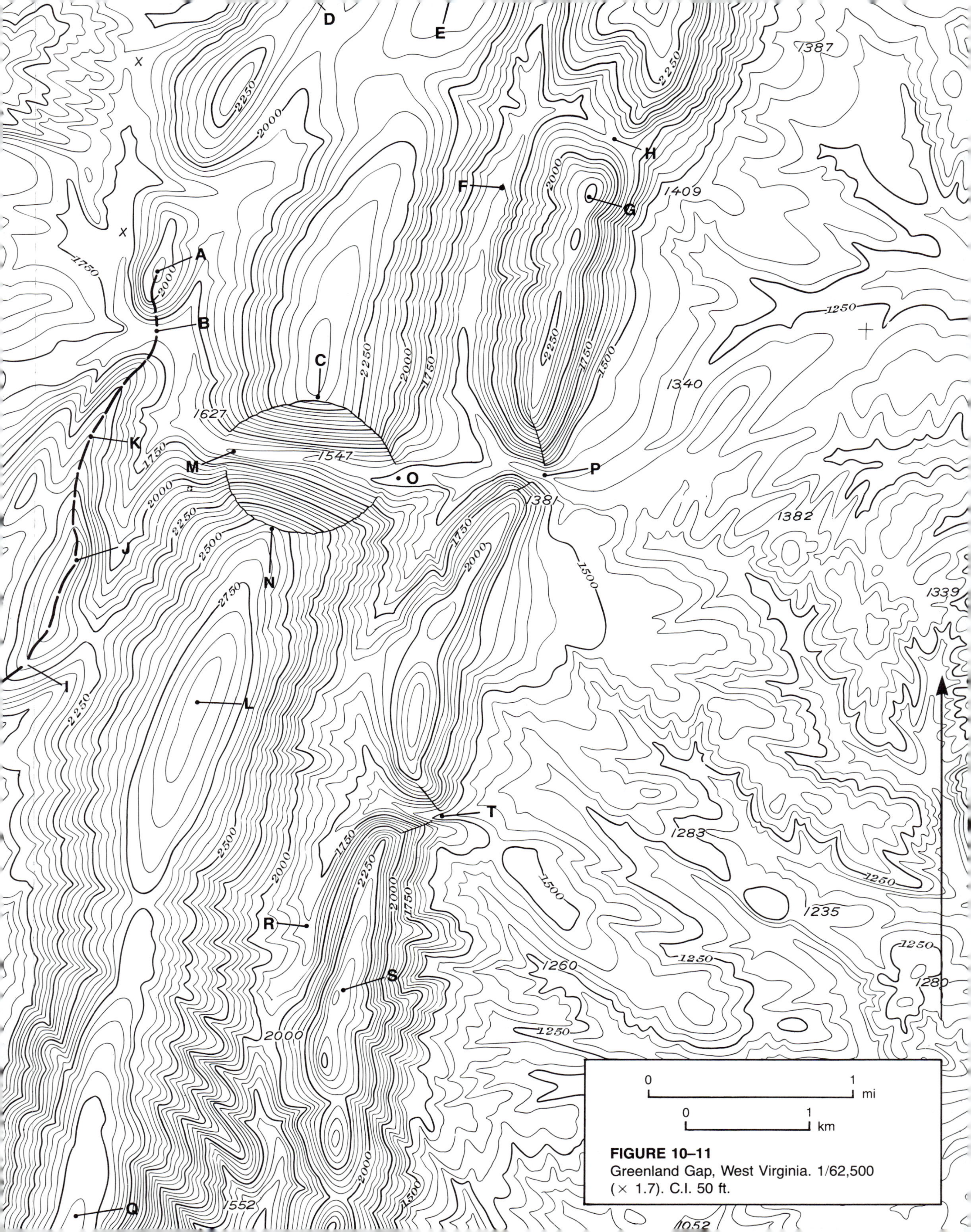

FIGURE 10–11
Greenland Gap, West Virginia. 1/62,500 (× 1.7). C.I. 50 ft.

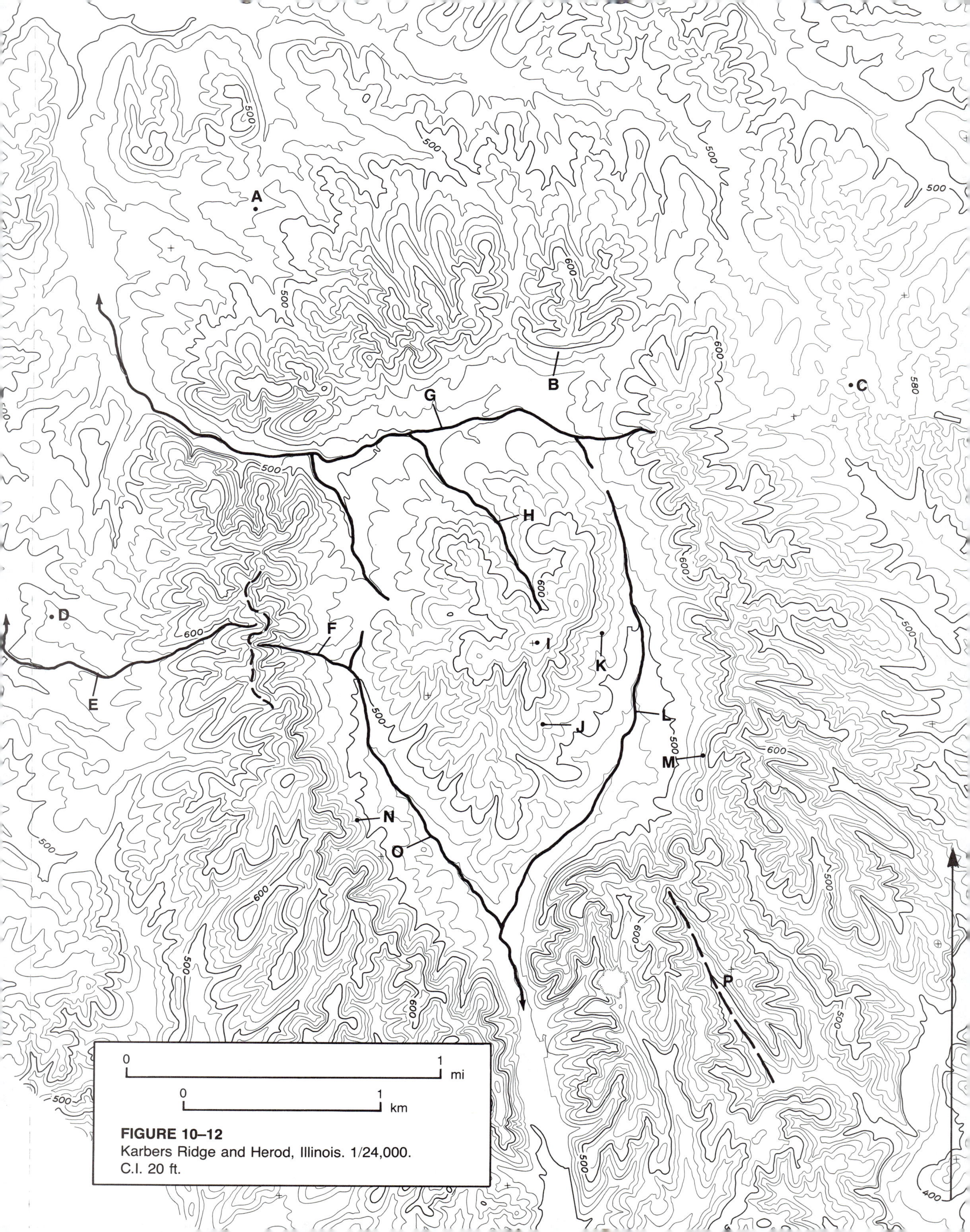

FIGURE 10–12
Karbers Ridge and Herod, Illinois. 1/24,000.
C.I. 20 ft.

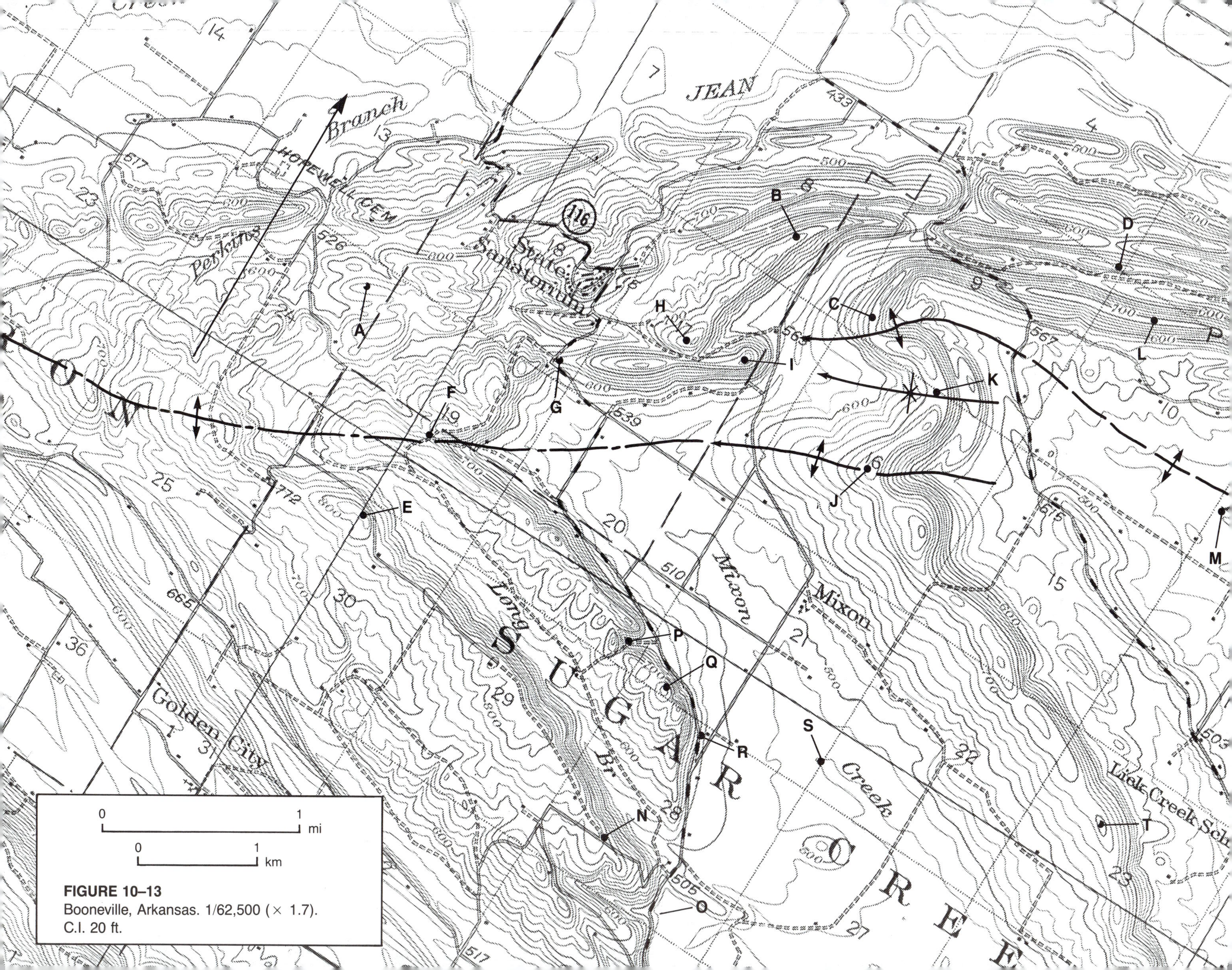

FIGURE 10–13
Booneville, Arkansas. 1/62,500 (× 1.7).
C.I. 20 ft.

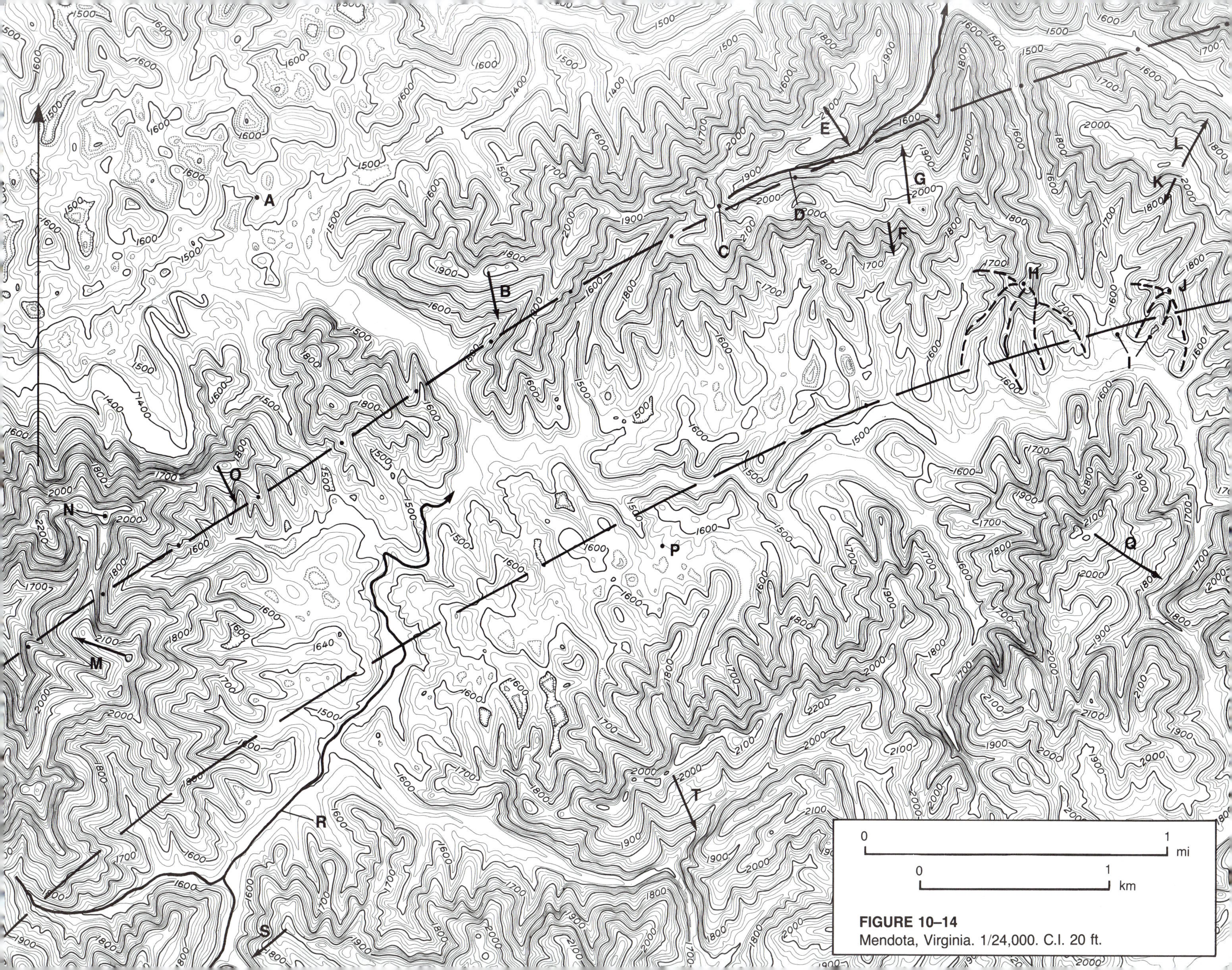

FIGURE 10–14
Mendota, Virginia. 1/24,000. C.I. 20 ft.

FIGURE 11–1
Kanab Point, Arizona. 1/62,500. C.I. 80 ft.

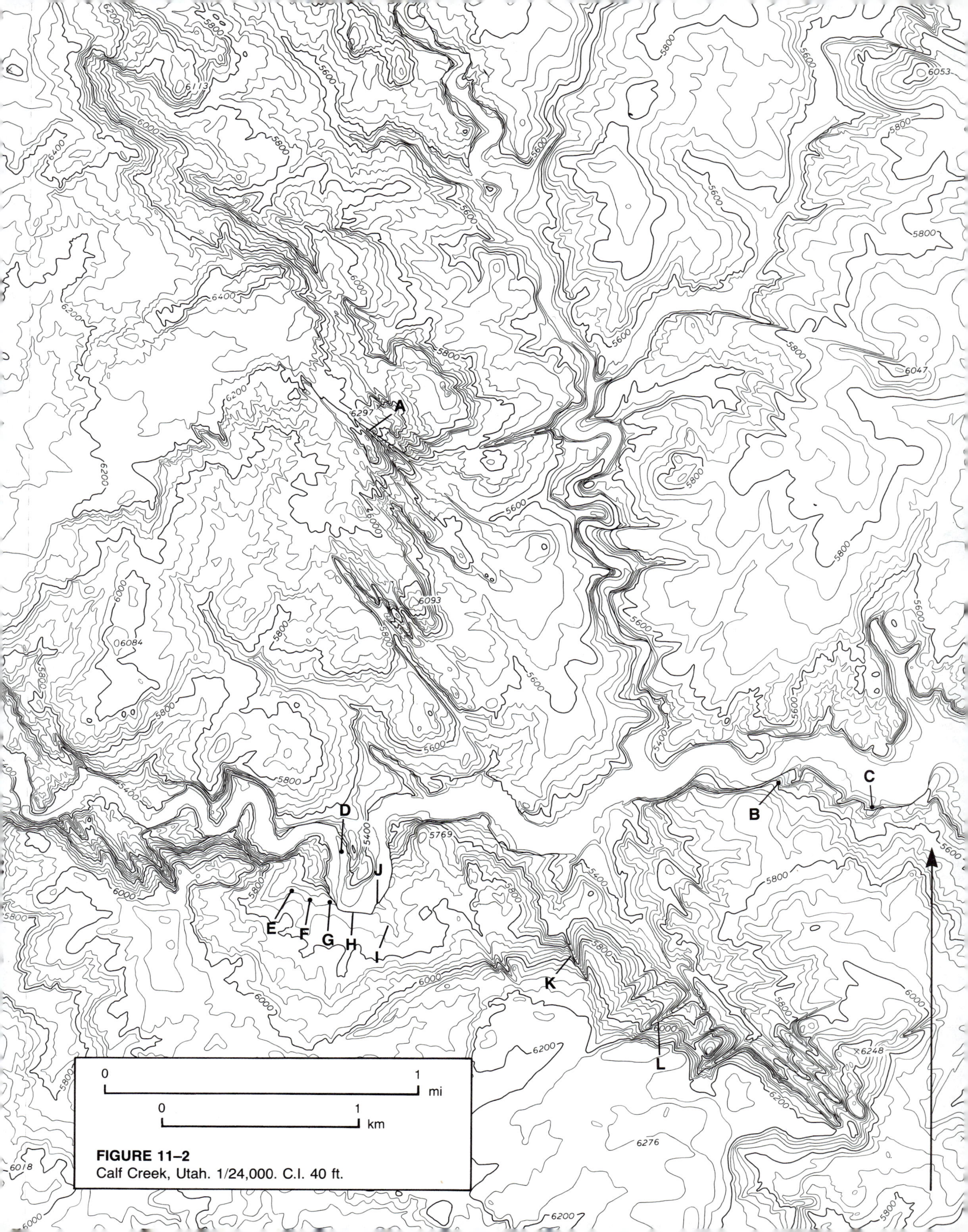

FIGURE 11–2
Calf Creek, Utah. 1/24,000. C.I. 40 ft.

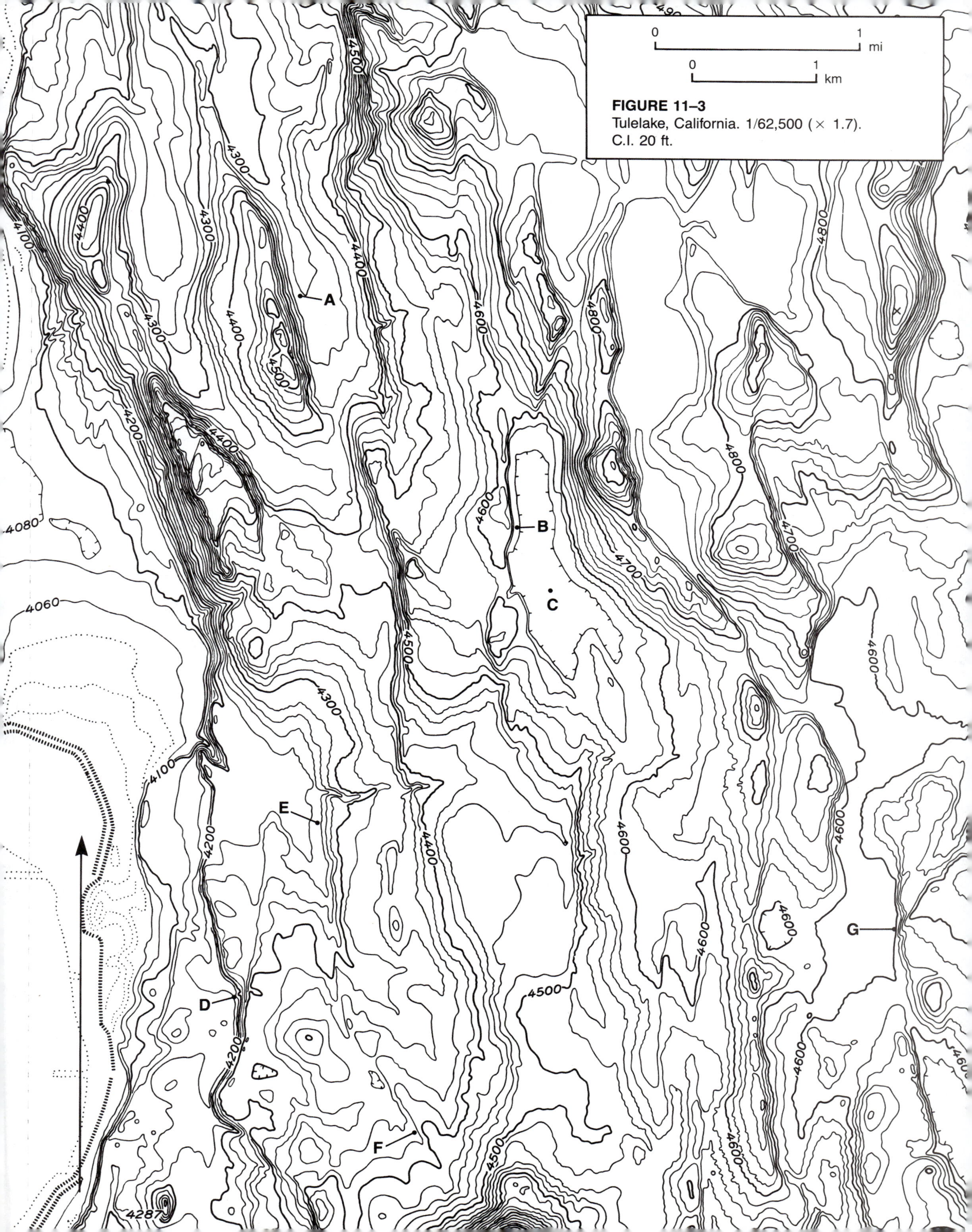

FIGURE 11–3
Tulelake, California. 1/62,500 (× 1.7).
C.I. 20 ft.

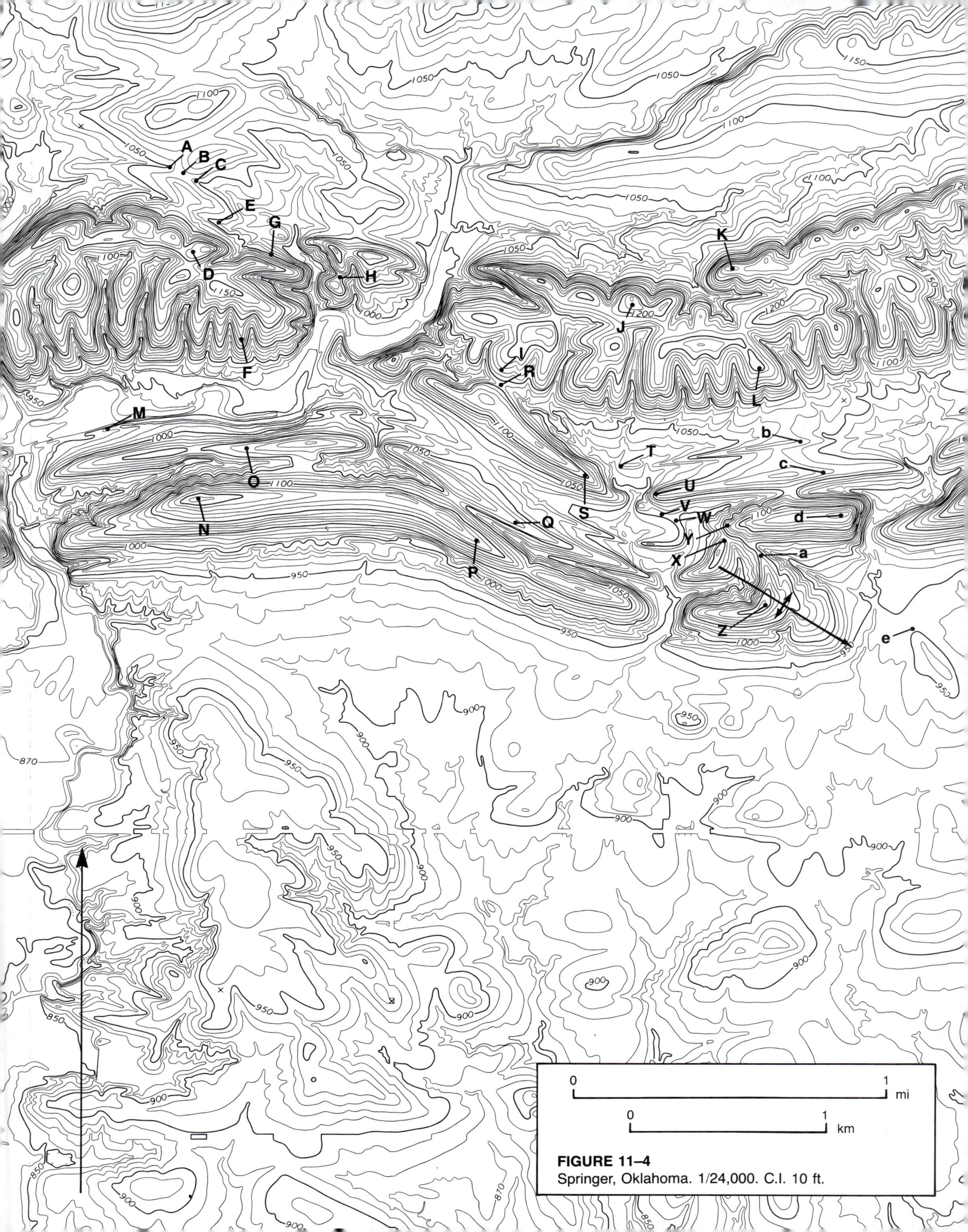

FIGURE 11–4
Springer, Oklahoma. 1/24,000. C.I. 10 ft.

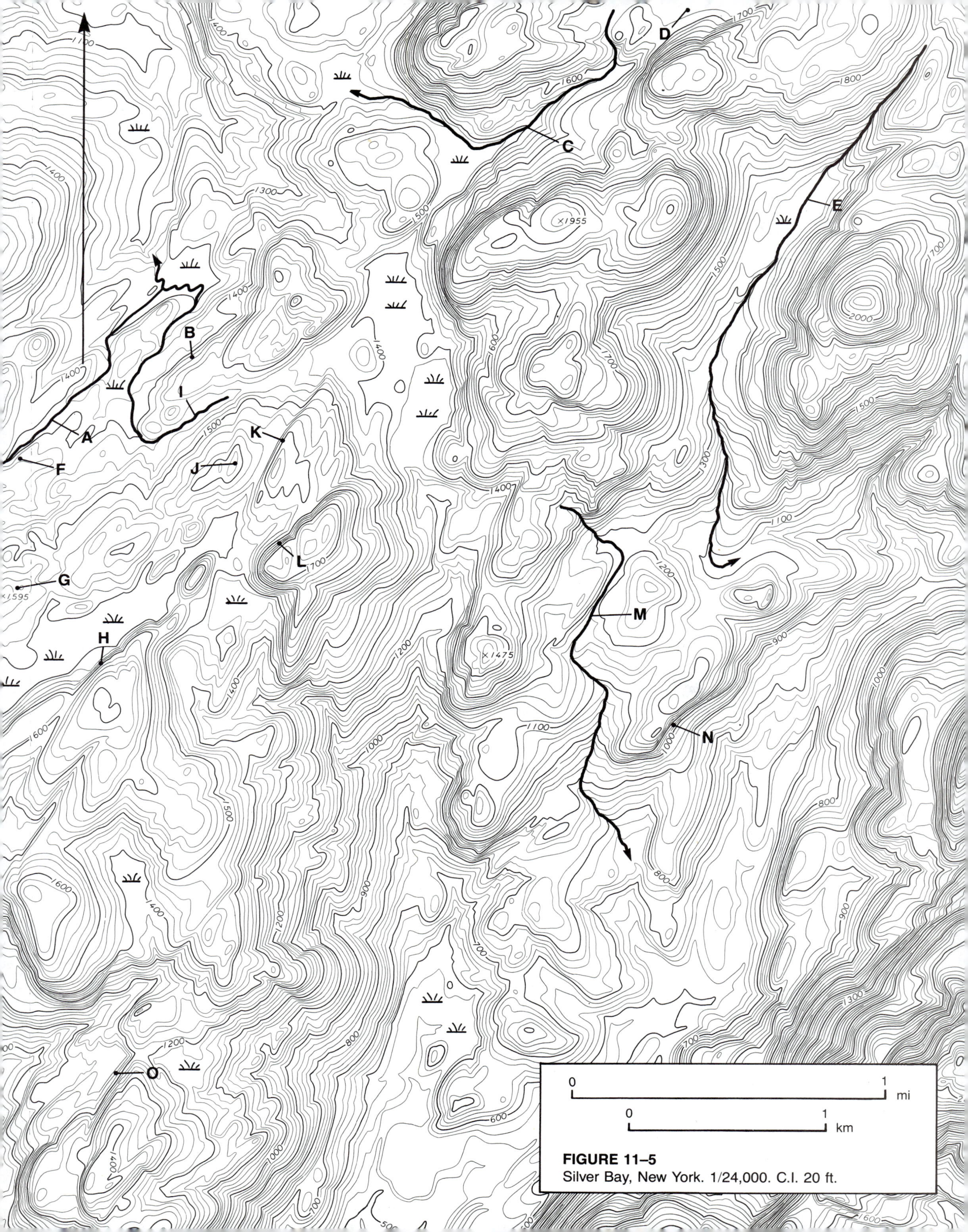

FIGURE 11–5
Silver Bay, New York. 1/24,000. C.I. 20 ft.

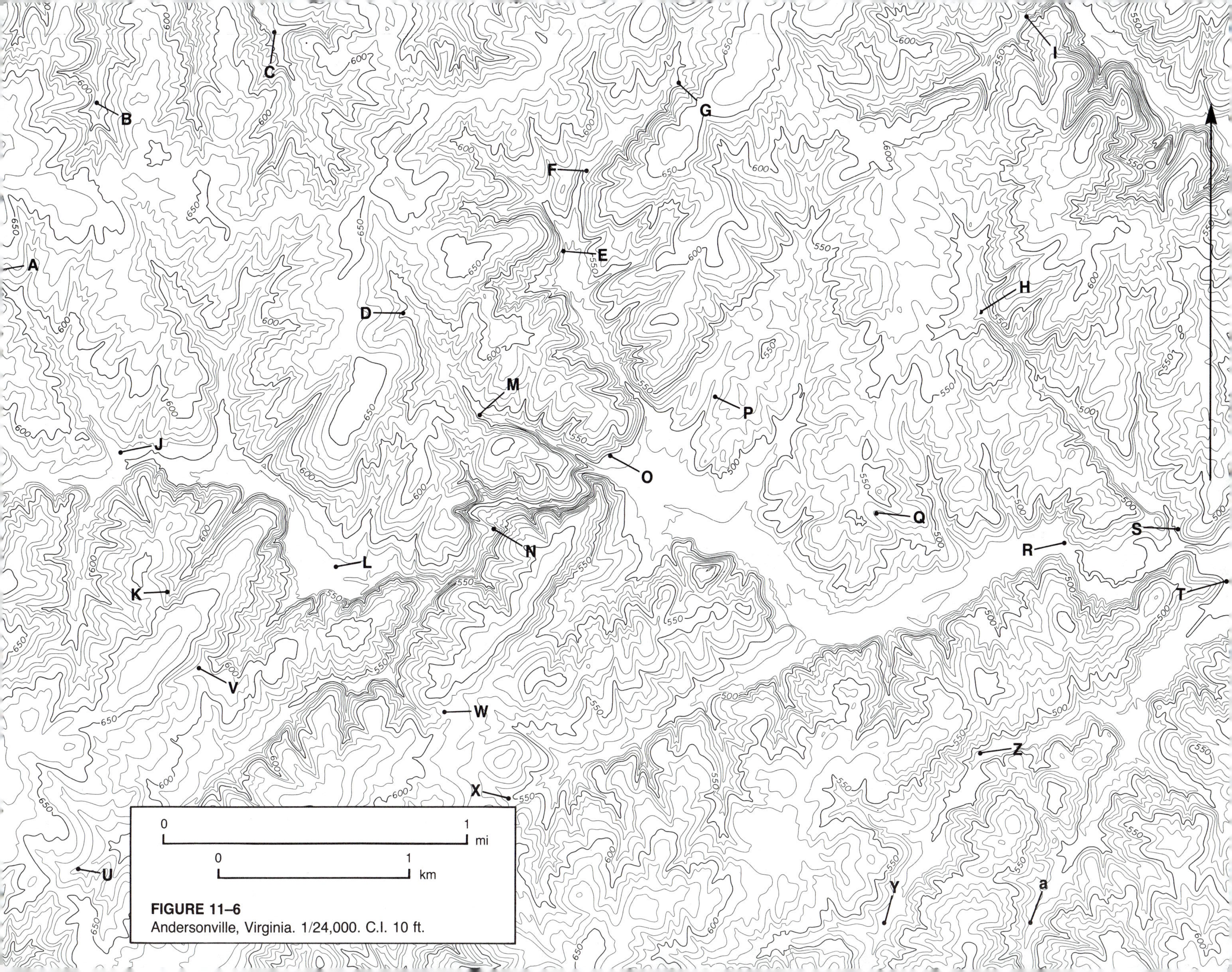

FIGURE 11–6
Andersonville, Virginia. 1/24,000. C.I. 10 ft.

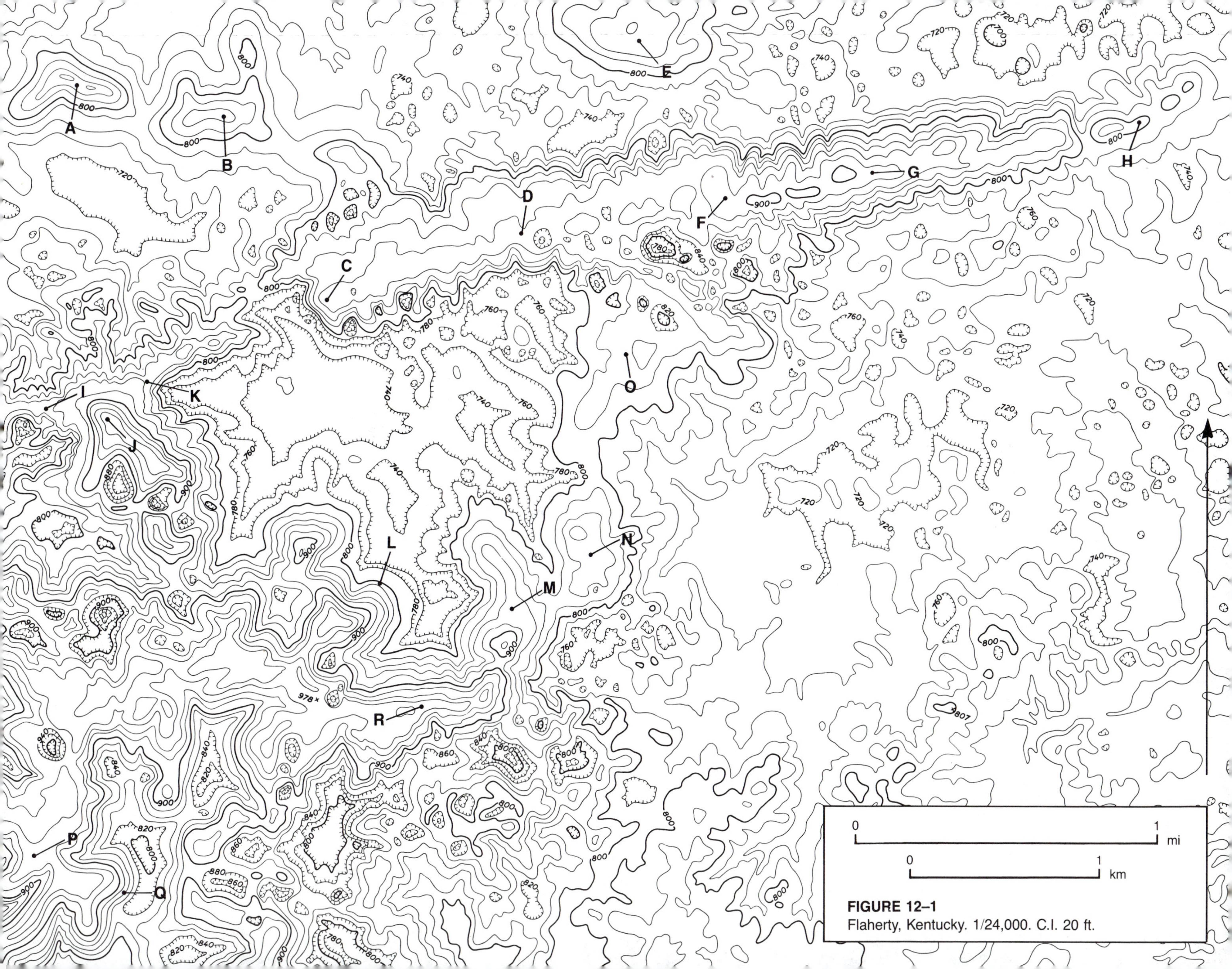

FIGURE 12–1
Flaherty, Kentucky. 1/24,000. C.I. 20 ft.

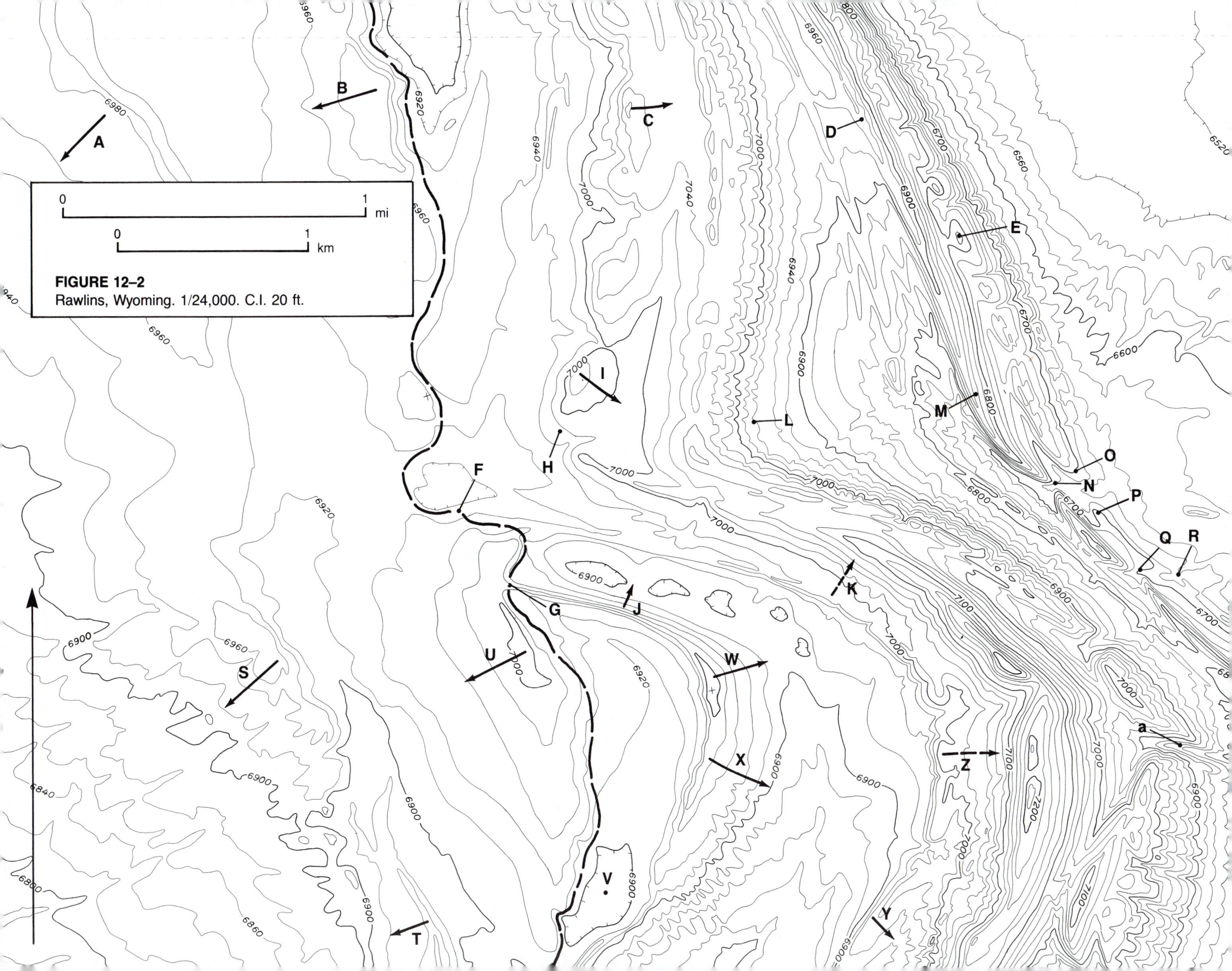

FIGURE 12–2
Rawlins, Wyoming. 1/24,000. C.I. 20 ft.

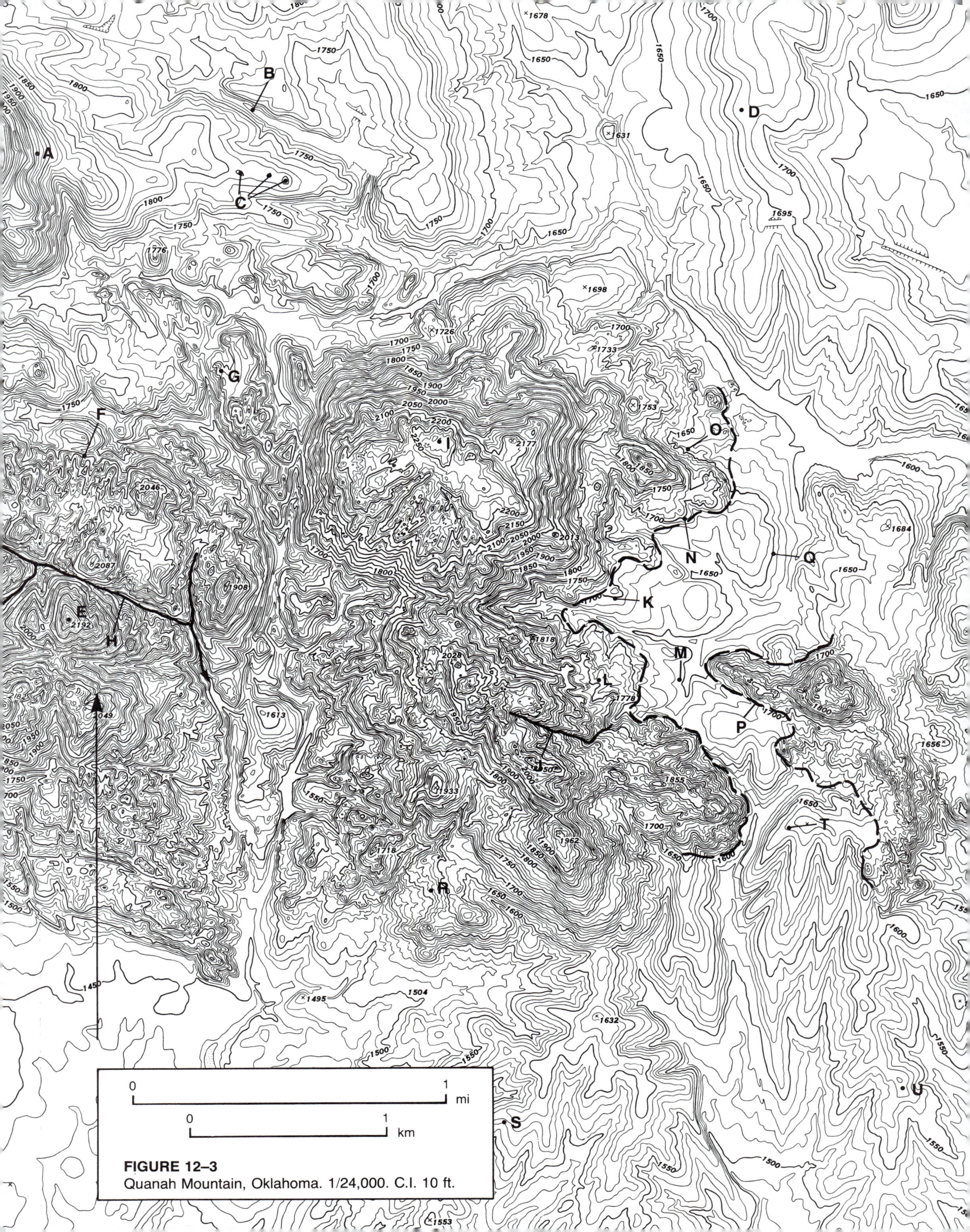

FIGURE 12–3
Quanah Mountain, Oklahoma. 1/24,000. C.I. 10 ft.

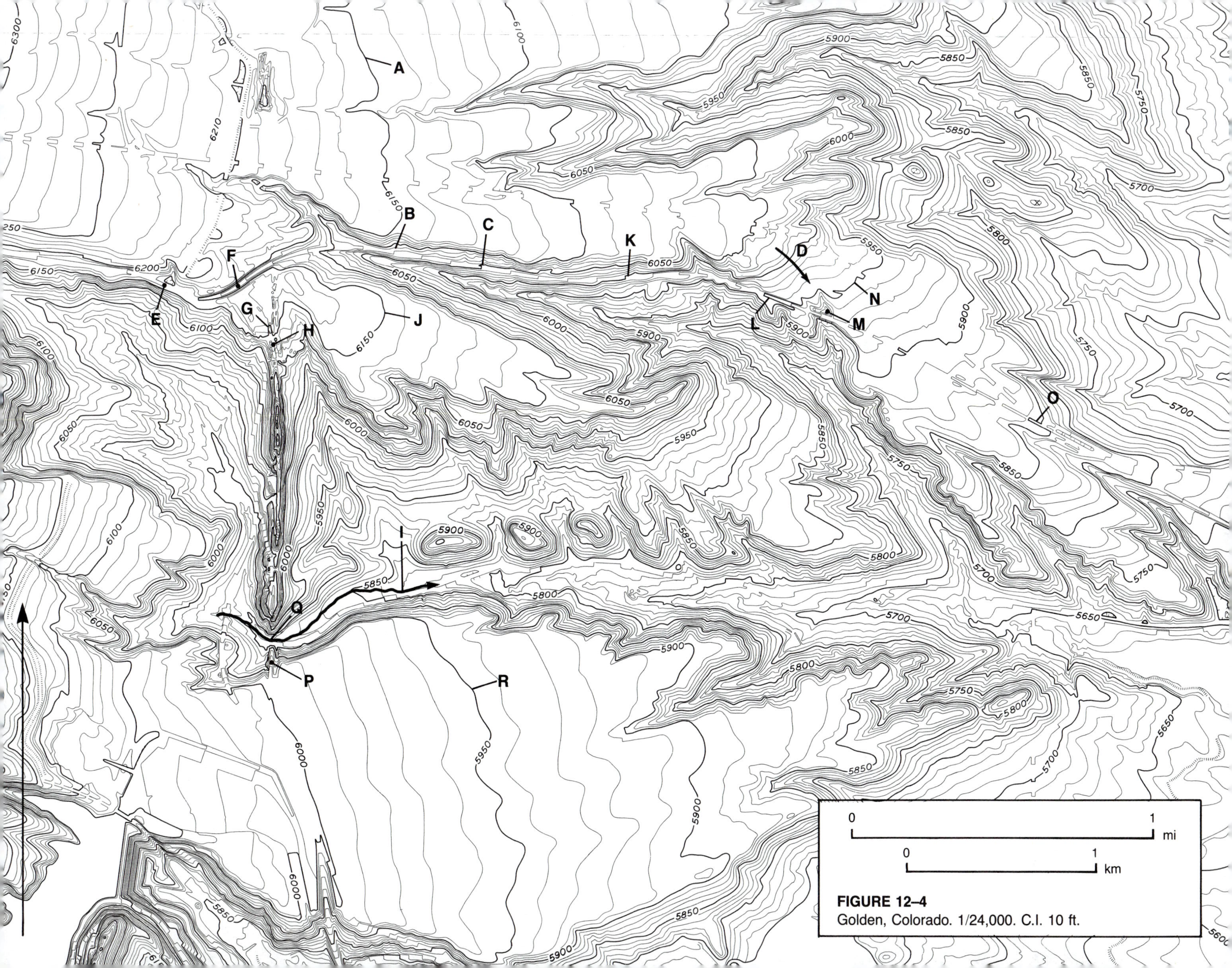

FIGURE 12–4
Golden, Colorado. 1/24,000. C.I. 10 ft.

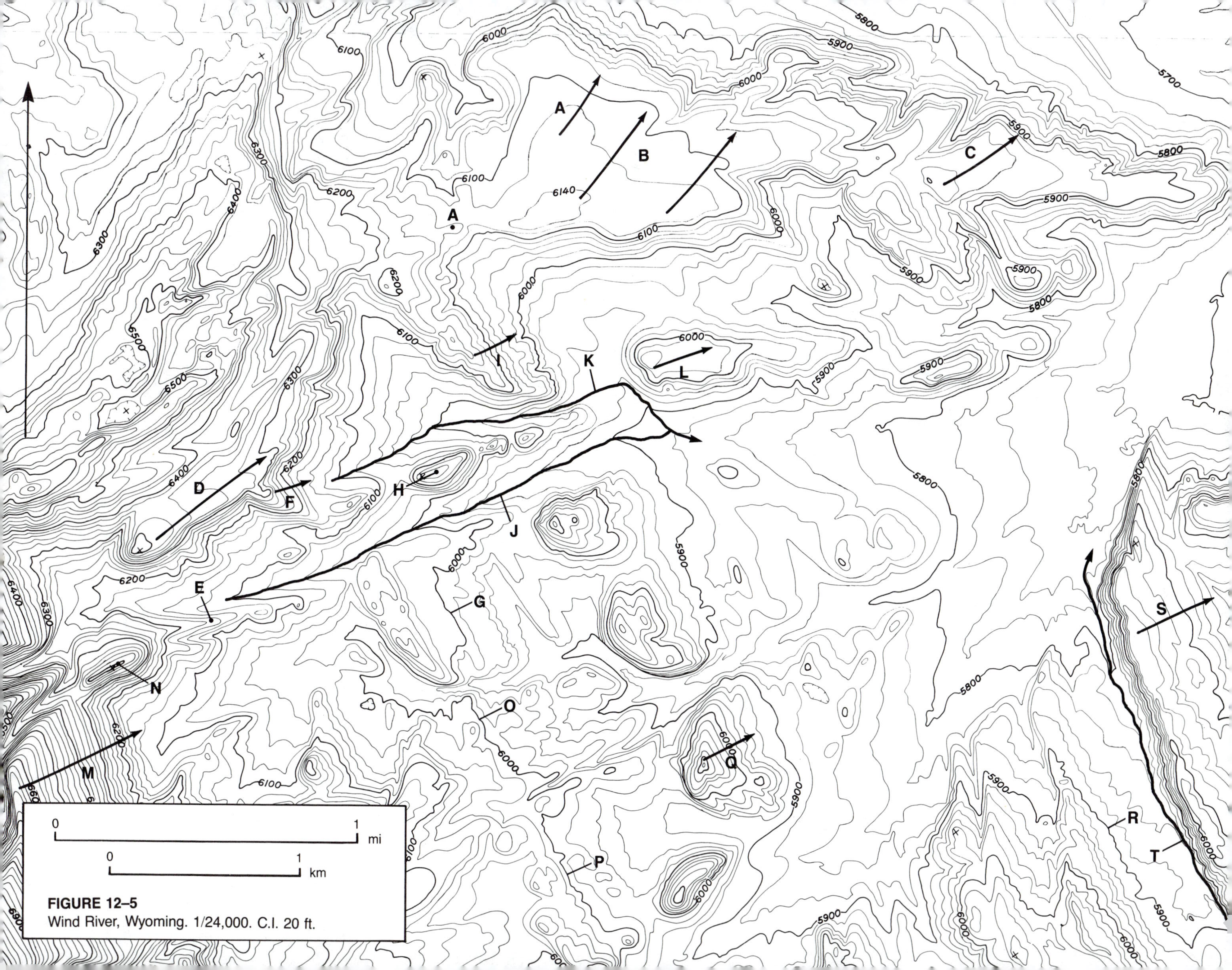

FIGURE 12–5
Wind River, Wyoming. 1/24,000. C.I. 20 ft.

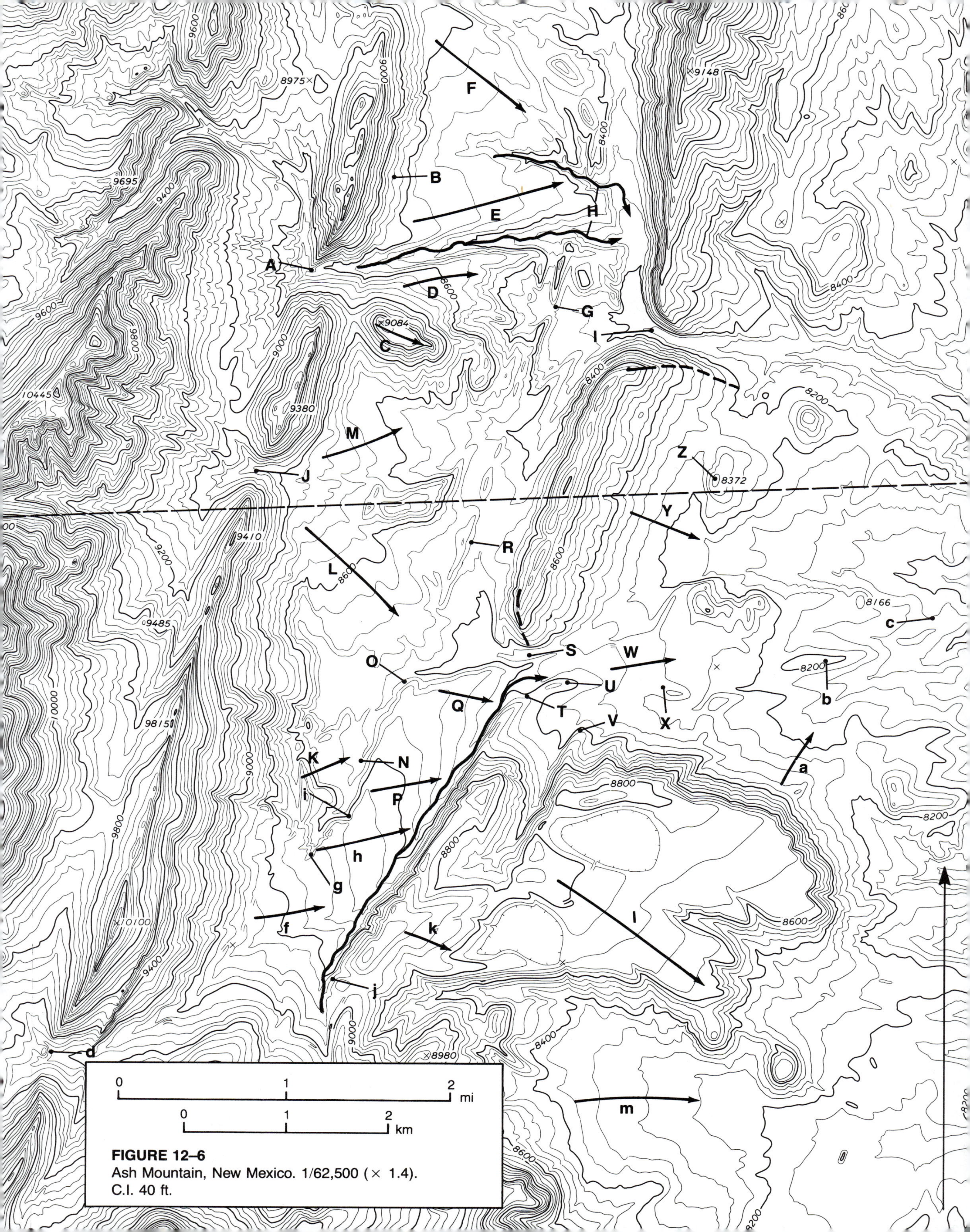

FIGURE 12–6
Ash Mountain, New Mexico. 1/62,500 (× 1.4).
C.I. 40 ft.

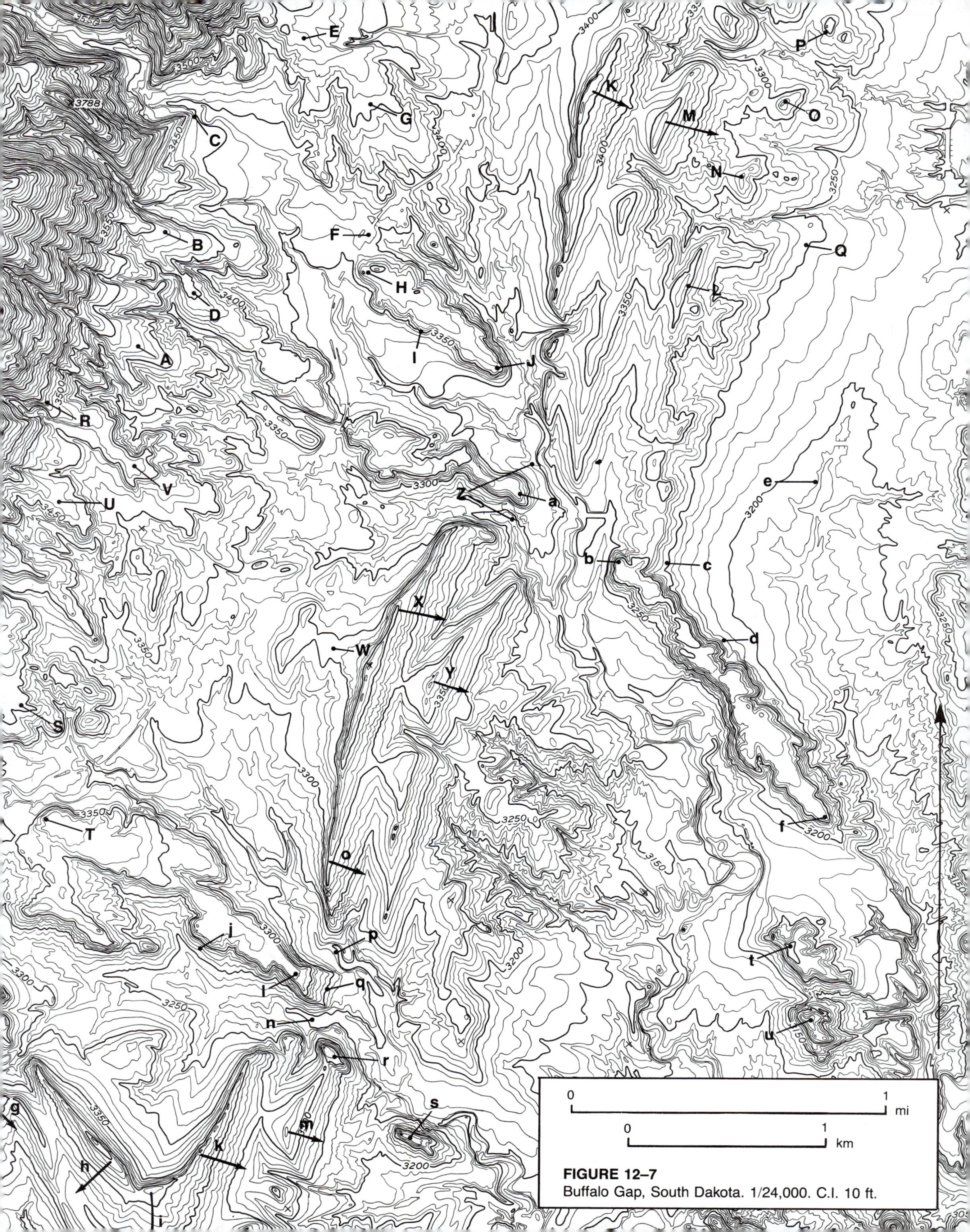

FIGURE 12–7
Buffalo Gap, South Dakota. 1/24,000. C.I. 10 ft.

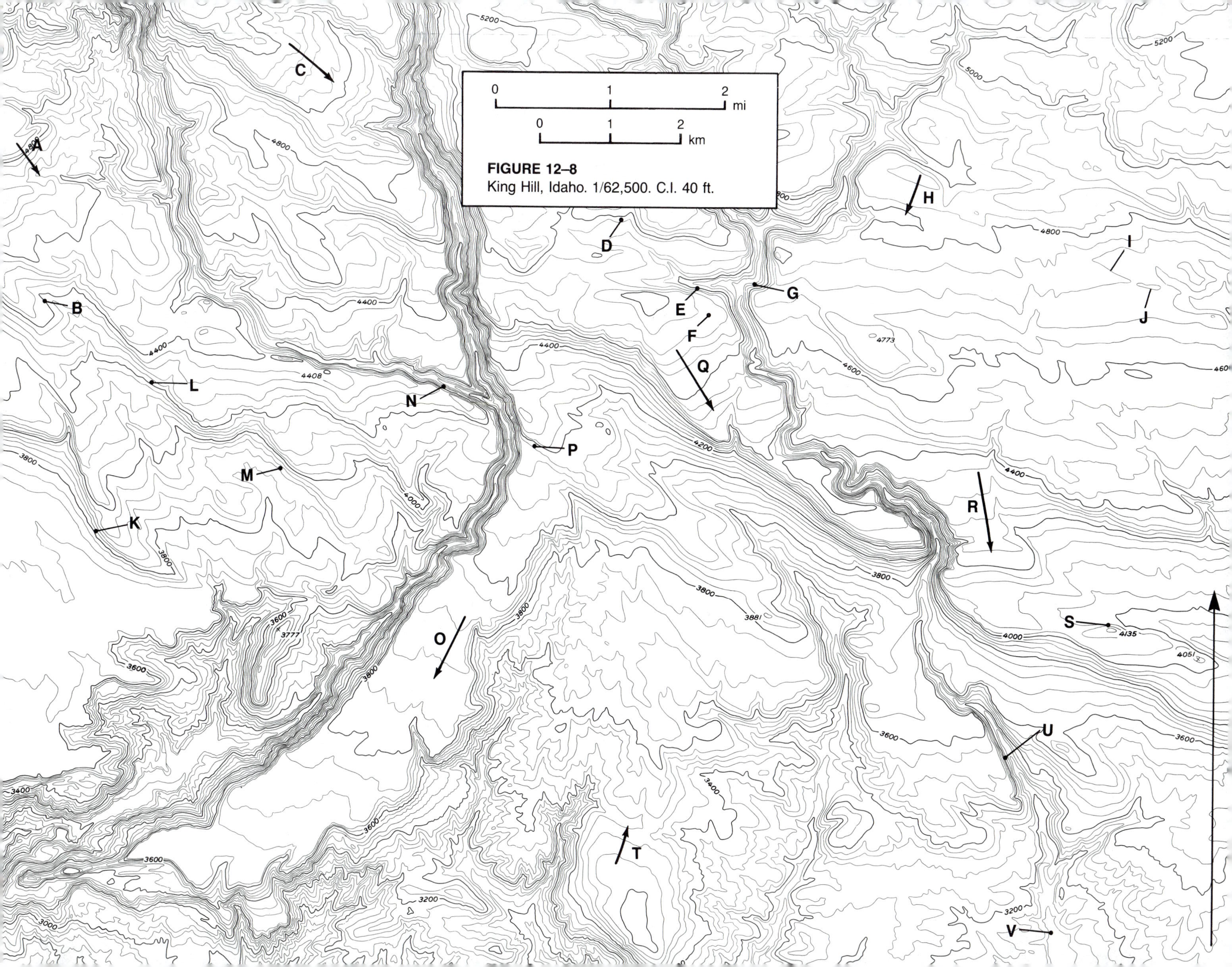

FIGURE 12–8
King Hill, Idaho. 1/62,500. C.I. 40 ft.

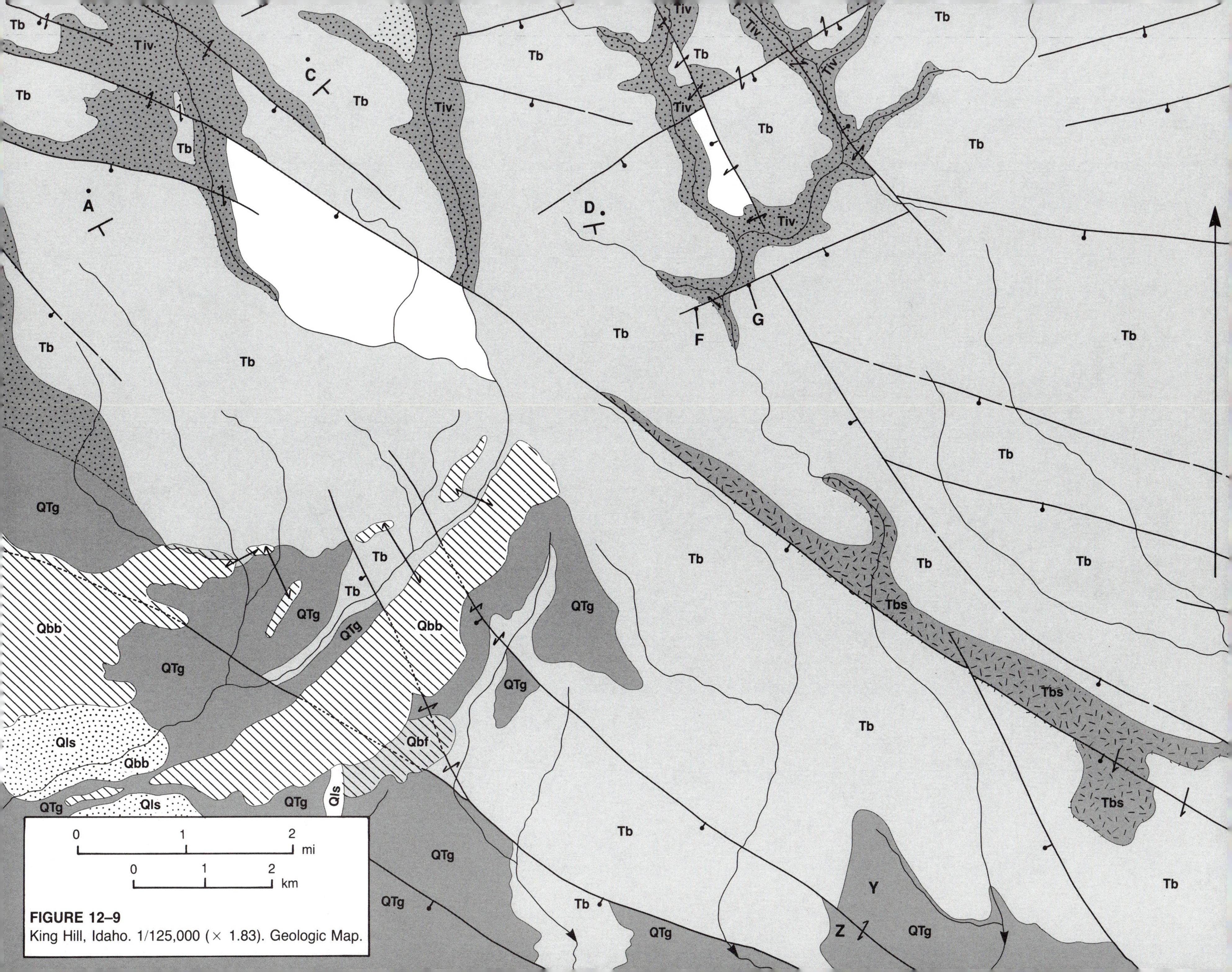

FIGURE 12–9
King Hill, Idaho. 1/125,000 (× 1.83). Geologic Map.

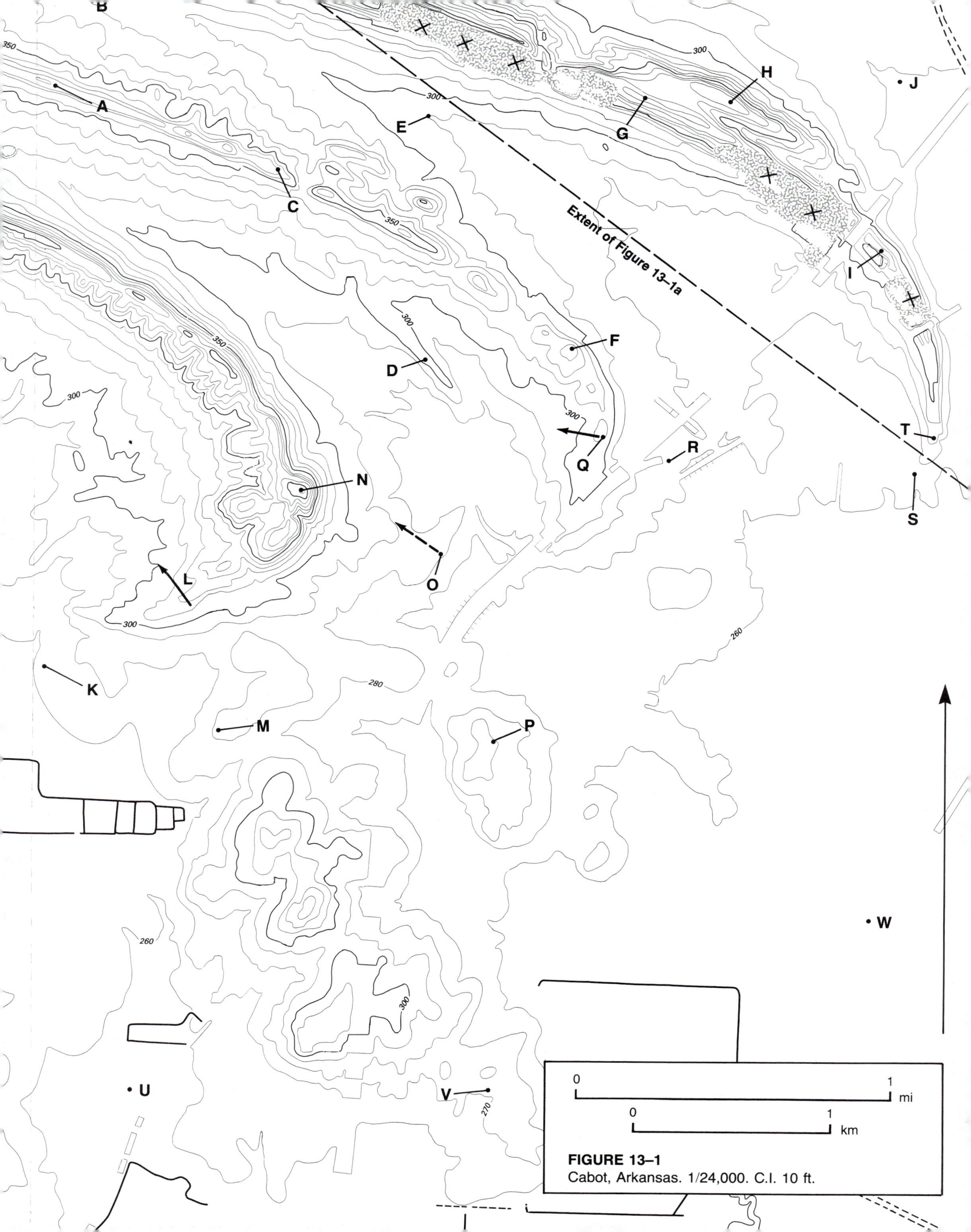

FIGURE 13–1
Cabot, Arkansas. 1/24,000. C.I. 10 ft.

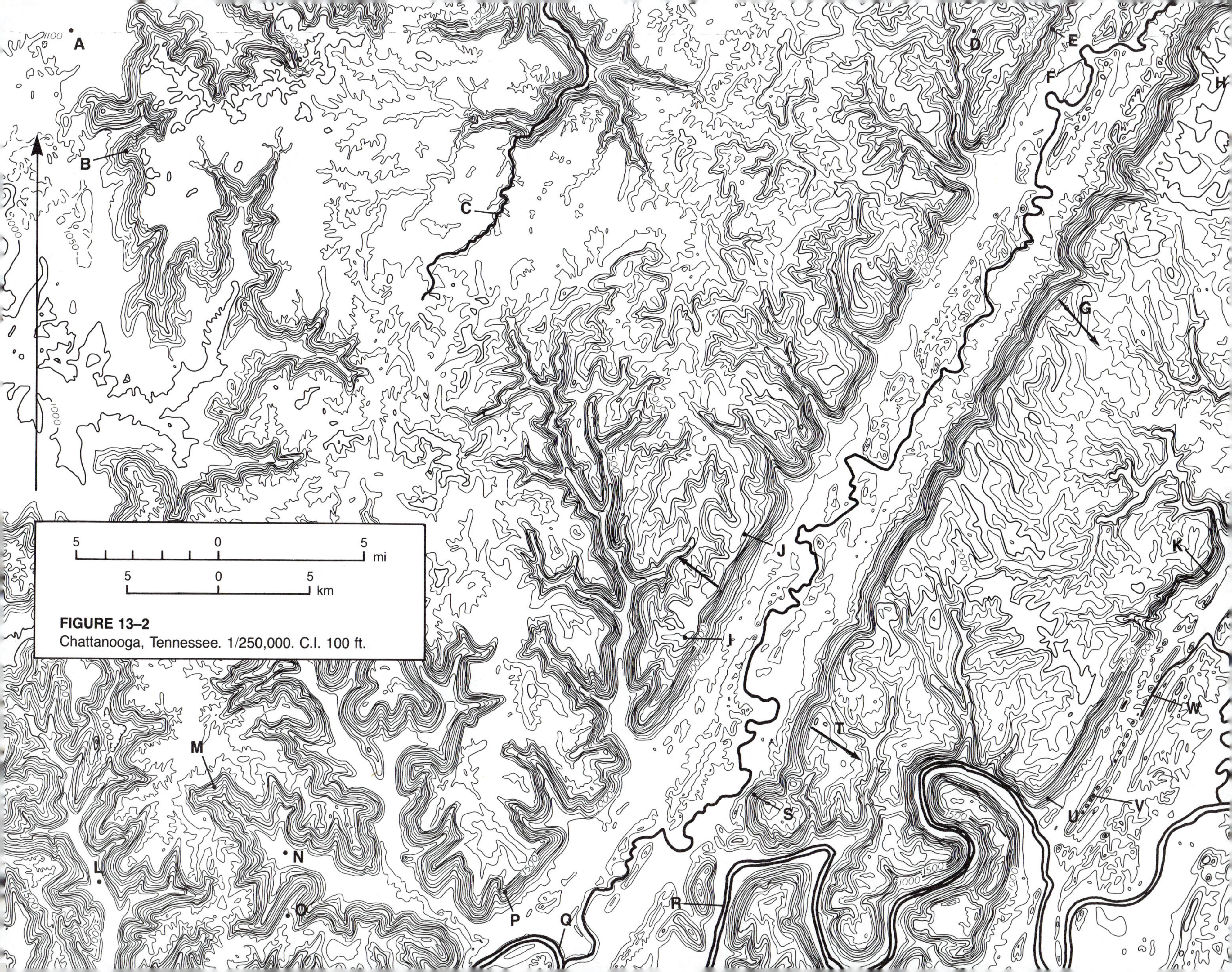

FIGURE 13–2
Chattanooga, Tennessee. 1/250,000. C.I. 100 ft.

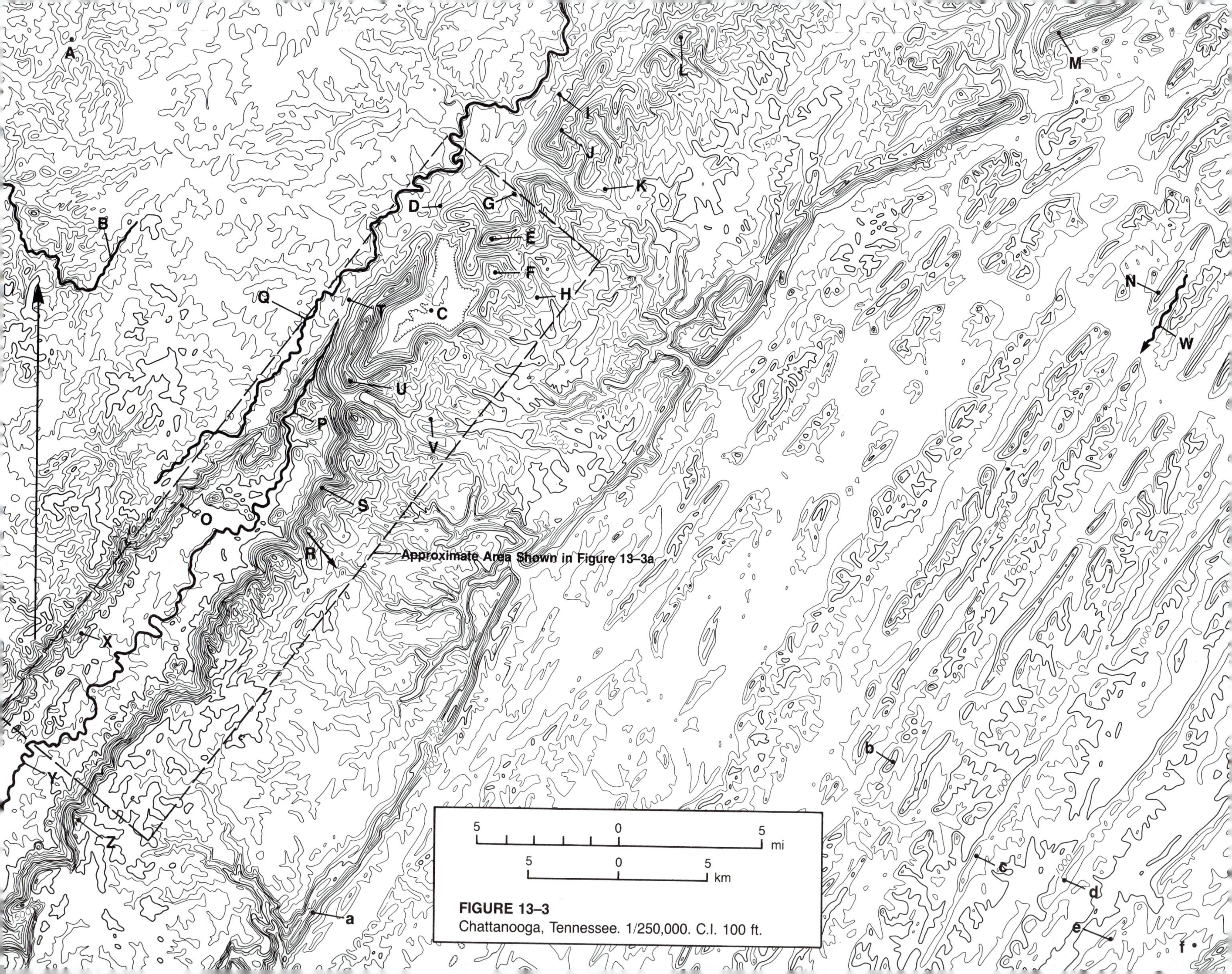

FIGURE 13–3
Chattanooga, Tennessee. 1/250,000. C.I. 100 ft.

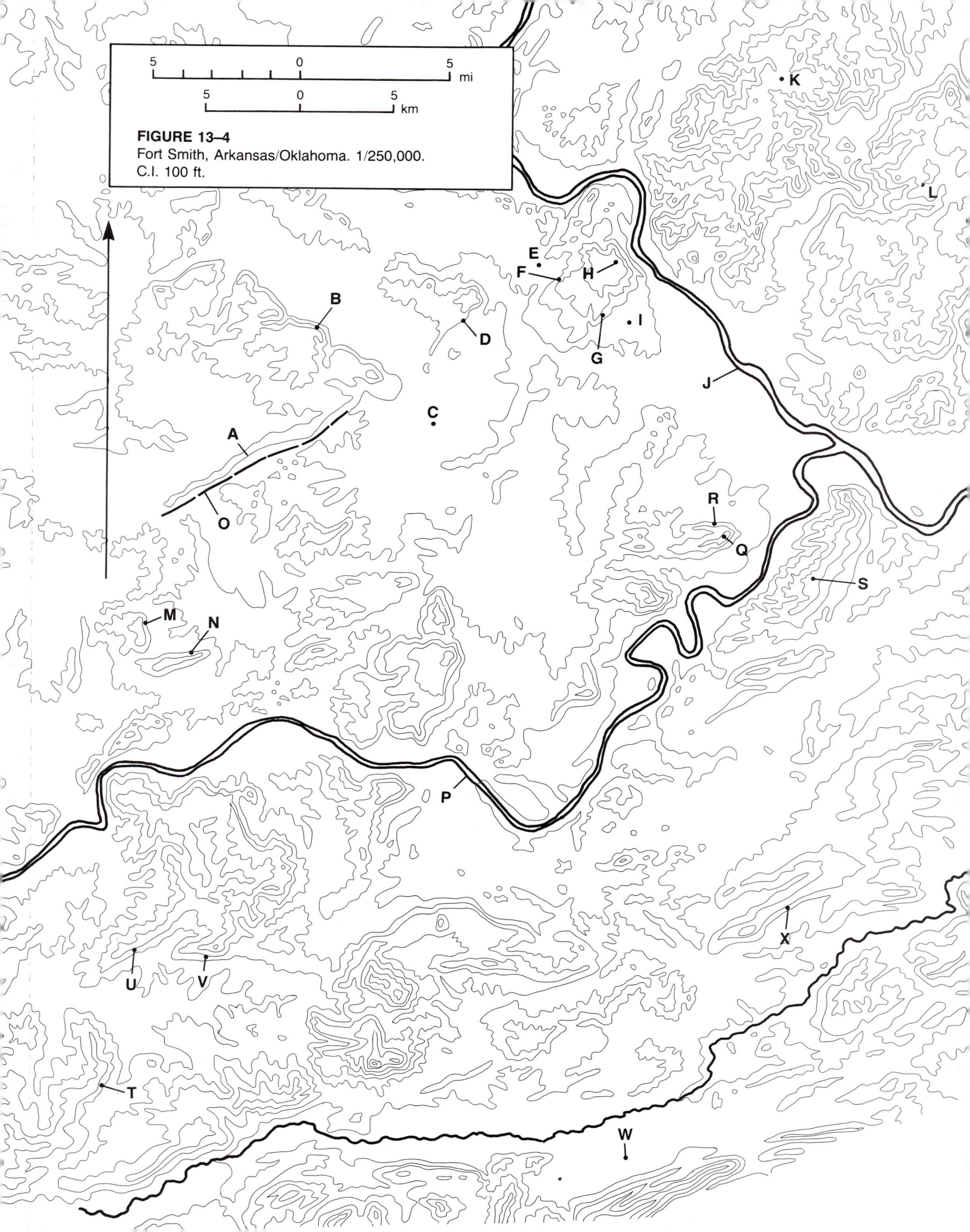

FIGURE 13–4
Fort Smith, Arkansas/Oklahoma. 1/250,000.
C.I. 100 ft.

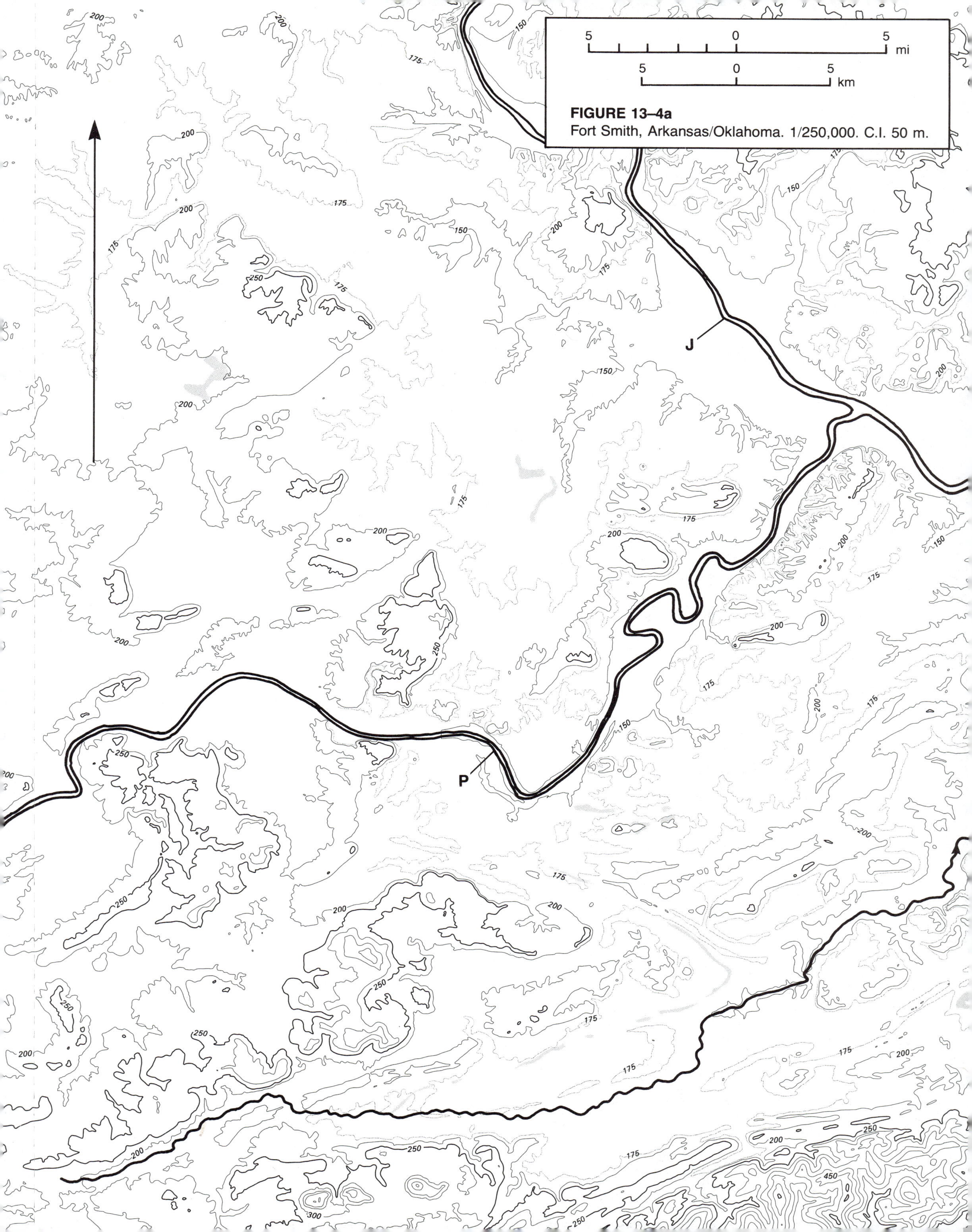

FIGURE 13–4a
Fort Smith, Arkansas/Oklahoma. 1/250,000. C.I. 50 m.

FIGURE 13–5
Fort Smith, Arkansas/Oklahoma. 1/250,000.
Geologic Map.

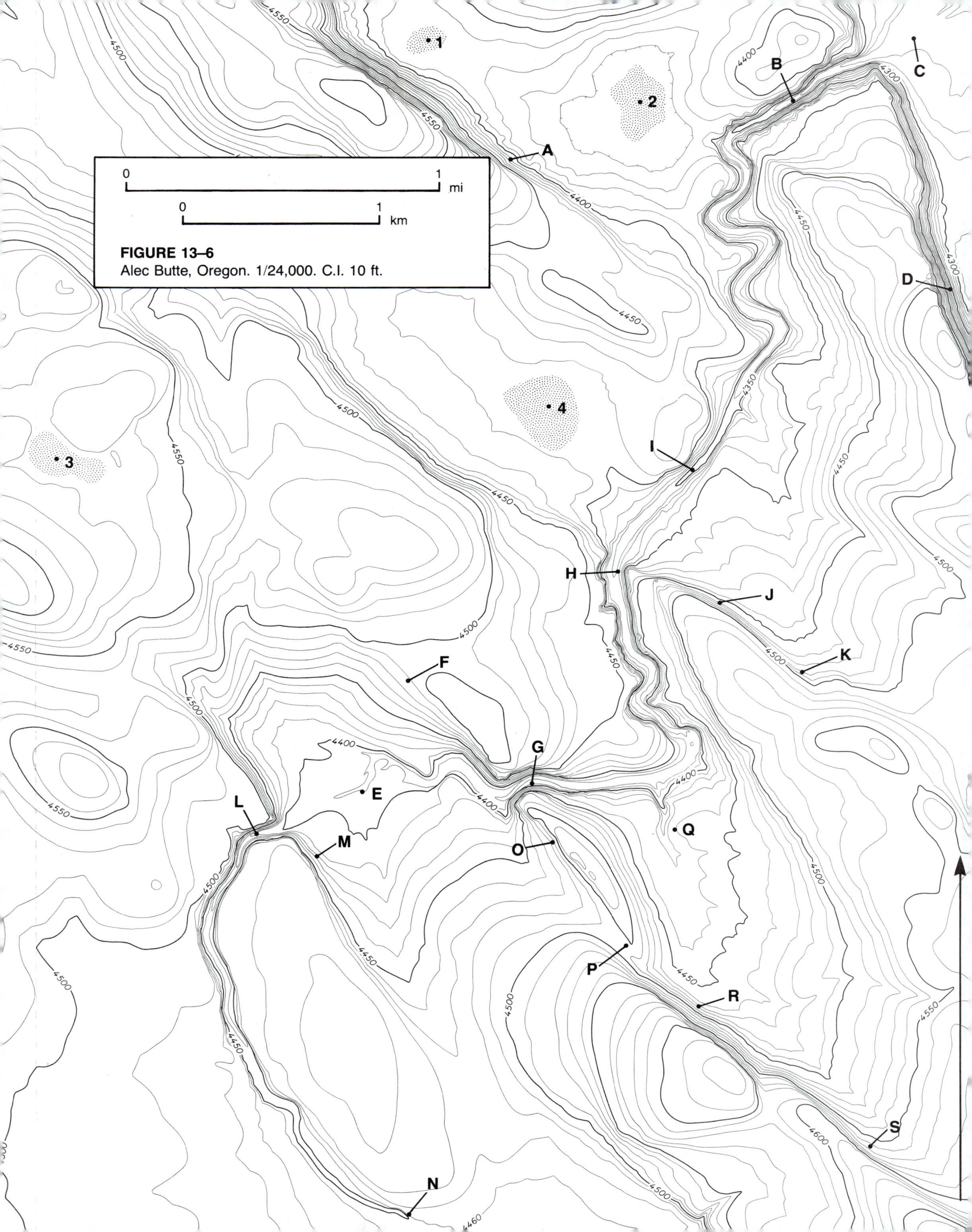

FIGURE 13–6
Alec Butte, Oregon. 1/24,000. C.I. 10 ft.

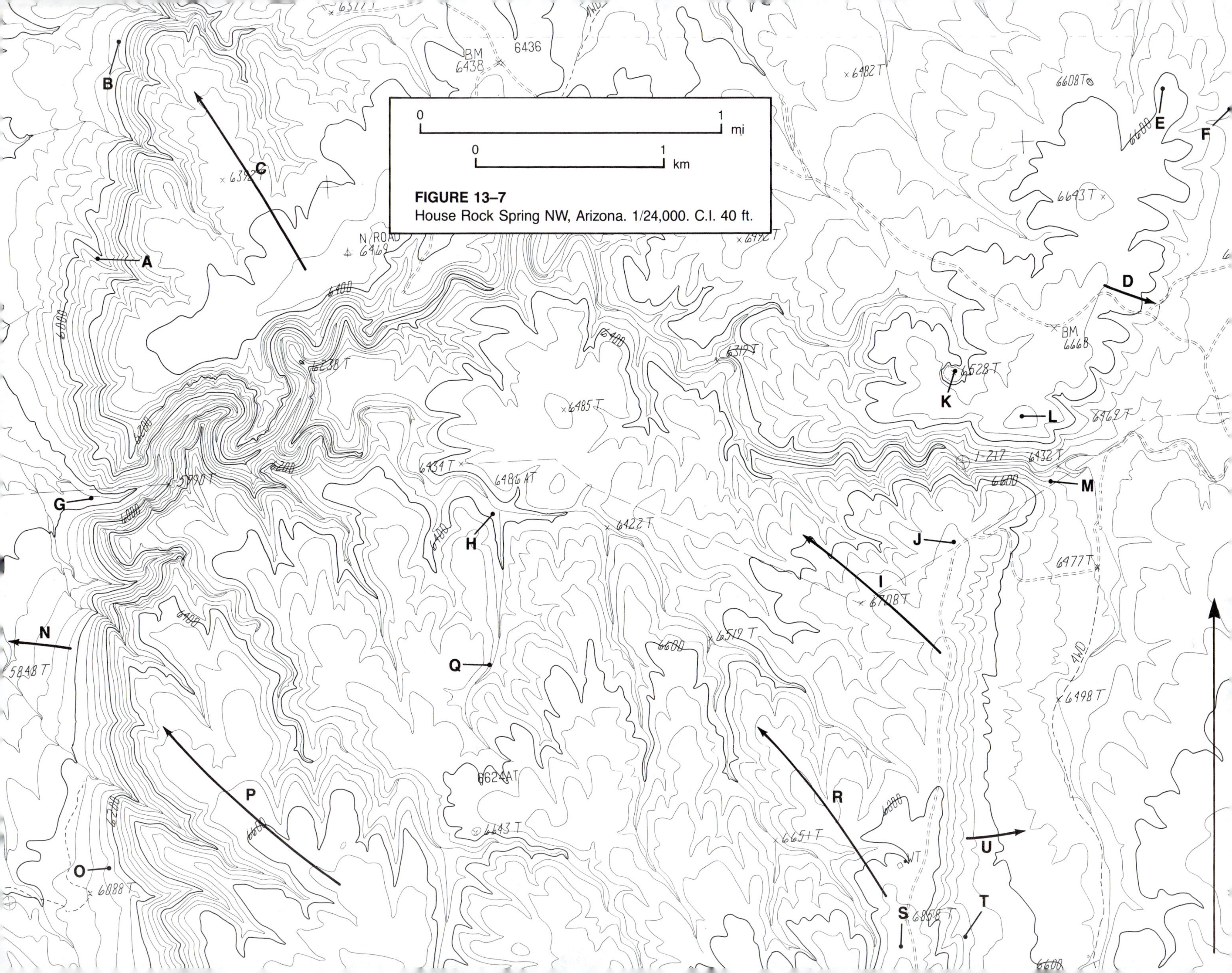

FIGURE 13–7
House Rock Spring NW, Arizona. 1/24,000. C.I. 40 ft.

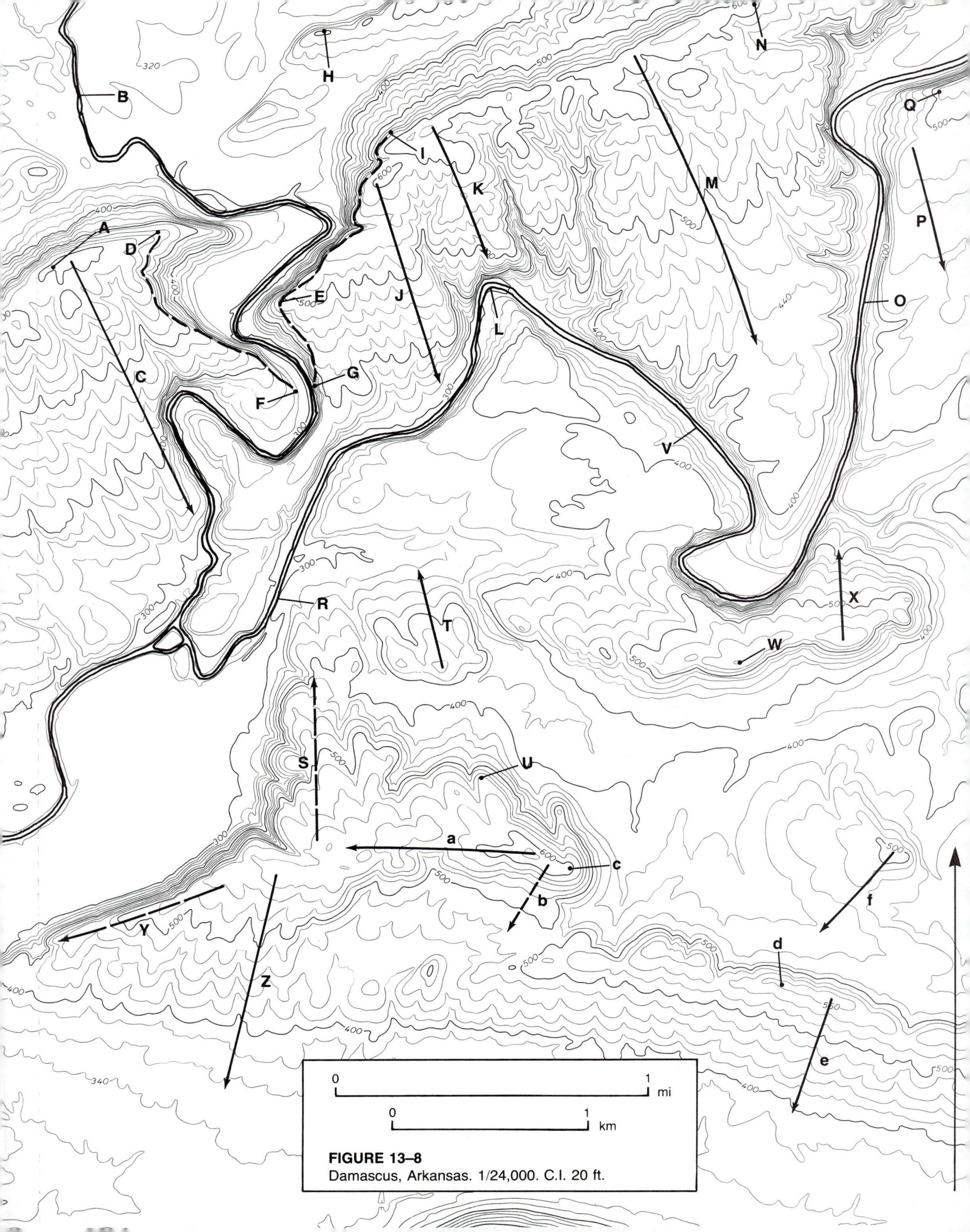

FIGURE 13–8
Damascus, Arkansas. 1/24,000. C.I. 20 ft.

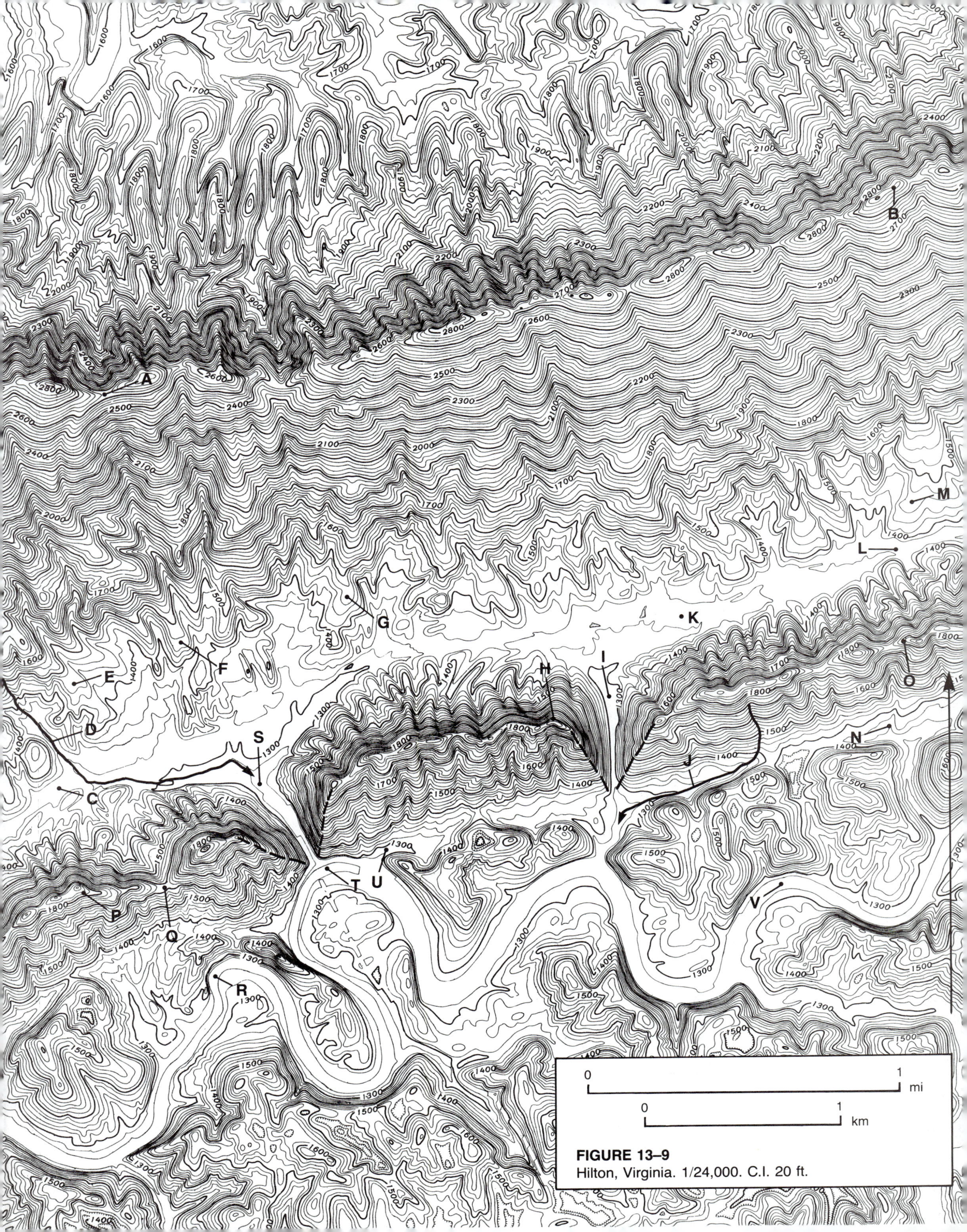

FIGURE 13–9
Hilton, Virginia. 1/24,000. C.I. 20 ft.

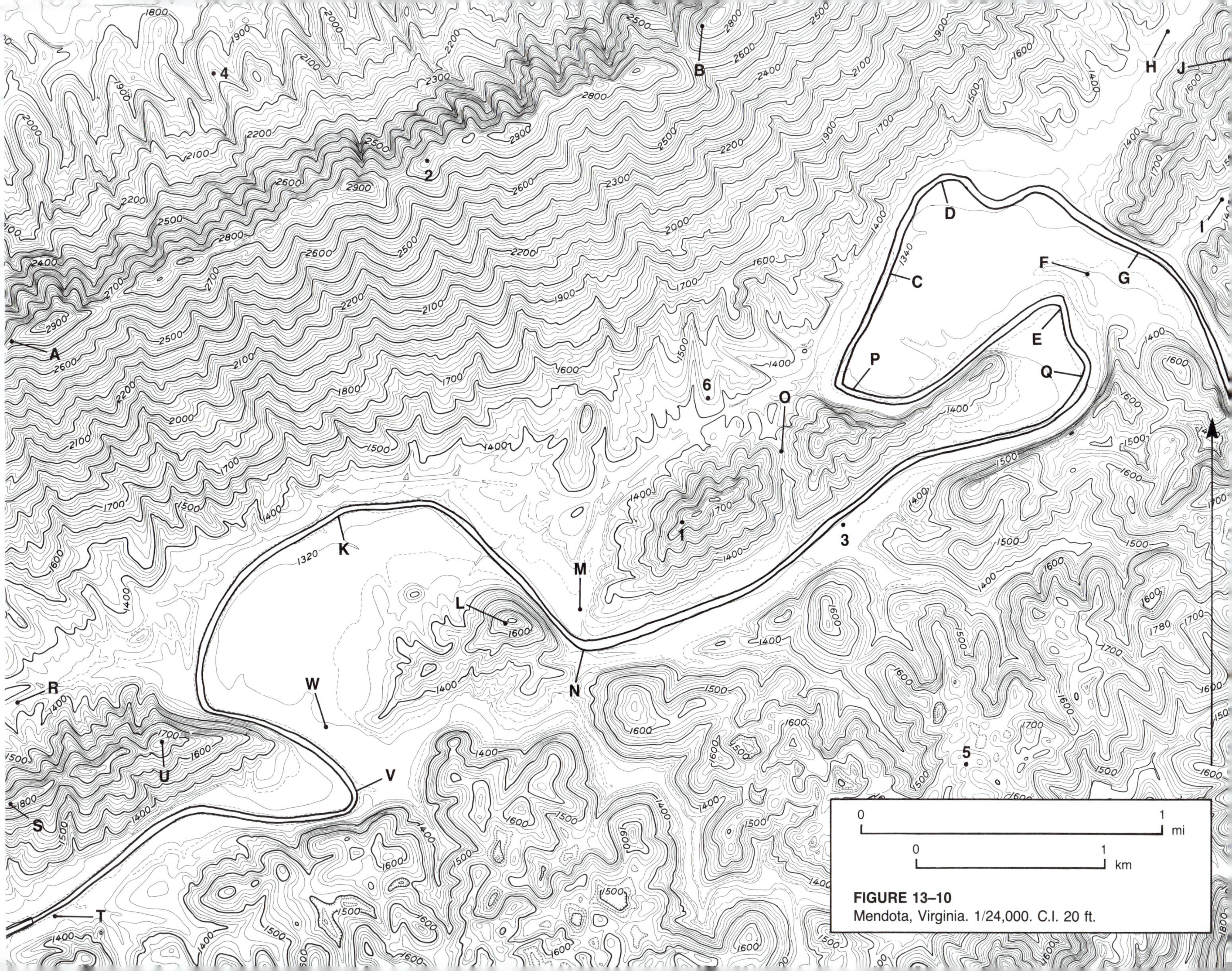

FIGURE 13–10
Mendota, Virginia. 1/24,000. C.I. 20 ft.

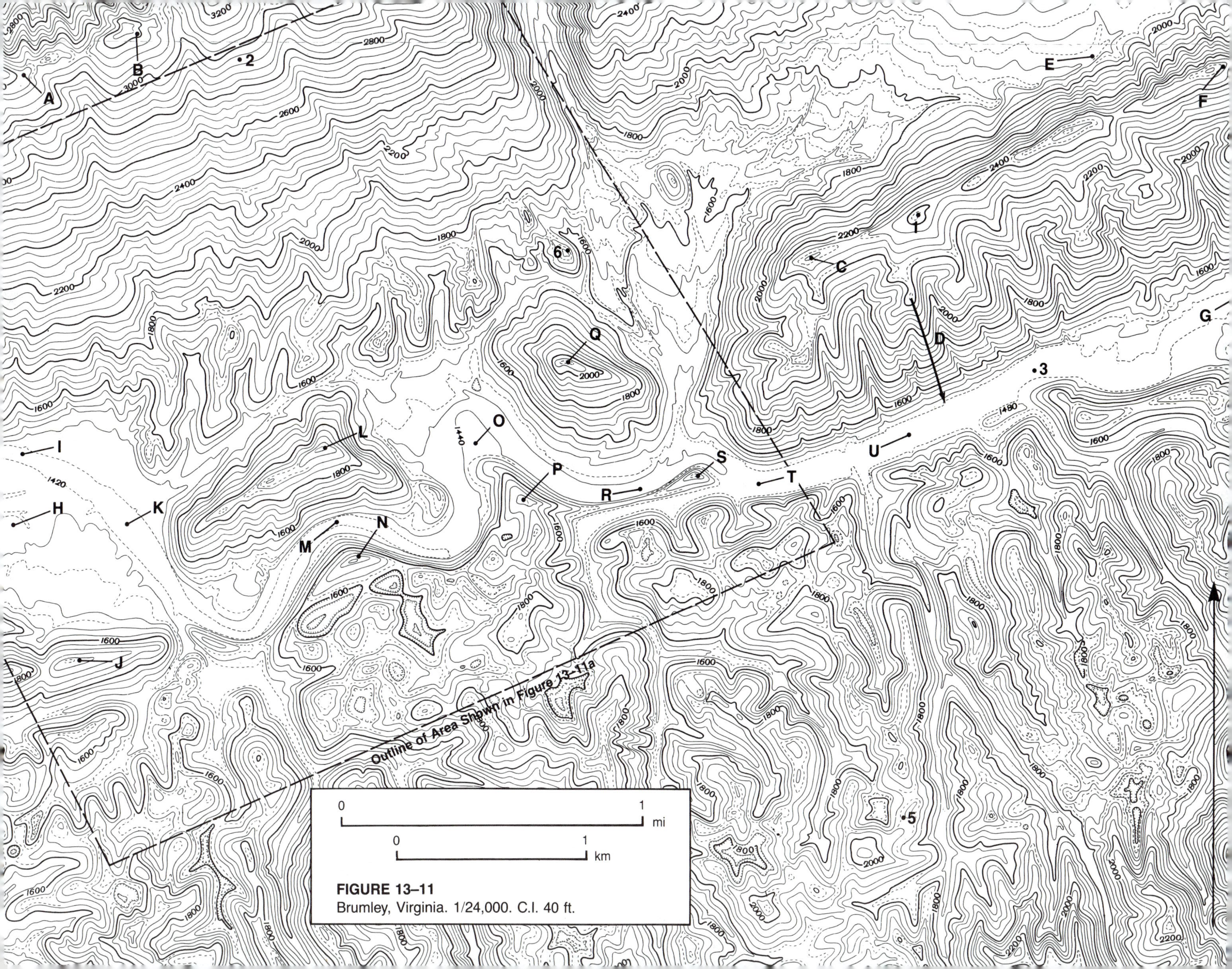

FIGURE 13–11
Brumley, Virginia. 1/24,000. C.I. 40 ft.

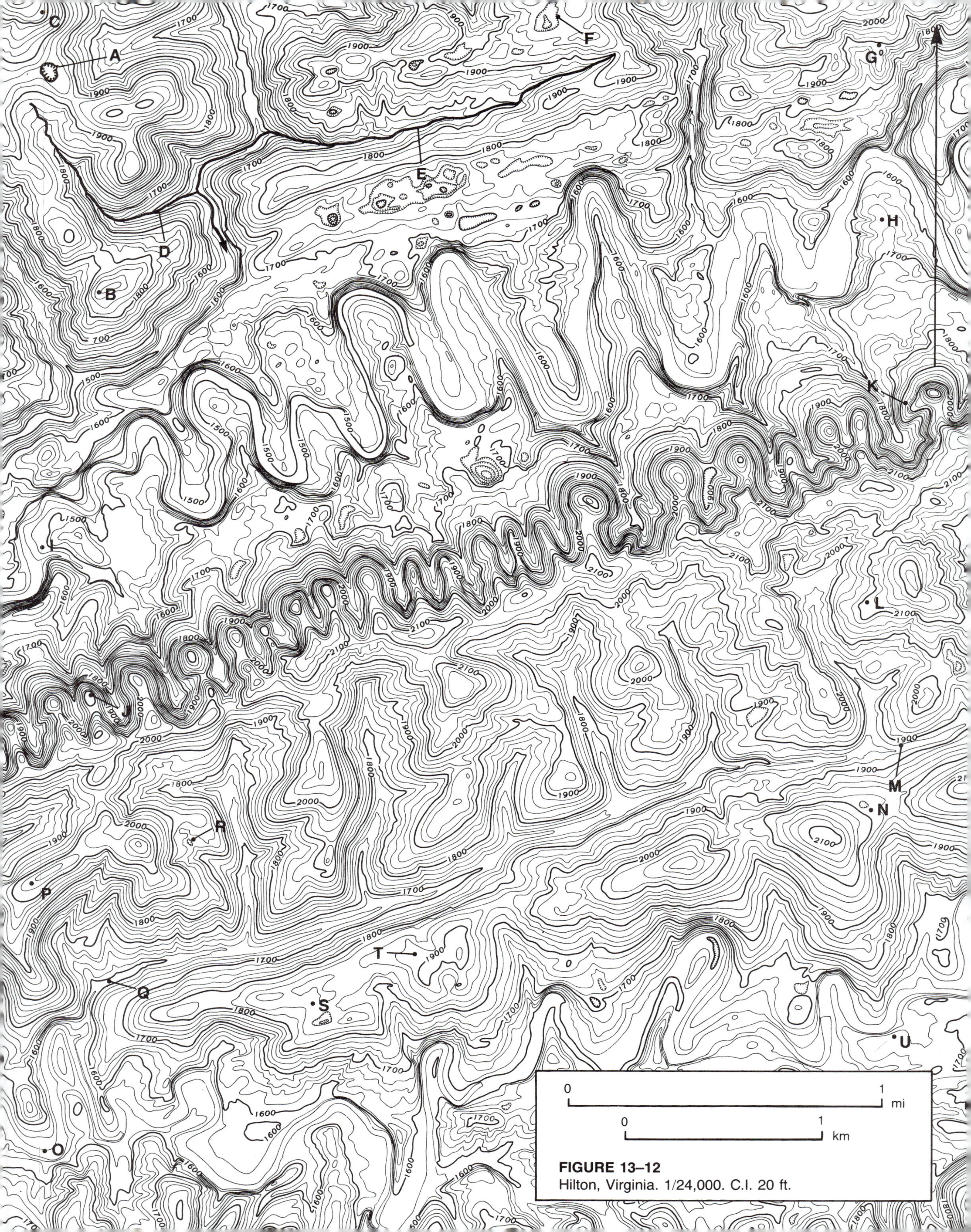

FIGURE 13–12
Hilton, Virginia. 1/24,000. C.I. 20 ft.

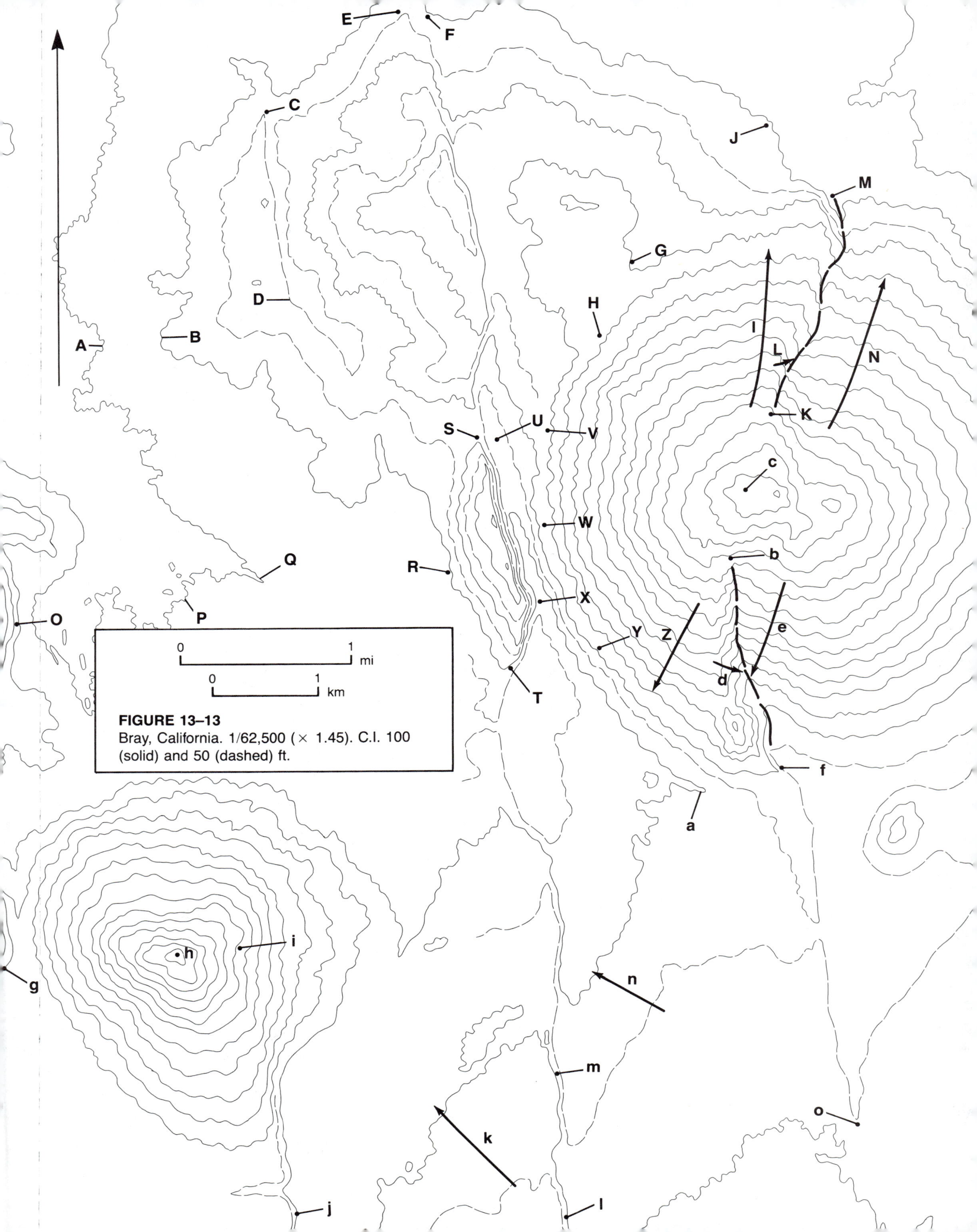

FIGURE 13–13
Bray, California. 1/62,500 (× 1.45). C.I. 100 (solid) and 50 (dashed) ft.

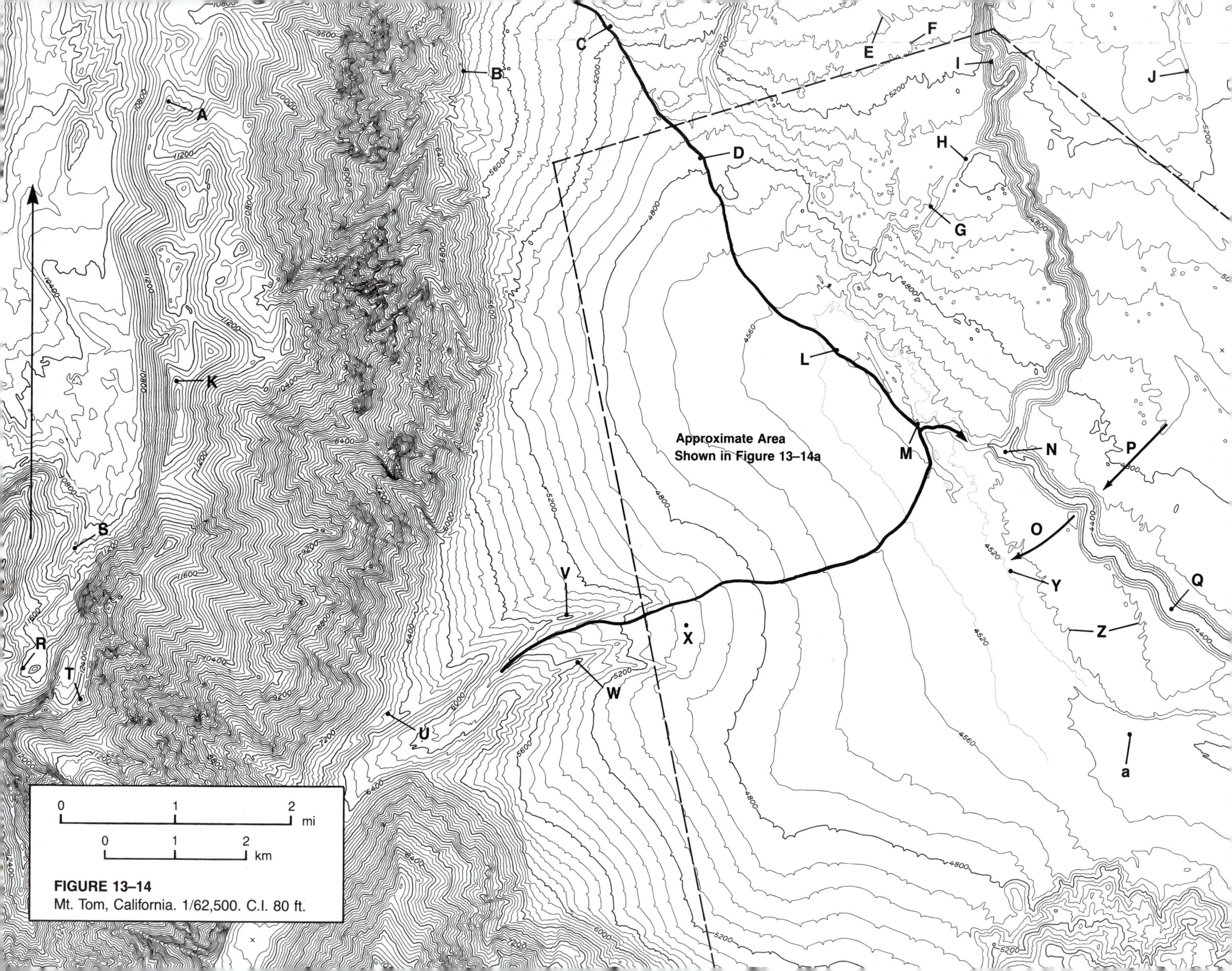

FIGURE 13–14
Mt. Tom, California. 1/62,500. C.I. 80 ft.

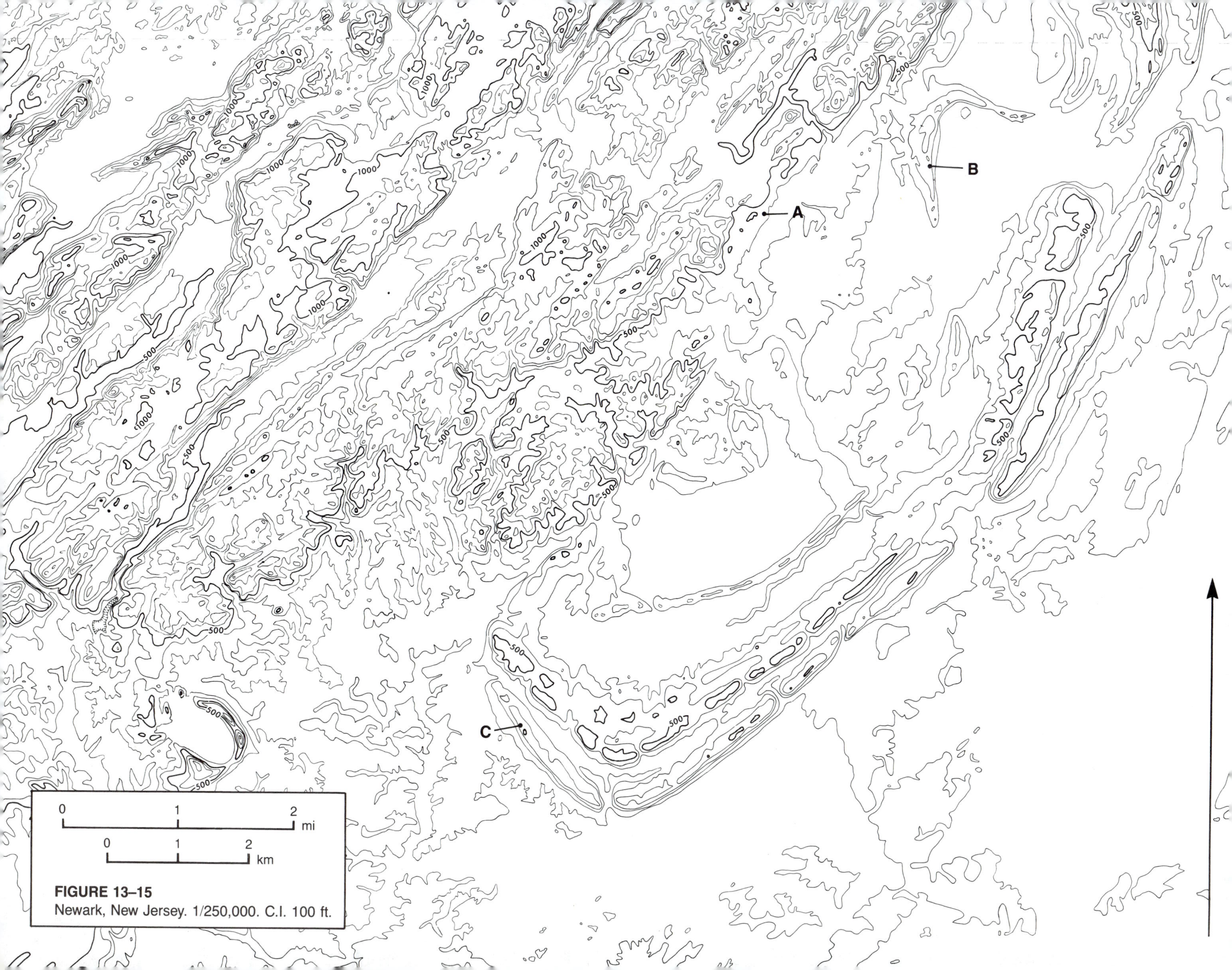

FIGURE 13–15
Newark, New Jersey. 1/250,000. C.I. 100 ft.

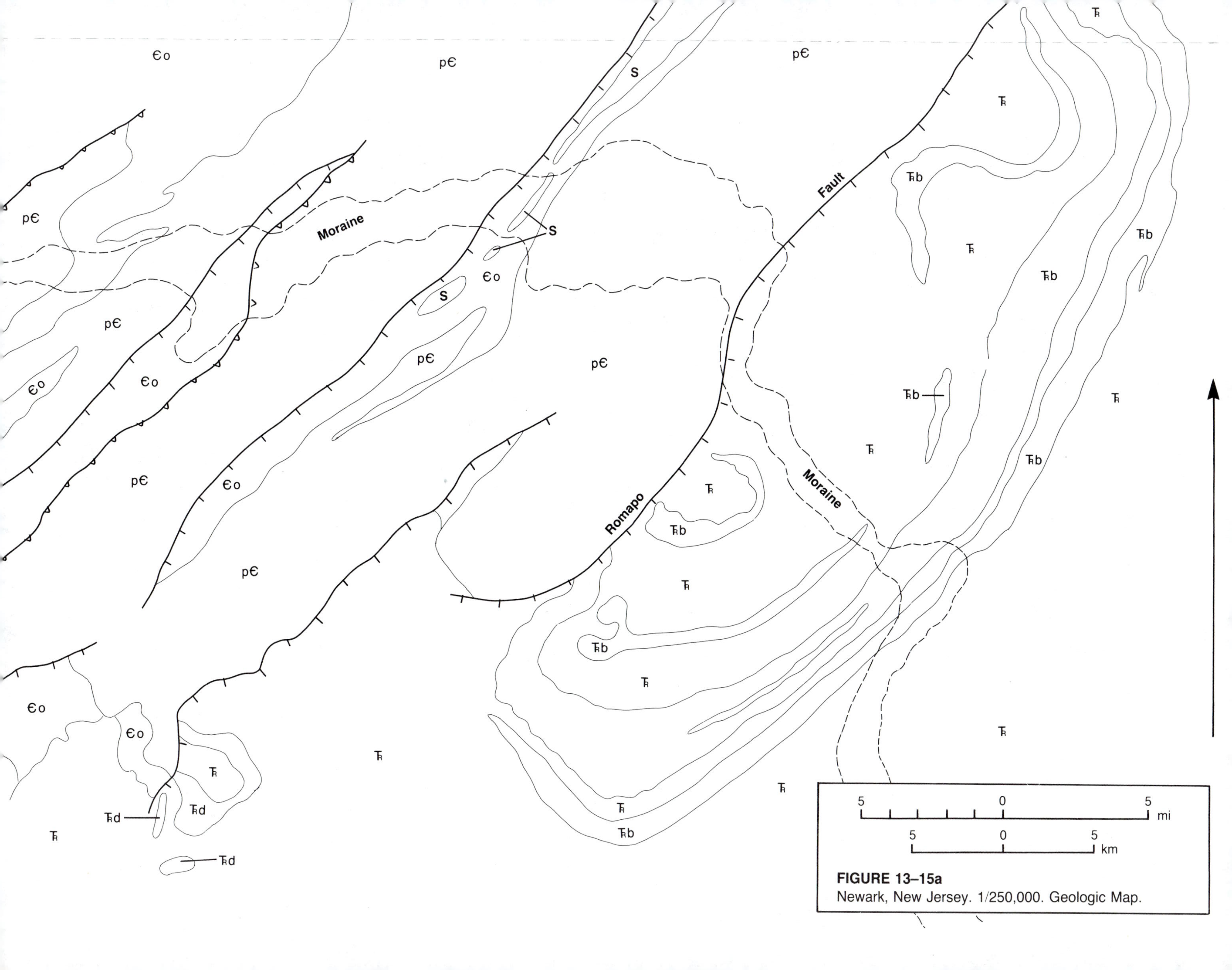

FIGURE 13–15a
Newark, New Jersey. 1/250,000. Geologic Map.

CENOZOIC

Period		Epoch		Age	Picks (Ma)
QUATERNARY		HOLOCENE			0.01
		PLEISTOCENE		CALABRIAN	1.6
TERTIARY	NEOGENE	PLIOCENE	L	PIACENZIAN	3.4
			E	ZANCLEAN	5.3
		MIOCENE	L	MESSINIAN	6.5
				TORTONIAN	11.2
			M	SERRAVALLIAN	15.1
				LANGHIAN	16.6
			E	BURDIGALIAN	21.8
				AQUITANIAN	23.7
	PALEOGENE	OLIGOCENE	L	CHATTIAN	30.0
			E	RUPELIAN	36.6
		EOCENE	L	PRIABONIAN	40.0
			M	BARTONIAN	43.6
				LUTETIAN	52.0
			E	YPRESIAN	57.8
		PALEOCENE	L	SELANDIAN: THANETIAN	60.6
				SELANDIAN: UNNAMED	63.6
			E	DANIAN	66.4

Magnetic polarity (Hist., Anom., Chron): anomalies 1, 2, 2A, 3, 3A, 4, 4A, 5, 5A, 5B, 5C, 5D, 5E, 6, 6A, 6B, 6C, 7, 7A, 8, 9, 10, 11, 12, 13, 15, 16, 17, 18, 19, 20, 21, 22, 23, 24, 25, 26, 27, 28, 29; chrons C1, C2, C2A, C3, C3A, C4, C4A, C5, C5A, C5B, C5C, C5D, C5E, C6, C6A, C6B, C6C, C7, C7A, C8, C9, C10, C11, C12, C13, C15, C16, C17, C18, C19, C20, C21, C22, C23, C24, C25, C26, C27, C28, C29. Age (Ma) scale 5–65.

MESOZOIC

Period	Epoch		Age	Picks (Ma)	Uncert. (m.y.)
CRETACEOUS	LATE			66.4	
			MAASTRICHTIAN	74.5	4
			CAMPANIAN	84.0	4.5
			SANTONIAN	87.5	
			CONIACIAN	88.5	2.5
			TURONIAN	91	
			CENOMANIAN	97.5	2.5
	EARLY		ALBIAN	113	4
			APTIAN	119	9
		NEOCOMIAN	BARREMIAN	124	9
			HAUTERIVIAN	131	8
			VALANGINIAN	138	5
			BERRIASIAN	144	5
JURASSIC	LATE		TITHONIAN	152	12
			KIMMERIDGIAN	156	6
			OXFORDIAN	163	15
	MIDDLE		CALLOVIAN	169	15
			BATHONIAN	176	34
			BAJOCIAN	183	34
			AALENIAN	187	34
	EARLY		TOARCIAN	193	28
			PLIENSBACHIAN	198	32
			SINEMURIAN	204	18
			HETTANGIAN	208	18
TRIASSIC	LATE		NORIAN	225	8
			CARNIAN	230	22
	MIDDLE		LADINIAN	235	10
			ANISIAN	240	22
	EARLY		SCYTHIAN	245	20

Magnetic polarity (Hist., Anom., Chron): anomalies 29, 30, 31, 32, 33; chrons C29, C30, C31, C32, C33; M0, M1, M3, M5, M10, M12, M14, M16, M18, M20, M22, M25, M29; RAPID POLARITY CHANGES. Age (Ma) scale 70–240.

PALEOZOIC

Period		Epoch	Age		Picks (Ma)	Uncert. (m.y.)
PERMIAN		LATE	TATARIAN		245	20
			KAZANIAN		253	20
			UFIMIAN		258	24
		EARLY	KUNGURIAN		263	22
			ARTINSKIAN		268	12
			SAKMARIAN			
			ASSELIAN		286	12
CARBONIFEROUS	PENNSYLVANIAN	LATE	GZELIAN	S.		
			KASIMOVIAN		296	10
			MOSCOVIAN	W.		
			BASHKIRIAN		315	20
	MISSISSIPPIAN	EARLY	SERPUKHOVIAN	N.	320	
			VISEAN		333	22
			TOURNAISIAN		352	8
DEVONIAN		LATE	FAMENNIAN		360	10
			FRASNIAN		367	12
		MIDDLE	GIVETIAN		374	18
			EIFELIAN		380	18
		EARLY	EMSIAN		387	28
			SIEGENIAN		394	22
			GEDINNIAN		401	18
SILURIAN		LATE	PRIDOLIAN		408	12
			LUDLOVIAN		414	12
		EARLY	WENLOCKIAN		421	12
			LLANDOVERIAN		428	8
ORDOVICIAN		LATE	ASHGILLIAN		438	12
			CARADOCIAN		448	12
		MIDDLE	LLANDEILAN		458	16
			LLANVIRNIAN		468	16
		EARLY	ARENIGIAN		478	16
			TREMADOCIAN		488	20
CAMBRIAN		LATE	TREMPEALEAUAN, FRANCONIAN, DRESBACHIAN		505	32
		MIDDLE			523	36
		EARLY			540	28
					570	

Age (Ma) scale 260–560.

PRECAMBRIAN

Eon	Era	Bdy. Ages (Ma)
PROTEROZOIC		570
	LATE	900
	MIDDLE	1600
	EARLY	2500
ARCHEAN	LATE	3000
	MIDDLE	3400
	EARLY	3800?

Age (Ma) scale 750–3750.

Compiled 1983

Published by: The Geological Society of America, Inc.
3300 Penrose Place, P.O. Box 9140
Boulder, Colorado 80301

MAP AND CHART SERIES MC-50

Decade of North American Geology. (Courtesy of the Geological Society of America, Inc.)